Communications
in Computer and Information Science   **1761**

Rationale

The CCIS series is devoted to the publication of proceedings of computer science conferences. Its aim is to efficiently disseminate original research results in informatics in printed and electronic form. While the focus is on publication of peer-reviewed full papers presenting mature work, inclusion of reviewed short papers reporting on work in progress is welcome, too. Besides globally relevant meetings with internationally representative program committees guaranteeing a strict peer-reviewing and paper selection process, conferences run by societies or of high regional or national relevance are also considered for publication.

Topics

The topical scope of CCIS spans the entire spectrum of informatics ranging from foundational topics in the theory of computing to information and communications science and technology and a broad variety of interdisciplinary application fields.

Information for Volume Editors and Authors

Publication in CCIS is free of charge. No royalties are paid, however, we offer registered conference participants temporary free access to the online version of the conference proceedings on SpringerLink (http://link.springer.com) by means of an http referrer from the conference website and/or a number of complimentary printed copies, as specified in the official acceptance email of the event.

CCIS proceedings can be published in time for distribution at conferences or as post-proceedings, and delivered in the form of printed books and/or electronically as USBs and/or e-content licenses for accessing proceedings at SpringerLink. Furthermore, CCIS proceedings are included in the CCIS electronic book series hosted in the SpringerLink digital library at http://link.springer.com/bookseries/7899. Conferences publishing in CCIS are allowed to use Online Conference Service (OCS) for managing the whole proceedings lifecycle (from submission and reviewing to preparing for publication) free of charge.

Publication process

The language of publication is exclusively English. Authors publishing in CCIS have to sign the Springer CCIS copyright transfer form, however, they are free to use their material published in CCIS for substantially changed, more elaborate subsequent publications elsewhere. For the preparation of the camera-ready papers/files, authors have to strictly adhere to the Springer CCIS Authors' Instructions and are strongly encouraged to use the CCIS LaTeX style files or templates.

Abstracting/Indexing

CCIS is abstracted/indexed in DBLP, Google Scholar, EI-Compendex, Mathematical Reviews, SCImago, Scopus. CCIS volumes are also submitted for the inclusion in ISI Proceedings.

How to start

To start the evaluation of your proposal for inclusion in the CCIS series, please send an e-mail to ccis@springer.com.

Dana Simian · Laura Florentina Stoica
Editors

Modelling and Development of Intelligent Systems

8th International Conference, MDIS 2022
Sibiu, Romania, October 28–30, 2022
Revised Selected Papers

Editors
Dana Simian 🆔
Lucian Blaga University of Sibiu
Sibiu, Romania

Laura Florentina Stoica 🆔
Lucian Blaga University of Sibiu
Sibiu, Romania

ISSN 1865-0929   ISSN 1865-0937 (electronic)
Communications in Computer and Information Science
ISBN 978-3-031-27033-8   ISBN 978-3-031-27034-5 (eBook)
https://doi.org/10.1007/978-3-031-27034-5

This Springer imprint is published by the registered company Springer Nature Switzerland AG
The registered company address is: Gewerbestrasse 11, 6330 Cham, Switzerland

Preface

This volume contains selected, peer-reviewed papers presented at the 8th International Conference on Modelling and Development of Intelligent Systems – MDIS 2022, which was held on October 28–30, 2022, in Sibiu, Romania. The conference was organized by the Research Center in Informatics and Information Technology and the Department of Mathematics and Informatics from the Faculty of Sciences, Lucian Blaga University of Sibiu (LBUS), Romania. In order to take into account the possible travel difficulties of some participants caused by the Covid 19 pandemic, this edition of the conference was organized in a hybrid format, allowing both on-site and online participation.

Intelligent systems modelling and development are currently areas of great interest, making an important contribution to solving the problems that today's society is facing. In this context, the importance of international forums allowing the exchange of ideas in topics connected to artificial intelligence is increasing globally.

The MDIS conferences aim to connect scientists, researchers, academics, IT specialists and students who want to share and discuss their original results in fields related to modelling and development of intelligent systems. The original contributions presented at MDIS 2022 ranged from concepts and theoretical developments to advanced technologies and innovative applications. The main topics of interest included but were not restricted to: machine learning, data mining, intelligent systems for decision support, natural language processing, robotics, swarm intelligence, metaheuristics and applications, adaptive systems, computational and conceptual models, pattern recognition, hybrid computation for artificial vision, modelling and optimization of dynamic systems, multiagent systems and mathematical models for the development of intelligent systems.

The 8th edition of MDIS was distinguished by a considerable extension of the geographical area of the authors, and brought together authors from 12 countries and four continents (Europe, Asia, North America and South America).

Four keynote speakers gave presentations, showing the high impact that research in artificial intelligence has or can have in all life areas. The invited keynote speakers and their lectures were:

Dan Cristea, A Technology of Deciphering Old Cyrillic-Romanian
Marcel Kyas, Autonomous Drones and Analogue Missions
Detlef Streitferdt, Inside Neural Networks
Milan Tuba, Recent Topics of Convolutional Neural Networks Applications

All submitted papers underwent a thorough single-blind peer review. Each paper was reviewed by at least 3 independent reviewers, chosen based on their qualifications and field of expertise.

This volume contains 21 selected papers from the total of 48 papers submitted to the conference. The papers in the volume are organized in the following topical sections: Intelligent Systems for Decision Support, Machine Learning, Mathematical Models for Development of Intelligent Systems, Modelling and Optimization of Dynamic Systems.

We thank all the participants for their interesting talks and discussions. We also thank the scientific committee members and all of the external reviewers for their help in reviewing the submitted papers and for their contributions to the scientific success of the conference and to the quality of this proceedings volume.

December 2022

Dana Simian
Laura Florentina Stoica

Organization

General Chair

Dana Simian   Lucian Blaga University of Sibiu, Romania

Scientific Committee

Kiril Alexiev   Bulgarian Academy of Sciences, Bulgaria
Anca Andreica   Babeş-Bolyai University, Romania
Christopher Auer   University of Applied Sciences Landshut,
                               Germany
Nebojsa Bacanin   Singidunum University, Serbia
Arndt Balzer   Technical University of Applied Sciences
                               Würzburg-Schweinfurt, Germany
Alina Bărbulescu   Ovidius University of Constanta, Romania
Lasse Berntzen   University of South-Eastern Norway, Norway
Charul Bhatnagar   GLA University, India
Florian Boian   Babeş-Bolyai University, Romania
Peter Braun   Technical University of Applied Sciences
                               Würzburg-Schweinfurt, Germany
Amelia Bucur   Lucian Blaga University of Sibiu, Romania
Steve Cassidy   Macquarie University, Australia
Camelia Chira   Babeş-Bolyai University, Romania
Nicolae Constantinescu   Lucian Blaga University of Sibiu, Romania
Ovidiu Cosma   Technical University of Cluj-Napoca, Romania
Cerasela Crişan   Vasile Alecsandri University of Bacău, Romania
Dan Cristea   Alexandru Ioan Cuza University, Romania
Gabriela Czibula   Babeş-Bolyai University, Romania
Daniela Dănciulescu   University of Craiova, Romania
Thierry Declerck   German Research Center for Artificial
                               Intelligence, Germany
Lyubomyr Demkiv   Lviv National Polytechnic University, Ukraine
Alexiei Dingli   University of Malta, Malta
Oleksandr Dorokhov   Kharkiv National University of Economics,
                               Ukraine
Tome Eftimov   Jožef Stefan Institute, Slovenia

George Eleftherakis   International Faculty of the University of
                             Sheffield, Greece
Călin Enăchescu   University of Medicine, Pharmacy, Science and
                             Technology of Târgu Mureş, Romania
Denis Enăchescu   University of Bucharest, Romania
Ralf Fabian   Lucian Blaga University of Sibiu, Romania
Stefka Fidanova   Bulgarian Academy of Sciences, Bulgaria
Ulrich Fiedler   Bern University of Applied Sciences, Switzerland
Martin Fränzle   Carl von Ossietzky University of Oldenburg,
                             Germany
Amir Gandomi   Michigan State University, USA
Dejan Gjorgjevikj   Ss. Cyril and Methodius University, Republic of
                             Macedonia
Teresa Gonçalves   University of Évora, Portugal
Andrina Granić   University of Split, Croatia
Katalina Grigorova   University of Ruse, Bulgaria
Axel Hahn   Carl von Ossietzky University of Oldenburg,
                             Germany
Piroska Haller   University of Medicine, Pharmacy, Science and
                             Technology of Târgu Mureş, Romania
Masafumi Hagiwara   Keio University, Japan
Tetiana Hladkykh   SoftServe, Ukraine
Daniel Hunyadi   Lucian Blaga University of Sibiu, Romania
Milena Karova   Technical University of Varna, Bulgaria
Sundaresan Krishnan Iyer   Infosys Limited, India
Marek Krótkiewicz   Wrocław University of Science and Technology,
                             Poland
Raka Jovanovic   Hamad Bin Khalifa University, Qatar
Saleema JS   Christ University, India
Adnan Khashman   European Centre for Research and Academic
                             Affairs, Cyprus
Wolfgang Kössler   Humboldt University of Berlin, Germany
Marcel Kyas   Reykjavik University, Iceland
Lixin Liang   Tsinghua University, China
Hans Peter Litz   University of Oldenburg, Germany
Suzana Loskovska   Ss. Cyril and Methodius University, Republic of
                             Macedonia
Vasyl Lytvyn   Lviv Polytechnic National University, Ukraine
Manuel Campos Martinez   University of Murcia, Spain
Gines Garcia Mateos   University of Murcia, Spain
Gerard de Melo   University of Potsdam, Germany
Matthew Montebello   University of Malta, Malta
Gummadi Jose Moses   Raghu Engineering College, India

Nicholas Muller	Technical University of Applied Sciences Wurzburg-Schweinfurt, Germany
Elena Nechita	Vasile Alecsandri University of Bacău, Romania
Eugénio Costa Oliveira	University of Porto, Portugal
Grażyna Paliwoda-Pękosz	Cracow University of Economics, Poland
Camelia Pintea	Technical University of Cluj-Napoca, Romania
Ivaylo Plamenov Penev	Technical University of Varna, Bulgaria
Horia Pop	Babeş-Bolyai University, Romania
Petrică Claudiu Pop Sitar	Technical University of Cluj-Napoca, Romania
Anca Ralescu	University of Cincinnati, USA
Mohammad Rezai	Sheffield Hallam University, UK
Abdel-Badeeh M. Salem	Ain Shams University, Egypt
Hanumat Sastry	University of Petroleum and Energy Studies, India
Willi Sauerbrei	University of Freiburg, Germany
Livia Sângeorzan	Transilvania University of Braşov, Romania
Vasile-Marian Scuturici	University of Lyon, France
Klaus Bruno Schebesch	Vasile Goldis Western University of Arad, Romania
Frank Schleif	Technical University of Applied Sciences Wurzburg-Schweinfurt, Germany
Soraya Sedkaoui	University of Khemis Miliana, Algeria
Francesco Sicurello	University of Milano-Bicocca, Italy
Andreas Siebert	University of Applied Sciences Landshut, Germany
Corina Simian	Whitehead Institute for Biomedical Research, USA
Dana Simian	Lucian Blaga University of Sibiu, Romania
Lior Solomovich	Kaye Academic College of Education, Israel
Srun Sovila	Royal University of Phnom Penh, Cambodia
Ansgar Steland	RWTH Aachen University, Germany
Laura Florentina Stoica	Lucian Blaga University of Sibiu, Romania
Florin Stoica	Lucian Blaga University of Sibiu, Romania
Detlef Streitferdt	Ilmenau University of Technology, Germany
Grażyna Suchacka	University of Opole, Poland
Ying Tan	Beijing University, China
Jolanta Tańcula	University of Opole, Poland
Mika Tonder	Saimaa University of Applied Sciences, Finland
Claude Touzet	Aix-Marseille University, France
Milan Tuba	Singidunum University, Serbia
Dan Tufiş	Romanian Academy, Romania
Alexander D. Veit	Harvard Medical School, USA
Anca Vasilescu	Transilvania University of Braşov, Romania

Sofia Visa	The College of Wooster, USA
Anca Vitcu	Alexandru Ioan Cuza University, Romania
Badri Vellambi	University of Cincinnati, USA
Xin-She Yang	Middlesex University London, UK

Sponsors

AUSY Technologies Romania

Fundatia Academia Ardeleana

Asociatia BIT

GSD

NTT DaTa
Trusted Global Innovator

NTT Data

Pan Food

ROPARDO

TopTech

wenglor
the innovative family

Wenglor

Plenary Lecture 1

A Technology of Deciphering Old Cyrillic-Romanian

Dan Cristea

Alexandru Ioan Cuza University Iasi, Romania
dcristea@info.uaic.ro

Abstract. Between the 16th century and the middle of the 19th, a unique Cyrillic alphabet has circulated on the territories of historical Romania, with slight variations in the shapes of graphemes or their phonetic values. As such, a huge bibliography of Cyrillic-Romanian texts has been accumulated in various libraries, while very few of these books have been transliterated by a small number of specialised linguists. The access of the large public of Romanian readers, interested in knowing these documents, is yet very restricted. This is why, an automatic deciphering of old Romanian documents from Cyrillic to Latin would be most welcome. I will present the DeLORo project (Deep Learning for Old Romanian), which aimed to build such a technology for printed and uncial Cyrillic-Romanian documents (not for manuscripts). In this talk I will describe the structure of DeLORo's data repository, which includes images of scanned pages, annotations operated collaboratively over them, and alignments between annotated objects in the images and (sequences of) decoded Latin characters. The primary data are used to train the deep learning recognition technology. Since the manual annotation process is very time consuming and the density of characters is highly non-uniform in documents, I also overview a strategy for data augmentation that exploits a collection of documents entirely transcribed by experts in other involvements than our project. Different phases of processing are applied over the images of pages, combining binarization and partial blurring operations with segmentation of the page image, detection of objects (as rows of text and characters) and labelling of characters. I will also show some results, as they will be reported at the end of the project (October 2022), for character detection and recognition, in comparison with other approaches.

Brief Biography of the Speaker: Dan Cristea is an emeritus professor of the Alexandru Ioan Cuza University of Iaşi (UAIC), Faculty of Computer Science, and still holds a part time position as a principal researcher at the Institute of Computer Science in the Iaşi branch of the Romanian Academy. He is a corresponding member of the Romanian Academy and a full member of the Academy of Technical Sciences of Romania. The main courses taught have been artificial intelligence, rule-based programming and techniques of natural language engineering. He is the initiator of the program of master studies in Computational Linguistics at UAIC. His research interests have been mainly related to discourse theories and applications (methods and techniques for anaphora resolution, discourse parsing, question answering, summarization), automatic config-uration of NLP architectures, language evolution, semantic representations of natural language, computerisation of dictionaries, lexical semantics. He was involved in the construction of resources for Romanian language and, recently, on computational stud-ies of the old Romanian language. Prof. Cristea has initiated and is the co-director of the EUROLAN series of Summer Schools on Human Language Technology (https:// eurolan.info.uaic.ro/2021/), which had 15 biennial editions since 1993, and co-initiated the series of international conferences on Linguistic Resources and Tools for Natural Language Processing, with its 17th edition this year (https://profs.info.uaic.ro/~consilr/ 2022/).

Plenary Lecture 2

Autonomous Drones and Analogue Missions

Marcel Kyas

Reykjavik University Reykjavik, Iceland
marcel@ru.is

Abstract. We report our involvement in an analog space mission testing unmanned aerial systems (UAS) intended for operating on Mars in Holuhraun, Iceland. Our goal in this mission was to test autonomous landing methods: the UAS has to find a suitable, safe landing site and land reliably. Iceland's lava fields, crevasses, and craters resemble the environment we expect on other planets. The environment allows us to experiment with revolutionary approaches to space exploration. At the same time, the mission allowed participating engineers, geologists, and computer scientists to express their views on mission design, equipment, and programming. Many components of such systems are mission and safety-critical. Their failure jeopardizes the mission and may result in tremendous financial loss. For example, the Perseverance launch is estimated to cost $2.9 billion. Consequently, the developers take utmost care to avoid all potential issues and follow strict coding standards like MISRA. Their goal to ensure predictable and deterministic behavior is at odds with revolutionary developments. We want to use machine learning for landing site identification and autonomous control of unmanned vehicles in planetary environments. Those systems are often unpredictable. Our position is that machine learning can be used safely in space missions and avionics, provided that they are shown not to jeopardize the system's operations or supervised by deterministic components or prioritized modularization.

Brief Biography of the Speaker: Marcel Kyas received his Ph.D. from the University of Leiden in 2006. In his dissertation "Verifying OCL specifications of UML models," he developed compositional, computer-aided verification methods for object-oriented real-time systems. After that, he became a PostDoc at the University of Oslo. He researched type systems for dynamically evolving distributed systems. He became an assistant professor at the Freie Universität Berlin. He studied indoor positioning systems and their empirical validation. He published competitive positioning methods and an indoor positioning test-bed. Since 2015, he has worked at Reykjavik University. He extends his work on indoor positioning systems to distributed systems. He works on autonomous unmanned aerial vehicles, especially landing in GPS-denied areas. His group participated in an analog mission to investigate equipment for geological sampling on Mars.

Plenary Lecture 3

Inside Neural Networks

Detlef Streitferdt

Technical University of Ilmenau Ilmenau, Germany
`detlef.streitferdt@tu-ilmenau.de`

Abstract. Machine learning became a prominent technology with artificial neural networks as its current and highly discussed model. Although the results of image analysis / recognition or speech detection are very promising, the analysis of neural networks itself and their behavior is still a very hard task due to the complexity of the models. Even a single software neuron puts high demands on the assessment of software quality aspects. Current models with thousands of interconnected neurons require by far more elaborated software tools and methods.

This talk gives a software engineering overview of the current state in analyzing neural networks within the software development life-cycle. It addresses the limits of using neural networks and emphasizes the corresponding pitfalls a software engineer has to cope with.

Brief Biography of the Speaker: Detlef Streitferdt is currently senior researcher at the Ilmenau University of Technology heading the research group Software Architectures and Product Lines since 2010. The research fields are the efficient development of software architectures and product lines, their analysis and their assessment as well as software development processes and model-driven development. Before returning to the University he was Principal Scientist at the ABB AG Corporate Research Center in Ladenburg, Germany. He was working in the field of software development for embedded systems. Detlef studied Computer Science at the University of Stuttgart and spent a year of his studies at the University of Waterloo in Canada. He received his doctoral degree from the Technical University of Ilmenau in the field of requirements engineering for product line software development in 2004.

Plenary Lecture 4

Recent Topics of Convolutional Neural Networks Applications

Milan Tuba

Singidunum University Belgrade, Serbia
tuba@ieee.org

Abstract. Artificial intelligence and machine learning algorithms have become a significant part of numerous applications used in various fields from medicine, and security, to agriculture, astronomy, and many more. In general, most of these applications require a classification algorithm, and often for the classification of digital images. Due to the wide need for classification methods and intensive study of the classification problem, numerous classification methods have been proposed and used. However, the convolutional neural networks have proven to be a far better method for certain classification problems and have brought some revolutionary changes in certain areas. Convolutional neural networks (CNNs) represent the type of deep artificial neural networks that, due to preserving spatial correlation in inputs, manage to significantly improve signal classification accuracy, especially of digital images. Using, creating and training CNN is a relatively simple task due to the various available software tools, but the problem with CNNs is finding the optimal configuration and architecture. Designing and tuning CNN represents a very challenging problem that should be dealt with in order to achieve the best possible results. The optimal CNN's configuration depends on the considered problem and one CNN that is good for one problem is not necessarily good for others. Finding the optimal configuration is not a simple task since there are numerous hyperparameters such as the number, type and order of layers, number of neurons in each layer, kernel size, optimization algorithm, padding, stride, and many others that should be fine-tuned for each classification problem. There is no unique efficient method for finding optimal values of CNNs' hyperparameters. A commonly used method for setting the CNN's configuration is to guess good starting values and estimate better values for the hyper-parameters (guesstimating). This method is simple but not the most efficient. Since this is an optimization problem, some recent studies tested different optimization metaheuristics such as swarm intelligence algorithms. Usage of swarm intelligence algorithms for finding CNNs' configuration can be time consuming but the improvement of the classification accuracy is significant. In this talk, the advantages and challenges of finding the optimal CNN configuration will be presented.

Brief Biography of the Speaker: Milan Tuba is the Vice Rector for International Relations, Singidunum University, Belgrade, Serbia and was the Head of the Department for Mathematical Sciences at State University of Novi Pazar and the Dean of the Graduate School of Computer Science at John Naisbitt University. He is listed in the World's Top 2% Scientists by Stanford University in 2020 and 2021. Prof. Tuba is the author or coauthor of more than 250 scientific papers (cited more than 5000 times, h-index 42) and editor, coeditor or member of the editorial board or scientific committee of a number of scientific journals and conferences. He was invited and delivered around 60 keynote lectures at international conferences.

He received B. S. in Mathematics, M. S. in Mathematics, M. S. in Computer Science, M. Ph. in Computer Science, Ph. D. in Computer Science from University of Belgrade and New York University. From 1983 to 1994 he was in the U.S.A. first at Vanderbilt University in Nashville and Courant Institute of Mathematical Sciences, New York University and later as Assistant Professor of Electrical Engineering at Cooper Union School of Engineering, New York. During that time he was the founder and director of Microprocessor Lab and VLSI Lab, leader of the NSF scientific projects and theses supervisor. From 1994 he was Assistant Professor of Computer Science and Director of Computer Center at University of Belgrade, from 2001 Associate Professor, Faculty of Mathematics, University of Belgrade, from 2004 also a Professor of Computer Science and Dean of the College of Computer Science, Megatrend University Belgrade. Prof. Tuba was the principal creator of the new curricula and programs at the Faculty of Mathematics and Computer Science at the University of Belgrade and later at John Naisbitt University where he was the founder and practically alone established a complete new school with bachelor, master and PhD program. He taught more than 20 graduate and undergraduate courses, from VLSI Design and Computer Architecture to Computer Networks, Operating Systems, Artificial Intelligence, Image Processing, Calculus and Queuing Theory.

His research interest includes nature-inspired optimizations applied to image processing, computer networks, and neural networks. He is a member of the ACM, IEEE, AMS, SIAM, IFNA, IASEI.

Contents

Intelligent Systems for Decision Support

Effective LSTM Neural Network with Adam Optimizer for Improving
Frost Prediction in Agriculture Data Stream 3
    Monika Arya and G. Hanumat Sastry

Gaze Tracking: A Survey of Devices, Libraries and Applications 18
    Edwin Cocha Toabanda, María Cristina Erazo, and Sang Guun Yoo

Group Decision-Making Involving Competence of Experts in Relation
to Evaluation Criteria: Case Study for e-Commerce Platform Selection 42
    Zornitsa Dimitrova, Daniela Borissova, Rossen Mikhov,
    and Vasil Dimitrov

Transparency and Traceability for AI-Based Defect Detection in PCB
Production 54
    Ahmad Rezaei, Johannes Richter, Johannes Nau, Detlef Streitferdt,
    and Michael Kirchhoff

Tasks Management Using Modern Devices 73
    Livia Sangeorzan, Nicoleta Enache-David, Claudia-Georgeta Carstea,
    and Ana-Casandra Cutulab

Machine Learning

A Method for Target Localization by Multistatic Radars 89
    Kiril Alexiev and Nevena Slavcheva

Intrusion Detection by XGBoost Model Tuned by Improved Social
Network Search Algorithm 104
    Nebojsa Bacanin, Aleksandar Petrovic, Milos Antonijevic,
    Miodrag Zivkovic, Marko Sarac, Eva Tuba, and Ivana Strumberger

Bridging the Resource Gap in Cross-Lingual Embedding Space 122
    Kowshik Bhowmik and Anca Ralescu

Classification of Microstructure Images of Metals Using Transfer Learning 136
    Mohammed Abdul Hafeez Khan, Hrishikesh Sabnis,
    J. Angel Arul Jothi, J. Kanishkha, and A. M. Deva Prasad

Generating Jigsaw Puzzles and an AI Powered Solver 148
  Stefan-Bogdan Marcu, Yanlin Mi, Venkata V. B. Yallapragada,
  Mark Tangney, and Sabin Tabirca

Morphology of Convolutional Neural Network with Diagonalized Pooling 161
  Roman Peleshchak, Vasyl Lytvyn, Oleksandr Mediakov,
  and Ivan Peleshchak

Challenges and Opportunities in Deep Learning Driven Fashion Design
and Textiles Patterns Development 173
  Dana Simian and Felix Husac

Feature Selection and Extreme Learning Machine Tuning by Hybrid Sand
Cat Optimization Algorithm for Diabetes Classification 188
  Marko Stankovic, Nebojsa Bacanin, Miodrag Zivkovic,
  Dijana Jovanovic, Milos Antonijevic, Milos Bukmira,
  and Ivana Strumberger

Enriching SQL-Driven Data Exploration with Different Machine Learning
Models 204
  Sabina Surdu

Mathematical Models for Development of Intelligent Systems

Analytical Solution of the Simplest Entropiece Inversion Problem 221
  Jean Dezert, Florentin Smarandache, and Albena Tchamova

Latent Semantic Structure in Malicious Programs 234
  John Musgrave, Temesguen Messay-Kebede, David Kapp,
  and Anca Ralescu

Innovative Lattice Sequences Based on Component by Component
Construction Method for Multidimensional Sensitivity Analysis 247
  Venelin Todorov and Slavi Georgiev

On an Optimization of the Lattice Sequence for the Multidimensional
Integrals Connected with Bayesian Statistics 264
  Venelin Todorov and Slavi Georgiev

Modelling and Optimization of Dynamic Systems

Numerical Optimization Identification of a Keller-Segel Model
for Thermoregulation in Honey Bee Colonies in Winter 279
  Atanas Z. Atanasov, Miglena N. Koleva, and Lubin Vulkov

Gradient Optimization in Reconstruction of the Diffusion Coefficient
in a Time Fractional Integro-Differential Equation of Pollution in Porous
Media 294
  Tihomir Gyulov and Lubin Vulkov

Flash Flood Simulation Between Slănic and Vărbilău Rivers in Vărbilău
Village, Prahova County, Romania, Using Hydraulic Modeling and GIS
Techniques 309
  Cristian Popescu and Alina Bărbulescu

Author Index 329

Intelligent Systems for Decision Support

Effective LSTM Neural Network with Adam Optimizer for Improving Frost Prediction in Agriculture Data Stream

Monika Arya[1]([⊠]) [iD] and G. Hanumat Sastry[2] [iD]

[1] Department of Computer Science and Engineering, Bhilai Institute of Technology, Durg, Chhattisgarh, India
arya.akshara@gmail.com
[2] School of Computer Science, University of Petroleum and Energy Studies, Dehradun, India
hsastry@ddn.upes.ac.in

Abstract. A country's economic progress would be impossible without the agriculture sector. The primary source of income for most countries is agricultural products and related enterprises. Farming is heavily influenced by weather factors like the amount of sunlight, kind of precipitation, temperature of the air, relative humidity, and wind speed, as well as fluctuations in these factors. The temperature directly influences the metabolic reactions that take place in plants. For example, freezing damages plants to the point where they cannot grow (frost). Farmers have traditionally monitored the weather in the spring by watching television, reading the newspaper, or following a detailed weather forecast.

Farm sustainability and output can be improved through smart agriculture technologies. Precision Farming (PF) can help farmers deal with various environmental issues instead of traditional agricultural approaches. It is possible to monitor farming conditions using sensors installed in the farmland area. This system requires predictive systems to increase yield. There is much interest in the field of prediction. For example, farmers can prevent crop frost by using anti-frost measures, and this research has developed a smart deep learning-based system for agricultural frost forecasting.

This paper uses Long-Short Term Memory (LSTM) neural networks for the time series prediction of low temperatures. Additionally, the LSTM model is combined with an Adam optimizer to intensify the prediction model's performance. The suggested approach is also compared with LSTM when combined with other optimizers. The findings indicate that the proposed model excels in base LSTM and LSTM with other optimizers.

Keywords: Precision farming · Artificial intelligence · Deep learning · Smart agriculture · LSTM · IoT

1 Introduction

The agriculture industry contributes significantly to a country's economic development [1]. In addition, most nations derive revenue from agricultural products and allied industries [2]. Therefore, one of the most basic and vital necessities for any country is the

D. Simian and L. F. Stoica (Eds.): MDIS 2022, CCIS 1761, pp. 3–17, 2023.
https://doi.org/10.1007/978-3-031-27034-5_1

protection and safety of agricultural commodities. Malnourishment has long been a concern in developing countries such as India and is strongly related to achieving food security. Farming is also affected by climatic elements such as sunlight and precipitation and changes in air temperature, relative humidity and wind [3]. In plants, the temperature has a direct effect on the biochemical activities that take place.

At 0 °C (frost), the climate becomes harsh for plant growth and development. Plants get severely damaged by freezing. Frost can have little or no effect on plants at times, but it can also have devastating effects on plants at other times. Only a few numbers of plants can withstand prolonged exposure to freezing temperatures. All new plant growth, flower buds, and freshly developed fruits can be wiped out if it's "strong". Temperatures that fall below freezing can harm or destroy many plants. "light frost" refers to temperatures between 28- and 32 °F, whereas "hard frost" refers to temperatures below 28 °F. The hard frost affects plants severely compared to light frost. The damage depends on the type of plant, the exposed tissue, and the temperature. Traditionally, farmers have kept tabs on the weather in the spring through television, newspaper, or a detailed weather forecast.

Smart agriculture technologies are powerful tools for enhancing farm sustainability and output [4]. However, even though crop yield losses have been minimized, conventional farming still has some drawbacks. Nutrient deficiencies, algorithm efficacy, inappropriate analysis, and the wrong parameter selection are all issues with contemporary farming practices and technology-based solutions. These issues have an impact on crop yields. Instead of traditional agricultural methods, Precision Farming (PF) can assist farmers in dealing with a wide range of problems related to the environment. PF is using AI to improve the quality and accuracy of the harvest as a whole [5]. In addition, AI technology aids in the detection of pests, plant diseases, and undernutrition on farms [6].

Daily, farms collect thousands of information about the temperature, soil, amount of water used, weather, etc. Then, in conjunction with the artificial intelligence and machine learning (ML) approach, this data is used to learn useful things like when to plant seeds, what crops to grow, which hybrid seeds can produce the most, and so on.

IoT solutions help farmers by ensuring high yields and profitability and protecting the environment. Using the Internet of Things (IoT), any device can communicate with a server through the Internet [7]. Data on agricultural areas can be gathered using IoT sensors. Sensors installed in the agricultural environment are perhaps utilized to monitor agricultural conditions. Soil moisture, temperature, and humidity can be measured with various sensors to avoid significant agricultural losses and increase agricultural output [8]. This technology allows farmers to monitor the actual health of the crops from a distance, eliminating the need for them to be on-site physically.

The sensor's readings are relayed to the server and entered into a database once they arrive [9]. Predictive systems are needed in this system to boost yield. Prediction is an intriguing area of research. Deep learning approaches offer a new framework for decision-making procedures frequently employed in numerous fields, including prediction and classification [10]. They are using decision-making technology in precision agriculture, which results in better productivity, lower expenses, and a more significant financial advantage for farmers.

Due to temperature changes and unexpected weather conditions, frost prediction has become one of the primary endeavors of agricultural market researchers and practitioners. As a result, frost prediction and forecasting are critical for decision-makers to establish anti-frost strategies and appropriate preventive measures. Frost prediction is volatile and influenced by various elements such as temperature, humidity, and season. Many algorithms have been used in Frost prediction and forecasting. However, the prediction accuracy needs to be enhanced. The prediction of time series is an uncertain modelling challenge.

This research proposes an intelligent deep-learning approach for predicting frost to enable farmers to protect the crop from frost and to use anti-frost strategies to safeguard the crop. The suggested model is intended to forecast low temperatures and is based on an LSTM model with an Adam optimizer. LSTM network is a sort of recurrent neural network (RNN) used in deep learning. LSTM, a superior architecture to conventional RNNs, enables the model to capture long-term dependencies well [11]. However, LSTM has had problems improving frost prediction accuracy. Many sectors of research and engineering use stochastic gradient-based optimization. Adam is a stochastic optimization approach that only needs first-order gradients and low memory. Adam, a method for effective stochastic optimization, improves the performance of an LSTM model by combining the strengths of two well-known optimization techniques, Adagrad and RMSprop. The model is built using data from an IoT architecture.

The remainder of the paper is structured as follows: Sect. 2 addresses the linked works and summarizes the methodology, datasets and evaluation metrics utilized in previous works. The methodology is explained in Sect. 3. Section 4 elaborates on the datasets, experiments and results and compares prediction errors. Finally, Sect. 5 surmised the research and looked at the future scope.

2 Literature Review

An LSTM neural network was compared for time series prediction in smart agriculture with the backpropagation technique in [12]. It was concluded that, in terms of the accuracy of predictions, LSTM has the potential to exceed Back-Propagation for an intelligent agriculture dataset. Deep Bid-LSTM was suggested in [13] to enhance moisture (SM) and soil electrical conductivity (SEC) forecasts using environmental information data, offering a valuable benchmark for irrigation and fertilization. A prediction model for crop recommendations, RDA-Bi-LSTM-EERNN that is optimized using RDA biases and weights was presented in [14]. The temperature prediction model was proposed in [15]. The model was built using an RNN called LSTM neural network. CLSTM networks were assessed for crop recognition in [16] using four datasets. LSTM neural networks, which forecast time series accurately, were used in [17] to create a reliable forecasting model. The LSTM model is associated with a smoothing approach for pre-processing data to boost prediction accuracy. To deal with the problems caused by the nonlinearity and non-stationarity of time series data, researchers in [18] developed a model combining the Seasonal-Trend decom-position using LOESS (STL) pre-processing approach and an attention mechanism based on LSTM to decompose seasonal trends. The most recent studies for frost prediction using deep learning approaches are listed in Table 1.

Table 1. List of the recent studies for frost prediction.

Author and Year	Methodology	Dataset	Evaluation parameters
Zhou et al. (2022) [19]	This paper presents an RNN technique consisting of the standard RNN, LSTM, and gated recurrent unit to optimize frost prediction	NSW and ACT of Australia	Mean Square Error (MSE)
Cadenas et al. [20] (2019)	This work proposes a method for regression based on the k-nearest neighbours and two methods for classification based on the FDTii and kNNimp	Data from various meteorological stations in the Region of Murcia were used for this work (south-east Spain)	Accuracy, MAE, MSE
Guillén et al. [21] (2021)	This work develops an LSTM-based deep learning model for frost forecasting	Data from various meteorological stations in the Region of Murcia were used for this work (south-east Spain)	Accuracy, quadratic error, determination coefficient R2
Castañeda-Miranda et al. [22] (2017)	In this work, a Multi-Layer Perceptron (MLP) ANN was used to predict temperature, which was then trained with backpropagation	Data is obtained from the central region of Mexico for this study	Determination coefficient R2
Diedrichs et al. [23] (2018)	In this work, the training data for the model was supplemented with SMOTE data to compensate for the paucity of frost data	DACC provided data from five meteorological stations for this study	Sensitivity, precision and F1

2.1 Motivation

The ADAM algorithm converges faster than the traditional stochastic gradient de-scent to find the best weights at every level. The first-order gradient descent modified ADAM optimization algorithm is suitable for models with many parameters and is easy to run on a computer. Optimizing LSTM with Adam makes the LSTM model perform even better.

2.2 Contribution

1. The main contribution of this work is to propose and implement an effective LSTM neural network with an Adam optimizer for improving frost prediction in the agriculture data stream.
2. The results of the proposed work are compared with LSTM and LSTM with other optimizers for frost prediction over significant parameters.
3. This study intends to boost crop yields, monitor crops in real time, and assist farmers in making better decisions for applying anti-frost techniques.

3 Proposed Work

Most of the frost prediction models that have been used in the past use either RNN, ML, or LSTM. Standard RNNs are sometimes challenging to train to handle issues necessitating the identification of long-term temporal dependencies. LSTM is better than RNN because it can keep information in its memory for a longer time than RNN. But it takes longer to train LSTMs. This is mainly because the LSTM layer works sequentially. The previous works used traditional Stochastic Gradient Decent (SGD) for optimization. Despite being easy to implement and effective, SGD may not have optimal convergence performance because it only employs a single learning rate overall gradient coordinate. As a result, SGD is more likely to get caught in a local minimum and converge more slowly [24]. Since its introduction as an extension to SGD, the Adam optimization technique has seen increasing use in deep learning projects, including computer vision and natural language processing. Adam employs adaptive learning rates and momentum to help it converge more quickly. Adam tries to fine-tune the learning rate for each parameter, while SGD keeps the rates constant across the board. The Adam optimizer produces superior results to conventional optimization methods, is faster to compute, and has fewer tuning parameters. As a result, the LSTM model and Adam optimizer are coupled in the suggested technique to enhance the prediction model's performance.

3.1 LSTM

LSTM functions similarly to an RNN cell. LSTM has a hidden state H referred to as "Short term memory". H(t −1) and H(t) represent the hidden state before timestamp t and at timestamp t, respectively. The cell state of LSTM referred to as "Long term memory", is indicated by C(t − 1) and C(t) for the previous and current timestamps. The LSTM is composed of three components.

The information from the previous timestamp is evaluated in the first part to see if it needs to be remembered or if it can be discarded. In the subsequent stage, the cell strives to gain new knowledge from the data it has acquired from the first phase. In the final stage, the acquired knowledge is communicated from the present timestamp to the following timestamp by the cell. The three LSTM cell components are referred to as gates. The Forget gate is the first component, the Input gate is the second, and the Output gate is the third. Figure 1 gives the details of the components of the LSTM cell [25].

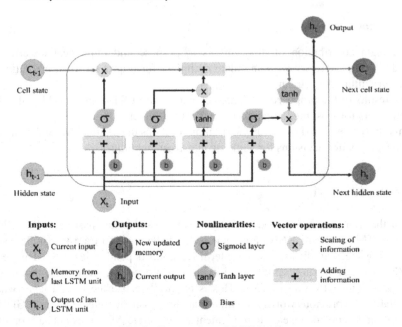

Fig. 1. Components of LSTM cell.

The equation for forget gate is given in Eq. 1.

$$f_t = \sigma \left(x_t * u_f + H_{t-1} * W_f \right) \tag{1}$$

where,

- x_t is the current input
- u_f is the weight of input
- H_{t-1} is the hidden state
- W_f is the weight matrix of the hidden state.

After that, a sigmoid function is applied that results in f_t. *where* $0 < f_t < 1$
The equation for the input gate is given in Eq. 2.

$$i_t = \sigma \left(x_t * u_i + H_{t-1} * W_i \right) \tag{2}$$

where,

- u_i is the weight matrix of input
- H_{t-1} is the hidden state
- W_i is the weight matrix of the input state

Again, a sigmoid function is applied that results in i_t. Where $0 < i_t < 1$.
The equation for the output gate is given in Eq. 3.

$$o_t = \sigma \left(x_t * u_o + H_{t-1} * W_o \right) \tag{3}$$

where,

- u_0 is the weight matrix of input
- H_{t-1} is the hidden state
- W_0 is the weight matrix of the input state

Due to the sigmoid function, the value o_t also lies between 0 and 1.

Optimizers are critical in boosting the accuracy of the LSTM. Numerous optimizer variations like Stochastic Gradient Descent (SGD), Nesterov accelerated Gradient, Adagrad, AdaDelta, RMSProp, and Adam can be employed. RMSProp, AdaDelta, and Adam are all algorithms that are very similar to each other. However, since Adam was found to perform slightly better than RMSProp, it is usually the best choice overall.

3.2 Adam Optimizer

Adam uses exponentially moving averages, produced based on the gradient evaluated on a current mini-batch, to calculate the moments. The formula to calculate the moments is given in Eq. 4 and Eq. 5:

$$p_t = m_1 * p_{t-1} + (1 - m_1).g_t \tag{4}$$

$$q_t = m_2 * q_{t-1} + (1 - m_2).g_t^2 \tag{5}$$

where,

- p and q are moving averages
- g is the gradient of the current batch
- m_1 and m_2 are hyperparameters

At the beginning of the first iteration, the values of the moving average vectors are all set to zero. Then, the equation given in Eq. 6 is used to do a weight update:

$$w_t = w_{t-1} - \eta \frac{\widehat{p_t}}{\sqrt{\widehat{q_t}} + \epsilon} \tag{6}$$

where,

- W is the weight of the model
- η is the size of steps that depends on iteration.

3.3 Proposed LSTM Neural Network with Adam Optimizer

The steps to develop the proposed LSTM model for frost prediction typically include:

1. Data pre-processing- For data, pre-processing normalization is done. Normalization is essential when dealing with attributes on multiple scales; otherwise, it may dilute the impact of an essential attribute (on a lower scale) due to another attribute's more significant scale values.
2. Initializing the parameters-Hyper parameters like Learning rate (LR) value, Number of epochs, Number of LSTM units, loss function, dropout rate and Maximum Epoch are initialized before training the model. The hyperparameter values set for the current work are given in Table 2.

Table 2. Hyperparameter values.

Hyperparameters	Value
Optimizer	Adam
Hyperparameter values	$m_1 = 0.9, m_2 = 0.999$
LR	$1 * 10^{-3}$
Number of epochs	100
Number of LSTM units	10
Loss function	Cross entropy
Dropout rate	0.2

3. Training the LSTM network using ADAM optimization- for training the proposed model, all the functions of gate units like input gate, forget gate and cell gate are calculated using Eqs. 1, 2 and 3. Finally, the linear activation function in the output layer is calculated with the formula:

$$f(x) = a.x \tag{7}$$

Subsequent optimization is done by Adam optimization for updating the system weight and bias. After that, repeat the above steps for a specified number of epochs and then stop.
4. Comparison of the results-The results are compared with the performance of LSTM using different optimizers using significant metrics.

Figure 2 gives the detailed workflow of the LSTM + Adam model.

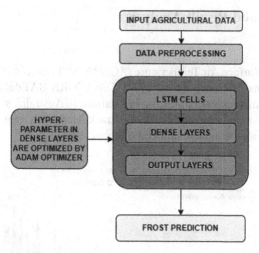

Fig. 2. Flowchart for the proposed work.

The algorithm for the proposed work is given below.

BEGIN
1. Initialize the hyperparameters of LSTM (refer to table 2)
2. Initialize weight W$_t$.
3. Set the maximum training time
4. Pre-process the input data
5. Input the training data
6. **REPEAT**
7.   i) Calculate the output: $y = LSTM(x)$;
8.   ii) Calculate the loss function $Loss = \frac{1}{T}\sum_{i=1}^{T}(y_i - y_i')^2$;
9.   iii) While W$_t$ do not converge, Do
10.   a) Calculate the gradient $g_t = \frac{\partial f}{\partial w}(x, w)$;
11.   b) Calculate $p_t = m_1 * p_{t-1} + (1 - m_1).g_t$;
12.   c) Calculate $q_t = m_2 * q_{t-1} + (1 - m_2).g_t^2$;
13.   d) Calculate $\widehat{p_t} = {p_t}/{(1 - m_t^2)}$;
14.   e) Calculate $\widehat{q_t} = {q_t}/{(1 - m_t^2)}$;
15.   f) Update the parameter $w_t = w_{t-1} - \eta\frac{\widehat{p_t}}{\sqrt{\widehat{q_t}}+\epsilon}$;
16.   iv) Return W$_t$
17. **UNTIL** (Loss meets the termination condition or training time reaches the maximum)
END

---

4 Experiment and Result Analysis

4.1 Dataset

This study uses the Surface Air Temperature (ACORN-SAT) dataset from the Australian Climate Observations Reference Network [24]. The ACORN-SAT dataset contains daily maximum and minimum temperatures for 112 stations in Australia, starting from 1910. The Bureau of Meteorology uses this dataset to track Australia's long-term temperature trends. Figure 3 shows the annual temperature from 1910 to 2020.

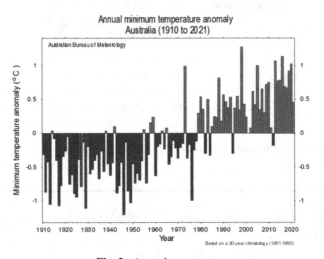

Fig. 3. Annual temperature

For training and testing, the proposed model only used 10 years (from 2010–2020) of data.

4.2 Results and Comparison

To assess the effectiveness of the suggested model following loss functions are used:

• Mean squared error

$$MSE = \frac{1}{T} \sum_{i=1}^{T} \left(\hat{x}_t - x_t \right)^2 \tag{8}$$

• Root mean squared error

$$RMSE = \sqrt{\frac{1}{T} \sum_{i=1}^{T} \left(\hat{x}_t - x_t \right)^2} \tag{9}$$

• Mean absolute error

$$MAE = \frac{1}{T} \sum_{i=1}^{T} \left| \hat{x}_t - x_t \right| \tag{10}$$

- Mean absolute percentage error

$$MAPE = \frac{1}{T} \sum_{i=1}^{T} \frac{|x_t - \hat{x}_t|}{|x_t|} * 100 \qquad (11)$$

The performance is better when the loss function is less.

The efficacy of the suggested model (LSTM + Adam) is compared with the LSTM model and LSTM with other optimizers like Stochastic Gradient Descent (SGD), Adagrad, and RMSProp. The results of the proposed model and LSTM with other optimizers are summarized in Table 3.

Table 3. Comparison of prediction error.

Year	Method	MSE	RMSE	MAE	MAPE
2010–2011	LSTM	1.43	3.67	3.54	4.13
	LSTM+SGD	0.89	2.88	2.10	3.87
	LSTM+Adagrad	0.78	2.76	3.41	3.71
	LSTM+RMSProp	0.63	2.56	1.83	3.62
	LSTM+Adam (proposed)	**0.42**	**2.43**	**1.59**	**3.54**
2011–2012	LSTM	1.33	4.68	4.34	3.43
	LSTM+SGD	0.79	3.88	3.10	4.87
	LSTM+Adagrad	0.68	2.65	4.42	3.91
	LSTM+RMSProp	0.62	3.56	2.83	4.62
	LSTM+Adam (proposed)	**0.32**	**2.05**	**2.39**	**2.54**
2012–2013	LSTM	1.43	2.63	4.34	5.33
	LSTM+SGD	0.79	1.17	4.15	4.68
	LSTM+Adagrad	0.68	1.08	3.56	4.75
	LSTM+RMSProp	0.53	1.06	2.83	4.62
	LSTM+Adam (proposed)	**0.22**	**0.98**	**2.59**	**3.63**
2013–2014	LSTM	4.43	6.67	5.29	6.13
	LSTM+SGD	0.89	1.60	2.10	3.87
	LSTM+Adagrad	0.78	1.71	3.41	3.71
	LSTM+RMSProp	0.63	2.18	1.83	3.62
	LSTM+Adam (proposed)	**0.42**	**1.02**	**1.59**	**3.54**
2014–2015	LSTM	1.23	2.87	2.52	3.53
	LSTM+SGD	0.79	2.52	1.65	2.47
	LSTM+Adagrad	0.68	2.47	1.62	2.31
	LSTM+RMSProp	0.43	2.32	1.54	1.62
	LSTM+Adam (proposed)	**0.32**	**2.03**	**1.54**	**1.34**
2015–2016	LSTM	5.43	6.67	5.54	6.13
	LSTM+SGD	3.89	5.88	4.10	5.87
	LSTM+Adagrad	2.78	4.76	3.41	5.71
	LSTM+RMSProp	2.63	4.56	2.83	4.62
	LSTM+Adam (proposed)	**1.42**	**3.43**	**2.59**	**4.54**

(continued)

Table 3. (*continued*)

Year	Method	MSE	RMSE	MAE	MAPE
2016–2017	LSTM	4.23	5.63	6.13	5.12
	LSTM+SGD	3.79	4.84	5.34	4.77
	LSTM+Adagrad	3.75	4.56	4.31	4.51
	LSTM+RMSProp	2.43	3.76	3.65	3.65
	LSTM+Adam (proposed)	**2.12**	**3.43**	**2.19**	**3.03**
2017–2018	LSTM	23.58	22.67	13.54	14.63
	LSTM+SGD	19.89	17.88	12.14	13.67
	LSTM+Adagrad	7.78	12.76	8.41	8.17
	LSTM+RMSProp	3.43	4.16	4.83	2.62
	LSTM+Adam (proposed)	**3.02**	**3.45**	**2.19**	**2.54**
2018–2019	LSTM	5.43	10.41	6.34	12.02
	LSTM+SGD	5.19	7.76	5.10	9.87
	LSTM+Adagrad	4.58	5.08	6.11	3.80
	LSTM+RMSProp	4.13	4.66	4.23	3.50
	LSTM+Adam (proposed)	**3.12**	**3.43**	**3.59**	**2.54**
2019–2020	LSTM	22.43	11.17	9.54	13.73
	LSTM+SGD	20.89	10.88	7.10	11.17
	LSTM+Adagrad	10.78	3.16	4.41	9.71
	LSTM+RMSProp	3.63	2.62	3.33	2.12
	LSTM+Adam (proposed)	**2.42**	**2.13**	**2.39**	**2.00**

The average results of the 10 years findings are summarized in Table 4.

Table 4. Average results of the 10 years findings.

Method	MSE	RMSE	MAE	MAPE
LSTM	5.26	6.04	5.28	6.61
LSTM+SGD	4.21	4.60	3.86	5.71
LSTM+Adagrad	2.83	3.13	3.85	4.68
LSTM+RMSProp	1.74	3.03	2.76	3.55
LSTM+Adam (proposed)	1.19	2.32	2.27	2.96

Figure 4 compares the average prediction error of 10 years, i.e., from 2010 to 2020.

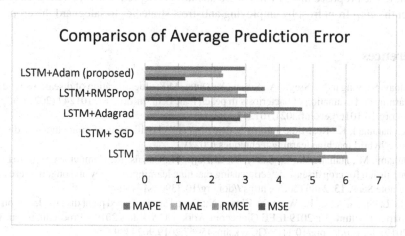

Fig. 4. Comparison of average prediction error.

The above graph denotes that the prediction errors of the proposed LSTM + Adam model for frost prediction outperform base LSTM and LSTM with other optimizers.

The proposed LSTM + Adam model for frost prediction is also compared with the recent and related techniques for frost prediction on parameters like MAE and MSE. Table 5 below presents the comparison.

Table 5. Comparison of proposed model with recent techniques for frost prediction.

Compared works	MSE	MAE	Technique used
Zhou et al. (2022) [24]	0.7813	NA	RNN technique consisting of the standard RNN, LSTM, and gated recurrent unit to optimize frost prediction was used
Cadenas et al. [20] (2019)	0.4870	0.5122	A method for regression based on the k-nearest neighbors was used
Proposed (LSTM+Adam)	1.19	2.27	LSTM with Adam optimizer was used

5   Conclusion and Future Scope

LSTM neural network has garnered substantial interest in language modelling, speech recognition, and natural language inference. It combines the advantages of both Ada-Grad, which is effective with sparse gradients, and RMSProp, which is effective in online and non-stationary environments. This research proposes an improved forecasting model based on Adam-optimized LSTM. The proposed approach is also compared with the performance of combining LSTM with other optimizers. Results concluded that the Adam optimizer reduces the LSTM model's prediction error and thus improves the prediction accuracy. The proposed model will enable the farmers to predict the frost accurately and protect the crop from frost by employing anti-frost strategies to safeguard the crop.

References

1. Yadav, S., Sengar, N., Singh, A., Singh, A., Dutta, M.K.: Identification of disease using deep learning and evaluation of bacteriosis in peach leaf. Ecol. Inform. **61**, 101247 (2021). https://doi.org/10.1016/j.ecoinf.2021.101247
2. Seetharaman, K.: Real-time automatic detection and classification of groundnut leaf disease using hybrid machine learning techniques (2022)
3. Ouhami, M., Hafiane, A., Es-Saady, Y., El Hajji, M., Canals, R.: Computer vision, IoT and data fusion for crop disease detection using machine learning: a survey and ongoing research. Remote Sens. **13**, 2486 (2021). https://doi.org/10.3390/rs13132486
4. Ale, L., Sheta, A., Li, L., Wang, Y., Zhang, N.: Deep learning based plant disease detection for smart agriculture. In: 2019 IEEE Globecom Work. GC Wkshps 2019 – Proceedings, pp. 1–6 (2019). https://doi.org/10.1109/GCWkshps45667.2019.9024439
5. Dyson, J., Mancini, A., Frontoni, E., Zingaretti, P.: Deep learning for soil and crop segmentation from remotely sensed data. Remote Sens. **11**, 7–9 (2019). https://doi.org/10.3390/rs11161859
6. Karar, M.E., Alsunaydi, F., Albusaymi, S., Alotaibi, S.: A new mobile application of agricultural pests recognition using deep learning in cloud computing system. Alex. Eng. J. **60**, 4423–4432 (2021). https://doi.org/10.1016/j.aej.2021.03.009
7. Maduranga, M.W.., Abeysekera, R.: Machine learning applications in IoT based agriculture and smart farming: a review. Int. J. Eng. Appl. Sci. Technol. **04**, 24–27 (2020). https://doi.org/10.33564/ijeast.2020.v04i12.004
8. Pang, H., Zheng, Z., Zhen, T., Sharma, A.: Smart farming: an approach for disease detection implementing iot and image processing. Int. J. Agric. Environ. Inf. Syst. **12**, 55–67 (2021). https://doi.org/10.4018/IJAEIS.20210101.oa4
9. Moso, J.C., Cormier, S., de Runz, C., Fouchal, H., Wandeto, J.M.: Anomaly detection on data streams for smart agriculture. Agric. **11**, 1–17 (2021). https://doi.org/10.3390/agriculture11111083
10. Magomadov, V.S.: Deep learning and its role in smart agriculture. J. Phys. Conf. Ser. **1399** (2019). https://doi.org/10.1088/1742-6596/1399/4/044109
11. Nguyen, T.T., et al.: Monitoring agriculture areas with satellite images and deep learning. Appl. Soft Comput. J. **95**, 106565 (2020). https://doi.org/10.1016/j.asoc.2020.106565
12. Suryo Putro S, B.C., Wayan Mustika, I., Wahyunggoro, O., Wasisto, H.S.: Improved time series prediction using LSTM neural network for smart agriculture application. In: Proceedings - 2019 5th International Conference on Science and Technology, ICST 2019, pp. 6–9 (2019). https://doi.org/10.1109/ICST47872.2019.9166401
13. Gao, P., et al.: Improved soil moisture and electrical conductivity prediction of citrus orchards based on IoT using deep bidirectional LSTM (2021)
14. Education, M., Mythili, K., Rangaraj, R., Coimbatore, S.: A Swarm based bi-directional LSTM-Enhanced elman recurrent neural network algorithm for better crop yield in precision agriculture. Turk. J. Comput. Math. Educ. (TURCOMAT) **12**, 7497–7510 (2021)
15. Guillén-Navarro, M.A., Martínez-España, R., Llanes, A., Bueno-Crespo, A., Cecilia, J.M.: A deep learning model to predict lower temperatures in agriculture. J. Ambient Intell. Smart Environ. **12**, 21–34 (2020). https://doi.org/10.3233/AIS-200546

16. De MacEdo, M.M.G., Mattos, A.B., Oliveira, D.A.B.: Generalization of convolutional LSTM models for crop area estimation. IEEE J. Sel. Top. Appl. Earth Obs. Remote Sens. **13**, 1134–1142 (2020). https://doi.org/10.1109/JSTARS.2020.2973602

17. Haider, S.A., et al.: LSTM neural network based forecasting model for wheat production in Pakistan. Agronomy **9**, 1–12 (2019). https://doi.org/10.3390/agronomy9020072

18. Yin, H., Jin, D., Gu, Y.H., Park, C.J., Han, S.K., Yoo, S.J.: STL-ATTLSTM: vegetable price forecasting using stl and attention mechanism-based LSTM. Agric. **10**, 1–17 (2020). https://doi.org/10.3390/agriculture10120612

19. Zhou, P., Feng, J., Ma, C., Xiong, C., Hoi, S., Weinan, E.: Towards theoretically understanding why SGD generalizes better than ADAM in deep learning. Adv. Neural Inf. Process. Syst. **33**, 21285–21296 (2020)

20. Cadenas, J.M., Garrido, M.C., Martínez-España, R., Guillén-Navarro, M.A.: Making decisions for frost prediction in agricultural crops in a soft computing framework. Comput. Electron. Agric. **175**, 105587 (2020). https://doi.org/10.1016/j.compag.2020.105587

21. Guillén, M.A., et al.: Performance evaluation of edge-computing platforms for the prediction of low temperatures in agriculture using deep learning. J. Supercomput. **77**(1), 818–840 (2020). https://doi.org/10.1007/s11227-020-03288-w

22. Castañeda-Miranda, A., Castaño, V.M.: Smart frost control in greenhouses by neural networks models. Comput. Electron. Agric. **137**, 102–114 (2017). https://doi.org/10.1016/j.compag.2017.03.024

23. Diedrichs, A.L., et al.: Prediction of Frost events using Bayesian networks and random forest to cite this version: HAL Id: hal-01867780 prediction of frost events using Bayesian networks and random forest (2019)

24. Zhou, I., Lipman, J., Abolhasan, M., Shariati, N.: Minute-wise frost prediction: an approach of recurrent neural networks. Array **14**, 100158 (2022). https://doi.org/10.1016/j.array.2022.100158

25. Ho, H.V., Nguyen, D.H., Le, X.H., Lee, G.: Multi-step-ahead water level forecasting for operating sluice gates in Hai Duong, Vietnam. Environ. Monit. Assess. **194**, 251 (2022)

Gaze Tracking: A Survey of Devices, Libraries and Applications

Edwin Cocha Toabanda[1,2], María Cristina Erazo[1,2], and Sang Guun Yoo[1,2(✉)] (iD)

[1] Departamento de Informática y Ciencias de la Computación, Escuela Politécnica Nacional,
Quito, Ecuador
{edwin.cocha,maria.erazo,sang.yoo}@epn.edu.ec
[2] Smart Lab, Escuela Politécnica Nacional, Quito, Ecuador

Abstract. Gaze tracking is a technological discipline that offers an alternative interaction between human and computer. A common solution based on gaze tracking has two elements: a hardware device to obtain data from user's gaze and a library to process the gathered data. This work makes a formal analysis of previous works in gaze tracking. It presents the most used devices and libraries in gaze tracking solutions in the last 5 years as well as the most important characteristics of each of them. To fulfill this purpose, a search for articles related with eye tracking solutions was carried out in different scientific databases. After that, a classification of results was made, as well as a review of them. In addition, the article offers the areas of IoT in which gaze tracking has had the most influence. Finally, various selection criteria for devices and libraries are offered. These criteria are based on features such as price, technical performance and compatibility.

Keywords: Gaze tracking · Eye tacking · IoT

1 Introduction

The Internet of Things (IoT) is a technology that is present in several disciplines which is rapidly evolving, not only in the field of computing and communication, but also in areas such as logistics, medicine, among others [1–4]. The main objective of IoT is to allow things to connect and communicate from anywhere at any time [5]. It has allowed people to easily interact with smart devices using their voice, body [1, 6] and web/mobile applications [7, 8]. However, this type of solutions excludes people with severe motor and speech disabilities since they cannot use the aforementioned interfaces due to their physical limitations. Within this group of people are those suffering from motor disability caused by Spinal Cord Injury (SCI) [9] and Amyotrophic Lateral Sclerosis (ALS), and their number is increasing by 50000 people each year [10]. Other diseases that produce this type of disabilities are Multiple Sclerosis, Cerebral Palsy, Cerebrovascular Attack, Muscular Dystrophy, among others [11–14].

Based on the information mentioned previously, it is possible to say that most IoT solutions are not fully inclusive since people with movement disabilities cannot access the most common interfaces to control IoT devices. In this situation, eye or gaze tracking

© The Author(s), under exclusive license to Springer Nature Switzerland AG 2023
D. Simian and L. F. Stoica (Eds.): MDIS 2022, CCIS 1761, pp. 18–41, 2023.
https://doi.org/10.1007/978-3-031-27034-5_2

appears as one of the complementary technologies to provide an inclusive interface [15]. Eye or gaze tracking is a technological discipline that has been gaining importance in several areas because it can facilitate the implementation of technological solutions that are used in daily life of people. An example of these solutions can be turning a light on or off through slight movements of user's eyes [1].

Gaze tracking is the process of following the eye movement to determine exactly where and how long a person is looking at [16]. There are different devices and algorithms that allow us to determine the position and direction of the gaze. Both hardware and software have been implemented by different research groups or companies related to this technology. It is also important to indicate that there are several solutions that make use of gaze tracking technologies e.g., those proposed in [17–22] which are used in different areas. In order to carry out this kind of implementations, different tools available in IoT and Gaze Tracking technologies are required. In [1, 23–27], systems composed by both cameras and gaze tracking algorithms were used. There are also other works such as [28–31] using infrared (IR) cameras for obtaining gaze data in dark environments. It was also found that a simple webcam and open-source libraries can be used for performing gaze tracking [18, 32–34]. Additionally, glasses containing a video camera were also used [15, 19, 21, 35–37].

Based on the implementations explained before, it can be observed that the essential tools for gaze tracking are: (1) a device that allows to obtain data of user's eye in real time and (2) an algorithm/library that processes the gathered data for determining the position of the user's gaze on the screen.

Due to the importance of eye/gaze tracking technology, the present work intends to carry out a study of different works developed in the last five years in this area. This study will analyze the different types of hardware and software (libraries) that can be used for the implementation of solutions based on gaze tracking and the different applications in which this technology can be applied. We believe that this work will be of great contribution to people who start their research in the area helping them to understand the situation and trend of eye/gaze tracking technology.

2 Methodology

The main objective of this work is to analyze the trend of hardware devices and libraries used in gaze tracking from a technical perspective and the applications in which they can be used. To meet this objective, a semi-cyclical research method was used, which is based on the Research-Action process [2, 38] and systematic literature review [39, 40]. The method is composed of three main phases, where each one contains its own activities that were developed during the proposed study (see Fig. 1).

1) Planning Phase: This phase is intended to minimize the scope, and at the same time, establish the criteria for selecting the previous works to be analyzed. For this, the following research questions were established for this study: (1) what are the IoT solutions that can be implemented using gaze tracking? and (2) what hardware devices and libraries are used in the implementation of those solutions?

The defined questions have the intention of knowing and reviewing the implementations of gaze tracking systems that allow controlling IoT devices from a technological perspective. Based on the research questions indicated previously, the following keywords were identified: "tracking", "eye", "gaze", "IoT", "devices", "hardware", "libraries", "solutions", "implementation". These keyworks allowed the establishment of preliminary search strings such as "gaze tracking solutions", "gaze tracking hardware", "gaze tracking libraries". Since the first search returned a very large number of results, it was necessary to use logical connectors (AND, OR and NOT) with the previously identified keywords to make the search more specific. The scientific databases where the search strings were executed were IEEE Xplore, Scopus, Science Direct and ACM Digital Library.

2) Review Phase: Throughout this phase, the search for the previously defined search was carried out. During the search, the date filter was applied in order to obtain recent solutions (after 2016) since the technology is constantly changing, as indicated in [2, 41]. From the total number of articles, the title and abstract were reviewed, in order to eliminate irrelevant articles and obtain articles that would allow answering the research questions. The articles obtained after performing this activity were retrieved and loaded into a well-known tool called Mendeley. This is a free tool that allows you to manage and share bibliographic references and research documents, find new references, documents and collaborate online [42–44]. Finally, the remaining research papers were analyzed to identify if they were focused on the research topic. Figure 2 shows the used article selection process, which was taken from [2].

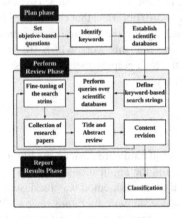

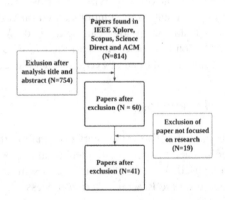

Fig. 1. Research methodology.   **Fig. 2.** Article selection process.

In the first activity of the article selection process, 814 articles were obtained that included relevant and non-relevant documents for our research. To eliminate non-relevant documents, the title and summary of each one of them were analyzed. When carrying out this activity, 754 articles were discarded. Then, a detailed review was carried out using the remaining 60 articles, which allowed discarding 19 articles that were not related to the research topic. At the end of the selection process, 41 articles were obtained, which were used to carry out the present study.

From the obtained articles, a list of the devices and libraries was drawn up. Since the resulting list did not contain enough elements to meet the goal of the research, so we proceeded to search the Internet for more devices and libraries. With this process, a list with 10 devices and 20 libraries was obtained, which are shown in Table 1.

After getting the list of devices and libraries, we proceeded to search for scientific articles in which the devices and libraries were used. This search was executed to answer the first research question i.e., to understand what IoT solutions can be implemented using gaze tracking. For this search, strings made up of the name of the library or device and keywords such as "gaze tracking" and "eye tracking" were used. The scientific databases where the queries were carried out were IEEE Xplore, Scopus, Science Direct and ACM Digital Library. Table 2 shows the obtained results.

Table 1. Devices and libraries used for gaze tracking.

Devices	Libraries	
EyeLink 1000 Plus	Eye Like	Tobii Pro-SDK
Tobii Pro Nano	WebGazer	Tobii Pro Glasses 3 API
Tobii Pro Fusion	OpenCV	SMI BeGaze
Smart Eye AI-X	Turker Gaze	SmartGaze
EyeTech VT3 Mini	Gaze Tracking	Eyeware – GazeSense
Eyegaze Edge Encore Camera	Pupil Core	openEyes
Eyegaze Edge Prime Camera	OGAMA	GazeParse
GazePoint 3 (GP3)	Gaze-Detection	GazePointer
SMI Eye Tracking Glasses	iMotions	GazePoint
Tobii Pro Glasses 3	OpenGazer	PyGaze (Webcam Eye Tracker)

To reduce the number of articles to be analyzed, it was necessary to follow the same process shown in Fig. 2. In the first activity, 1111 articles including both articles relevant and not relevant to the research were obtained. In order to get a set of relevant documents, the title and abstract of each one of them were analyzed, making possible to reduce this number discarding a total of 1001 articles. The next activity was to completely review the remaining 110 articles in order to obtain only those that actually made use of gaze tracking devices and libraries. Finally, after the process of elimination and classification, 70 articles were obtained which were used for this work.

3) Result Phase: In this phase, all findings and results were documented. The results of this phase were used to develop the following section that corresponds to the main part of this work. In addition, an analysis and classification of gaze tracking devices and libraries was carried out. This was done to understand the trend of using the different technological components within gaze tracking solutions to control IoT devices.

Table 2. Search result of gaze tracking applications.

Search string	Number of documents
IEEE Xplore Library Digital	
((<<device>>) AND ("gaze tracking" OR "eye tracking"))	1
((<<library>>) AND ("gaze tracking" OR "eye tracking"))	69
Scopus	
(<<device>> AND "gaze tracking" OR "eye tracking")	24
(<<library>> AND "gaze tracking" OR "eye tracking")	111
Science Direct	
(<<device>> AND "gaze tracking" OR "eye tracking")	256
(<<library>> AND "gaze tracking" OR "eye tracking")	354
ACM Digital Library	
(<<device>> AND "gaze tracking" OR "eye tracking")	67
(<<library>> AND "gaze tracking" OR "eye tracking")	229

3   Study of Gaze Tracking Solutions

Gaze tracking technology can make possible the creation of new interface allowing people to interact with systems and devices in a different way, making it more inclusive for people with severe motor and speech disabilities.

A common gaze tracking-based solution makes use of two main elements: a device used for capturing gaze data in real time and a library to process the captured data and determine gaze location. In this sense, there are different types of devices and libraries which can be used for gaze tracking, which are going to be detailed in this section.

3.1   Types of Devices

There are many devices that are used in gaze tracking. To meet the goal of gaze tracking, the devices make use of different components such as video cameras, infrared cameras, and brainwave sensors. They also make use of different methods, being the most common the detection of the dark pupil and bright pupil [45]. These devices work with different sampling frequencies that depend on the characteristics of the hardware. Precision and accuracy are the other differential characteristics. Some devices offer an acceptable margin of error, even when the user makes head movements. Additionally, depending on the used components, the equipment works in different lighting conditions. The weight and size also vary for each device, and they are compatible with different operating systems.

Among the different features, gaze tracking devices can be classified by the way they interact with the user, and from this point of view, devices are classified in: screen-based devices, head-mounted devices, and head-stabilized devices [46].

Screen-Based Devices: The main characteristic of this type of devices is that they are located in front of the user, either above or below the screen where the interaction interface is located. The data captured by these devices to perform gaze tracking is focused on the center of the pupil, reflection of the cornea and position of the head. Solutions developed with this kind of devices are extensive; among them, we can find solutions to control IoT devices [23], interact with a computer [35], evaluate user experiences [17], among others. In the market, the developers can find the following devices of this category. The list of this kind of devices can be found in Table 6. Notice that some values on Table 6 and after on Table 7 were labeled as N/S (Not Specified) which means that the documentation associated with a library did not provide the information about the characteristic.

Head-Mounted Devices: This kind of devices is generally used in the form of glasses. The main components that make up this device are a camera located in the visual path of the eyes and an additional camera that is responsible for recording the field of vision. The application of these devices in the real world is very wide. They can be used to make purchases in stores, for sports training, for simulations, among others [46]. For example, the reference [57] proposed the creation of an alert system to detect driver fatigue, using eye tracking glasses. In [58], a research work was carried out on the analysis of pilot's attention during takeoff and landing procedures. On the other hand, in [58], this kind of device was used to collect training data for neural networks with eye-tracking glasses. Head-mounted eye tracking systems are simple and comfortable to use, and they are less invasive to people than other devices that give little freedom of movement and cause user fatigue. However, this type of device has some limitations, such as the movement of the eyes outside the field of view captured by the cameras can cause errors, as well as the tracking of the gaze when sunlight generates excessive brightness [46]. The list of this kind of devices can be found in Table 6.

Head-Stabilized Devices: Some research areas, such as the field of Functional Magnetic Resonance, require precision and accuracy. To do this, the participant's head must be stabilized at the time of carrying out the experiment, leaving the user's comfort in the background. For these areas of research, there are head-stabilized gaze tracking devices. These devices offer greater precision in exchange for minimizing the good user experience. An important limitation of these devices is that they cannot be used in real-world scenarios [46] or in solutions that can facilitate the performance of daily life tasks. A representative device of this category is the EyeLink Plus 1000. The list of this kind of devices can be found in Table 6.

3.2 Types of Libraries

In the world of gaze tracking, there is a wide variety of libraries that can be used. These libraries allow processing the data flow coming from the devices and determining the position and direction of the gaze [73]. These gaze tracking tools are compatible with different programming languages and operating systems. During the review phase, it was possible to determine that there are open source and commercial libraries. Additionally, it was also found that some of the libraries are only compatible with specific eye-tracking devices.

Table 7 shows a summary of the gaze tracking libraries with the following information: capability, references using the library, type of license, installation simplicity, supported programming languages, supported Operating System (OS), availability of technical assistance and update date. For capability, the following labels were defined: w, o, sd, p, g, b and s. The library was labeled with "w" if it is compatible with webcams, "o" if it is compatible with other eye tracking devices, "sd" if it is compatible with only a specific device, "p" if it can do pupil tracking, "g" if it can do gaze tracking, "b" if it detects blinking and "s" if it detects saccades. For supported OS, labels W, M and L were defined for Windows, Mac and Linux respectively.

3.3 Types of Applications

Gaze tracking is being applied in different areas of knowledge. The classification of these areas is extensive, so in this work, the application of gaze tracking in IoT have been studied. A formal classification of IoT areas is obtained from [113]. The result of classifying the analyzed work by application areas is shown in Table 6.

Table 3. Application areas of gaze tracking.

Area	References	Number of references
Health & Wellness	[1, 15, 18, 19, 21, 22, 25, 27, 31, 32, 36, 60, 67, 70–72, 79, 84, 92, 94, 97, 102, 104, 106]	24
Psychology	[26, 47, 49–51, 53, 58–60, 63, 64, 69, 76, 85–87, 103, 107, 108, 111]	20
HCI	[15, 21–24, 28, 30, 33, 35–37, 54, 65, 71, 78, 79, 89, 95, 99]	19
Automotive industry	[57, 58, 80, 107–109]	6
Smart home	[1, 27, 31, 97]	4
Education	[20, 61, 103]	3

1) Health & Wellness: Within the health area, there are applications that can help to obtain a better diagnosis of diseases related to vision and others designed to help people with motor disabilities. For example, in [1], a system was created to control household appliances through eye tracking using a Tobii device, aimed at older adults and people with special needs. Another application aimed at helping this group of people was developed in [19]; in this work, the authors controlled a wheelchair through gaze tracking. Gaze tracking in the health area has been used in people with language problems, motor disabilities, people with spinal cord injuries, amyotrophic lateral sclerosis, patients with frontotemporal dementia, etc. (see Table 3).

2) Psychology: In the field of psychology, different applications have been developed to determine human behavior through facial expressions as well as to aid in the diagnosis of mental illness. In the recognition of facial expressions, several elements of the face are analyzed, such as the mouth, lips, but especially the eyes. Through the analysis of these elements, it is possible to identify common facial expressions of people when they are exposed to different stimuli [26]. In [59], it was possible to differentiate patients who had schizophrenia by measuring eye movements, duration of gaze fixations, and difference in gaze position in relation to a reference point. Additionally, in [47], measurements of user's eyes were analyzed to try to determine patients with delirium at Massachusetts General Hospital.

3) Human-Computer Interface: In human-computer interface (HCI), gaze tracking has been used to create new interaction interfaces that are manipulated through eye movement. These interfaces allow users to access the Internet and even email service using gaze tracking [36]. These two technologies (HCI and gaze tracking) can be very useful in museums, where people can interact with artworks [23]. There are also applications used in the daily activities of users, such as writing a message with a virtual keyboard [91].

4) Automotive Industry: In the automotive industry, gaze tracking is being used to create a safer driving environment. For example, smart cars have implemented alerts when the driver is in a state of sleepiness; in other words, the vehicle system detects prolonged blinking and send an alert to the user to avoid traffic accidents [57]. In [109], the visual attention that users have in response to collisions when driving semi-autonomous vehicles was evaluated.

5) Smart Home: In this application area, gaze tracking has made it possible to implement smart homes that can be controlled through eye movements. Inside the houses, there are several IoT devices that can be controlled with the gaze tracking of the user, among the most common controlled devices are lights, fans, doors, TVs, doorbells, and radio [27, 97]. In most of cases, these types of solutions were created for people with motor and speech disabilities [1, 31].

6) Education: In the area of education, gaze tracking was used in applications focused on evaluating learning in students and generating inclusive education solutions. For example, in [72], an eye tracker was used to improve the attention of students. Gaze tracking in education also allows the student's participation in learning situations and have feedback in real time [20].

It is important to do an honorific mention to other areas where eye tracking has been widely applied, such as in video games. Some video games released recently are compatible with eye tracking technology. Eye tracking is used to control the game's video camera or perform basic gaming tasks like opening the inventory [114]. Eye tracking could also work together with other technologies areas. For example, in [115], eye tracking, gaming and psychology were combined to study the behavior of video game players and found out differences between cognitive styles from players and non-players. In [116], this combination allows training visual attention in people with attention-deficit/hyperactivity disorder through a video game.

4 Discussion

In the present study, a classification per year of the selected research articles was made. As shown in Fig. 3, there was a growing trend in 2016, while in 2017 and 2018 there was a significant decrease. For 2019 and 2020, an increase of works can be observed again. A possible factor of works reduction in 2020 and 2021 could have been the COVID pandemic since many people and companies were affected and most of the resources were assigned to such research topic [117].

4.1 Types of Devices

Figure 4 indicates that the most used devices for gaze tracking are GazePoint 3 (GP3) with 31%, followed by WebCams with 29% and EyeLink 1000 Plus with 17%. Within the commercial devices, GazePoint 3 is the cheapest one and offers technical characteristics like other higher cost devices (see Table 6). These characteristics could be the influencing factors for it to be used in many research works. Something to highlight is the popularity of webcams, which despite not being devices specifically oriented to gaze tracking, they are widely used for this purpose. This may be because these devices are cheap and readily available. The EyeLink 1000 Plus, on the other hand, has a 2000 Hz sample rate and is the only one of the reviewed devices that belongs to the head-stabilized category. Devices of this type offer greater accuracy and precision, which are two essential factors in this field of research. This may be why the EyeLink 1000 Plus is quite popular in the field of gaze tracking despite its high cost. In Fig. 4, it can be seen that there are devices that are not mentioned in any article, among those are Eyegaze Edge Encore Camera and Eyegaze Edge Prime Camera. The absence of these devices within the research articles may be due to the fact that they are not easy to acquire, since they are part of commercial equipment for people with disabilities [118].

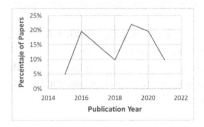

Fig. 3. Research works trend about gaze tracking.

4.2 Types of Libraries

From the libraries mentioned in the reviewed articles, OGAMA was used in 20%, while OpenCV, iMotions, Pupil Core and WebGazer were used in 18%, 15%, 11% and 9%, respectively (see Fig. 5).

OGAMA is a software that is compatible with different gaze tracking devices, such as Tobii Pro, Eye Tech, GazePoint, etc., which makes many researchers show interest in the use of this library, adding to the fact that it is an open-source software. However, it also has some disadvantages such as that it only runs on Windows and that it only supports infrared webcams. On the other hand, iMotions is a library developed for different purposes (i.e.

Electroencephalogram, Facial Expression, Eye Tracking) becoming a multidisciplinary library. Something in common between OGAMA and iMotions is that they have been in the field of gaze tracking for several years. The iMotions platform was founded in 2005, while OGAMA in 2007. Another important software is OpenCV, which is the oldest library since it has been operational since 2000. This software is compatible with different programming languages; this means that many developers can integrate this library into different gaze tracking projects on their own. One limitation of this library is that it is generic computer vision software, and it is not specifically focused on gaze tracking, so if an application needs to be created for this purpose, the programmer must add a significant amount of additional code. Unlike the previous libraries, Pupil Core and WebGazer are relatively new libraries; the former appearing in 2014 and the latter in 2016. On the one hand, Pupil Core is a specific platform for gaze tracking, and it is used for different research purposes, such as gaze estimation, pupillometry and egocentric vision, which makes it a very useful tool in this area of research. However, it has a big limitation which is it compatibility only with the Pupil Core eye tracking device. On the other hand, WebGazer is compatible with webcams, which can be ideal for research projects with few resources; in addition, it is easy to install and it is open source, making it very popular in different gaze tracking projects.

In Fig. 5, it can also be seen that other libraries such as Turker Gaze, Open Gazer, SmartGaze, etc., have not been used in research articles; this may be because, in most cases, the libraries have not received updates in the last five years.

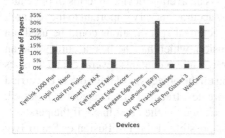

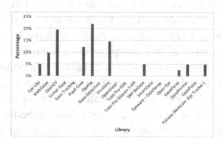

Fig. 4. Types of devices.

Fig. 5. Type of libraries and their use in different research articles.

4.3 Types of Applications

By reviewing the 76 selected research articles, it was possible to see that gaze tracking is used in different areas being the most popular areas health and well-being, psychology, and HCI with 32%, 26%, and 25% respectively (see Table 3).

Within the area of health and well-being, most of the technological solutions have been created with the purpose of helping people with special needs, for example, solutions for people with movement disabilities and sometimes with speech disabilities. Those solutions created a new inclusive option of interaction with different devices using gaze tracking.

It is known that, in a social interaction, nonverbal communication provides the 93% of the information. And from those 93%, the area of the face is the one that reveals the most amount of information (about 53%). This is the reason why the information revealed by the face of a person has been extensively studied in Psychology [119]. Among the different data revealed by the face, the data delivered by gaze tracking is also studied in psychology. In this area, gaze tracking is helping in the diagnosis of mental illnesses and in the recognition of different facial expressions in real time.

Other areas where the gaze tracking is used widely is HCI. Gaze tracking has opened the door for implementing easy-to-use interfaces and a positive user experience. Gaze tracking has allowed creating solutions with interfaces in which the user can interact with his eyes creating a novel and inclusive experience for people with motor and speech disabilities. This type of solutions goes beyond being an alternative, but a necessity in the present. Gaze tracking has made it possible to develop applications with more inclusive interfaces and is therefore widely used in HCI.

5 Criteria for Selecting Devices and Libraries

Criteria for Selecting Devices: It is important to indicate that the choice of a gaze tracking device will depend on the application in which it is going to be used. If the accuracy of the device does not drastically affect the results of the research or if there are few resources for the project, a low-cost device such as a webcam can be used. In this aspect, it can be seen that the GazePoint 3 is an average device in terms of price and features (see Table 6). On the other hand, if the accuracy is very important to the research project beyond cost, a device with the best technical specifications could be used; an example of this group could be Tobii Pro Fusion, which has the best accuracy and an excellent sample rate (see Table 6). In general, Table 6 contains several characteristics that can be used as selection criteria depending on the objectives of the work to be carried out. For example, if a device that offers the greatest freedom of movement is required, SMI Eye Tracking Glasses might be the best choice; on the other hand, if the person is required to be as far away from the screen as possible, Tobii Pro Fusion or Smart Eye AI-X are the ideal candidates.

Criteria for Selecting Libraries: Unlike gaze tracking devices, selecting a library can be more complex. For this reason, to provide a selection criterion for gaze tracking libraries, a method was proposed that consisted of selecting the largest number of features from Table 7 and assigning them a weight, which is detailed in Table 4. These weights stemmed from authors subjective preferences and were not obtained using a systematic and intersubjective process. In this table, the highest weight (35%) was given to Degree of Capability, because this feature determines compatibility with the different gaze tracking devices that were previously studied; in addition, it allows knowing if the library performs gaze tracking or only eye tracking, as well as if it could detect blinking and saccades. The values of this feature are high if it has the labels w, o, p, g, b, and s; medium if it has five, four labels excluded "sd" label, or three labels included the "w" label; and low if it has less than three labels or it has the "sd" label. The second most important characteristic considered in this work was the number of research articles, since it indicates the recognition that the library has within scientific research works. The assigned weigh for this feature was 30%. Additionally, it is important to know the conditions of use of the library, if the library can be freely used or distributed, or if it requires payment for its use, i.e., the type of license; this feature was assigned 15% of the total weight. Since there are virtual machines that allow any operating system to be simulated, the OS Compatibility feature would not be a limitation, however, performance would be lower than a native OS, so 10% was assigned to this feature. The values for this feature are high, medium and low, which means the library is compatible with three, two and one OS, respectively. Installation simplicity allows knowing how easy or difficult it is to install the library; once the installation is complete, regardless of the degree of complexity, this feature will no longer be important, so it was only assigned a weight of 10%. The labels for this are high, medium, and low. It was labeled with high if it is a stand-alone library; with medium if the installation is via command line interface (CLI); with low if the library requires dependencies, and additionally its installation is via CLI. Finally, the Support is a feature that allows you to know the recent updates of the library for its correct operation; however, a library that hasn't received updates in the last few years might still work fine. For this reason, technical support would not be a limitation when choosing or testing a library, which is why it has been assigned 5%. Together, all these features add up to 100%. The results of evaluating the libraries with the proposed method are shown in Table 5. It is important to indicate that the weights indicated above are appropriate for the development projects that the authors will carry out in the future. This means that it is a single example of how a library selection criterion could be generated according to the needs of each development project.

In Table 4, to weight the high, medium and low values, it was decided to assign 46%, 34% and 20% respectively. The percentage was taken from the total value of

Table 4. Ponderation criteria for the features.

Feature	Value	Ponderation
Degree of capability 35%	High	16
	Medium	12
	Low	7
Number of articles 30%	Number	Number * 3
Type of license 15%	Open source	7
	Freeware	5
	Commercial	3
Operating system compatibility 10%	High	5
	Medium	3
	Low	2
Installation simplicity 10%	High	5
	Medium	3
	Low	2
Technical support 5%	Yes	5
	No	0

the characteristic. In a similar way, the values of the License Type characteristic were weighted. To obtain the weighting of the number of articles in each library, the number of articles was multiplied by 3 to guarantee that the library with the most articles obtains a weighting of 30%. Finally, for the Support characteristic, a binary classification was made where YES takes all the weight (5%) and NO has a weight of zero.

As can be seen in Table 5, OpenCV ranks first, followed by OGAMA, iMotions and WebGazer. However, this ranking could be variable depending on the projects condition. For example, a project would require the usage of a webcam or have budget limitation for buying a commercial library, in which the ponderation should be changed.

Table 5. Libraries evaluation based on their features.

Library	Degree of capability	# of articles	Type of license	Installation simplicity	OS compatibility	Tech. support	Score
OpenCV	16	24	7	3	5	5	60
OGAMA	12	30	7	2	2	5	58
iMotions	16	18	3	5	–	5	47
WebGazer	12	12	7	3	5	5	44
Pupil Core	7	15	7	2	5	5	41
PyGaze	12	6	7	5	5	0	35
GazePointer	12	6	5	5	2	0	30
Gaze Tracking	12	0	7	5	0	5	29
GazeParse	12	3	7	2	5	0	29
Turker Gaze	12	0	7	5	5	0	29
Eyeware – GazeSense	12	0	3	5	3	5	28
SMI BeGaze	7	6	3	5	2	5	28
Eye Like	7	6	7	2	5	0	27
Tobii Pro-SDK	7	0	5	5	5	5	27
openEyes	12	0	7	3	5	0	27
Tobii Pro Glasses 3 API	7	0	5	3	5	5	25
GazePoint	12	0	5	2	5	0	24
OpenGazer	12	0	7	2	3	0	24
Gaze-Detection	7	0	7	3	3	0	20
SmartGaze	7	0	3	5	–	5	20

Table 6. Gaze tracking devices.

Device	Type of device	References using device	Screen size (inches)	Accuracy (grades)	Sampling Frequency (Hz)	Distance user-device	Head movement range	Head movement angle (grades)	Price (USD)
Tobii Pro Fusion	Screen based	[47, 48]	19″	0.3	120	45 cm–85 cm	65cm: 40 × 25 cm 80cm: 45 × 30 cm	29.36–31.59 H 20.55–21.01 V	$13900
Tobii Pro Nano	Screen based	[49–51]	24″	0.3	60, 120, 250	50 cm–80 cm	35 cm × 30 cm	23.63–34.99 H 20.55–30.96 V	$10000 +
Smart Eye AI-X	Screen based	None	24″	0.5	60	45 cm–85 cm	None	N/S	$3450
EyeTech VT3 Mini	Screen based	[20, 52]	22″	0.5	200	50 cm–70 cm	31.5 × 22.5 cm, 20 × 5 cm y 9 × 4 cm	16.07–21.97 H 8.53–11.86 V	$4000
Eyegaze Edge Encore Camera	Screen based	None	22″	0.45	50	40 cm–73 cm	17.8 × 11.9 × 7.6cm	13.66–23.98 H 9.26–16.57 V	$7500
Eyegaze Edge Prime Camera	Screen based	None	19″	0.45	60	43 cm–83 cm	7.6 × 6.4 × 6 cm	5.23–10.02 H 4.40–8.46 V	$7500
GazePoint 3 (GP3)	Screen based	[53–56] [64–70]	24″ [19] 1366 × 768 [21]	0.5	60	50 cm–80 cm 65 cm [16]	24 × 11 × 15 cm [15] 25 × 11 × 15 cm [17]	17.01–26.10 H 7.83–12.40 V	$845
Webcam	Screen based	[18, 19, 27, 29, 32–35, 71, 72]	23″	N/S	60	28.5 cm–60 cm	22–39 horizontal degrees	22–39 H	$15+
SMI Eye Tracking Glasses	Head mounted	[57]	N/S	0.5	120	N/S	80 × 60 degrees	80 H 60 V	$11900
Tobii Pro Glasses 3	Head mounted	[58]	N/S	0.6	100	N/S	N/S	N/S	$10000+
EyeLink 1000 Plus	Head stabilized	[59–63]	24″	0.5	2000	40 cm–70 cm	N/A	N/S	$10000+

Table 7. Gaze tracking libraries.

Library	Degree of capability	References using the library	Type of license	Installation	Supported programming languages	Supported OS	Technical support	Last update
Eye Like	w, p	[74, 75]	Open source	Stand-alone	C++, C	W, M, L	No	2019-03-31
WebGazer	w, o, p. g, b	[17, 76, 77, 95]	Open source	Install via CLI	JavaScript	W, M, L	Yes	2022-11-16
OpenCV	w, o, p, g, b, s	[78–80, 96–100]	Open source	Install via CLI	C++, C, Python, Java	W, M, L	Yes	2022–02-03
Turker Gaze	w, p, g, b, s	None	Open source	Need dependencies, Install via CLI	JavaScript	W, M, L	No	2016-04-06
Gaze Tracking	w, o, p. g, b	None	Open source	Need dependencies, Install via CLI	Python	W, L	Yes	2022-01-19
Pupil Core	sd, p, b	[81–83, 101, 102]	Open source	Stand-alone	Python, C++	W, M, L	Yes	2022-01-31
OGAMA	w, g, p, b, s	[53, 64, 84–86, 103–107]	Open source	Stand-alone	C#, C++	W	Yes	2021-03-30
Gaze-detection	w, p	None	Open source	Install via CLI	JavaScript	W, M, L	No	2021-06-24
iMotions	w, p, o, g, b, s	[87–90, 108] [109]	Commercial	Need dependencies, Install via CLI	N/S	W	Yes	N/A
OpenGazer	w, g, p, b	None	Open source	Stand-alone	C++, Python	L, M	No	2010-11-16
Tobii Pro-SDK	sd, g, p, b	None	Freeware	Need dependencies, Install via CLI	Python, Matlab y.NET	W, M, L	Yes	2021-09-15

(continued)

Table 7. (*continued*)

Library	Degree of capability	References using the library	Type of license	Installation	Supported programming languages	Supported OS	Technical support	Last update
Tobii Pro Glasses 3 API	sd, g, p	None	Freeware	Install via CLI	Python, Matlab y.NET	W, M, L	Yes	2020
SMI BeGaze	o, p, g, b, s	[110, 111]	Commercial	Need dependencies, Install via CLI	Proprietary language	W	Yes	2020
SmartGaze	o, p, g,	None	Commercial	Need dependencies, Install via CLI	N/S	N/S	Yes	N/S
Eyeware-GazeSense	w, o, p, g	None	Commercial	Need dependencies, Install via CLI	Just user interface	W, L	Yes	N/S
openEyes	w, p, g	None	Open source	Install via CLI	Matlab	W, M, L	No	N/S
GazeParse	w, o, p, g	[91]	Open source	Stand-alone	Python, C++	W, M, L	No	2021-03-22
GazePointer	w, o, p	[92, 93]	Freeware	Need dependencies, Install via CLI	N/S	W	No	2017
GazePoint	w, o, p, g	None	Freeware	Stand-alone	Proprietary commands, HTML	W, M, L	No	2018
PyGaze (Webcam Eye Tracker)	w, o, p, b, s	[94, 112]	Open source	Need dependencies, Install via CLI	Python	W, M, L	No	2016

6 Conclusions

By reviewing the previous works, it was possible to determine that gaze tracking technology has contributed significantly to the development of different technological solutions in multiple areas, and this trend has been thanks to the creation of different devices and libraries.

There are several gaze tracking devices that are available on the market. Depending on the budget and characteristics of the project, it is possible to choose from cheap devices but with little precision features, to high-cost devices with high-precision features that include their own gaze tracking algorithms. Similarly, there are different library options, from open-source projects to commercial software; there are also libraries that are compatible with webcams or specific gaze tracking devices.

The selection of the device and library will depend on the characteristics of the project to be developed. If only the position of the pupil is required, cheap devices and easy-to-use libraries can be used. While, if more precise applications are required, for example in situations where traffic accidents are to be avoided or to determine if a person suffers from mental illness, more precise devices and libraries should be chosen.

Gaze tracking is a discipline that easily couples with other areas of the IoT to generate different solutions, and this has been proven with multiple applications from different areas, such as Health and wellness, Psychology and HCI. It is in the field of health and well-being that gaze tracking has contributed the most, serving as a support tool for a better diagnosis of diseases. In addition, it is improving the quality of life for many people, especially those with special needs. Also, in psychology, gaze tracking has been used to diagnose mental illness and analyze human behavior. And as we've seen throughout this study, by merging gaze tracking with HCI, you can create technology solutions with more inclusive interfaces.

Although most of gaze tracking applications are still limited to scientific and academic prototypes, it is expected that through the different devices and libraries, real solutions can be generated for different types of users in the future.

References

1. Klaib, F.K., Alsrehin, N.O., Melhem, W.Y., Bashtawi, H.O.: IoT smart home using eye tracking and voice interfaces for elderly and special needs people. J. Commun. **14**(7), 614–621 (2019)
2. Barriga, J.J., et al.: Smart parking: a literature review from the technological perspective. Appl. Sci. **9**(21), 4569 (2019)
3. Zhaoa, W., Yi, L.: Research on the evolution of the innovation ecosystem of the internet of things: a case study of Xiaomi (China). Proc. Comput. Sci. **199**, 56–62 (2022)
4. Humayuna, M., Jhanjhi, N., Alsayat, A., Ponnusamy, V.: Internet of things and ransomware: evolution, mitigation and prevention. Egypt. Inform. **22**(1), 105–117 (2021)
5. Siddesh, G., Manjunath, S., Srinivasa, K.: Application for assisting mobility for the visually impaired using IoT infrastructure. In: 2016 International Conference on Computing, Communication and Automation (ICCCA), pp. 1244–1249 (2016)
6. Alanwar, A., Alzantot, M., Ho, B., Martin, P., Srivastava, M.: SeleCon: scalable IoT device selection and control using hand gestures. In: 2017 IEEE/ACM Second International Conference on Internet-of-Things Design and Implementation (IoTDI), pp. 47–58 (2017)

7. Tedla, T.B., Davydkin, M.N., Nafikov, A.M.: Development of an internet of things based electrical load management system. J. Phys. Conf. Ser. 1886 (2021)
8. Kirsh, I., Ruser, H.: Phone-pointing remote app: using smartphones as pointers in gesture-based IoT remote controls. In: Stephanidis, C., Antona, M., Ntoa, S. (eds.) HCII 2021. CCIS, vol. 1420, pp. 14–21. Springer, Cham (2021). https://doi.org/10.1007/978-3-030-78642-7_3
9. WHO: Spinal cord injury. https://www.who.int/news-room/fact-sheets/detail/spinal-cord-injury. Accessed 22 Feb 2022
10. A. N. Today. https://alsnewstoday.com/how-common-is-als/. Accessed 22 Feb 2022
11. Metz, C., Jaster, M., Walch, E., Sarpong-Bengelsdorf, A., Kaindl, A., Schneider, J.: Clinical phenotype of cerebral palsy depends on the cause: is it really cerebral palsy? A retrospective study. J. Child Neurol. 37(2), 112–118 (2021)
12. Randolph, A., Petter, S., Storey, V., Jackson, M.: Context-aware user profiles to improve media synchronicity for individuals whit sever motor disabilities. Inf. Syst. J. 32(1), 130–163 (2021)
13. El-Kafy, E.M.A., Alshehri, M.A., El-Fiky, A.A.R., Guermazi, M.A.: The effect of virtual reality-based therapy on improving upper limb functions in individuals with stroke: a randomized control trial. Front. Aging Neurosci. 13, 1–8 (2021)
14. Yayıcı, Ö., Taşkıran, C., Genç Sel, Ç., Aksoy, A., Yüksel, D.: Clinical features and quality of life in duchenne and becker muscular dystrophy patients from a tertiary center in Turkey. J. Curr. Pediatr. 19, 15–22 (2021)
15. Tarek, N., et al.: Morse glasses: an IoT communication system based on Morse code for users with speech impairments. Computing 104(4), 789–808 (2021). https://doi.org/10.1007/s00607-021-00959-1
16. Klaib, A., Alsrehin, N., Melhem, W., Bashtawi, H., Magableh, A.: Eye tracking algorithms, techniques, tools, and applications with an emphasis on machine learning and internet of things technologies. Expert Syst. Appl. 166, 114037 (2021)
17. Papoutsaki, A., Sangkloy, P., Laskey, J., Daskalova, N., Huang, J., Hays, J.: WebGazer: scalable webcam eye tracking using user interactions. In: Proceedings of the Twenty-Fifth International Joint Conference on Artificial Intelligence (IJCAI-16), pp. 3839–3845 (2016)
18. Veerati, R., Suresh, E., Chakilam, A., Ravula, S.P.: Eye monitoring based motion controlled wheelchair for quadriplegics. In: Anguera, J., Satapathy, S.C., Bhateja, V., Sunitha, K.V.N. (eds.) Microelectronics, Electromagnetics and Telecommunications. LNEE, vol. 471, pp. 41–49. Springer, Singapore (2018). https://doi.org/10.1007/978-981-10-7329-8_5
19. HemaMalini, B.H., Supritha, R.C., Venkatesh, N.K., Vandana, R., Yadav, R.: Eye and voice controlled wheel chair. In: 2020 IEEE Bangalore Humanitarian Technology Conference (B-HTC), pp. 1–3 (2020)
20. Carroll, M., et al.: Automatic detection of learner engagement using machine learning and wearable sensors. J. Behav. Brain Sci. 10(3), 165–178 (2020)
21. Wankhede, K., Pednekar, S.: Aid for ALS patient using ALS specs and IOT. In: 2019 2nd International Conference on Intelligent Autonomous Systems (ICoIAS), pp. 146–149 (2019)
22. Pai, S., Bhardwaj, A.: Eye gesture based communication for people with motor disabilities in developing nations. In: 2019 International Joint Conference on Neural Networks (IJCNN), pp. 1–8 (2019)
23. Dondi, P., Porta, M., Donvito, A., Volpe, G.: A gaze-based interactive system to explore artwork imagery. J. Multimodal User Interfaces 16(1), 55–67 (2022)
24. Su, Z., Zhang, X., Kimura, N., Rekimoto, J.: Gaze+Lip: rapid, precise and expressive interactions combining gaze input and silent speech commands for hands-free smart TV control. In: ETRA 2021 Short Papers: ACM Symposium on Eye Tracking Research and Applications, pp. 1–6 (2021)

25. Arias, E., Lópiz, G., Quesada, L., Guerrero, L.: Web accessibility for people with reduced mobility: a case study using eye tracking. In: Di Bucchianico, G., Kercher, P. (eds.) Advances in Design for Inclusion. Advances in Intelligent Systems and Computing, vol. 500, pp. 463–473. Springer, Cham (2016). https://doi.org/10.1007/978-3-319-41962-6_41
26. Abdallah, A.S., Elliott, L.J., Donley, D.: Toward smart internet of things (IoT) devices: exploring the regions of interest for recognition of facial expressions using eye-gaze tracking. In: 2020 IEEE Canadian Conference on Electrical and Computer Engineering (CCECE), pp. 1–4 (2020)
27. Bissoli, A., Lavino-Junior, D., Sime, M., Encarnação, L., Bastos, T.: A human-machine interface based on eye tracking for controlling and monitoring a smart home using the internet of things. Sensors **19**(4), 859 (2019)
28. Brousseau, B., Rose, J., Eizenman, M.: SmartEye: an accurate infrared eye tracking system for smartphones. In: 2018 9th IEEE Annual Ubiquitous Computing, pp. 951–959 (2018)
29. Brunete, A., Gambo, E., Hernando, M., Cedazo, R.: Smart assistive architecture for the integration of IoT devices, robotic systems, and multimodal interfaces in healthcare environments. Sensors **21**(6), 1–25 (2021)
30. Chandra, S., Sharma, G., Malhotra, S., Jha, D., Prakash, A.: Eye tracking based human computer interaction: applications and their uses. In: Proceedings of 2015 International Conference on Man and Machine Interfacing, pp. 1–5 (2016)
31. Heravian, S., Nouri, N., Behnam, M., Seyedkashi, S.M.H.: Implementation of eye tracking in an IoT-based smart home for spinal cord injury patients. Mod. Care J. **16**(4), 1–8 (2019)
32. Chew, M.T., Penver, K.: Low-cost eye gesture communication system for people with motor disabilities. In: Proceedings of 2019 IEEE International Instrumentation and Measurement Technology Conference, pp. 1–5 (2019)
33. Robal, T.: Spontaneous webcam instance for user attention tracking. In: Proceedings of 2019 Portland International Conference on Management of Engineering and Technology (PICMET), pp. 1–8 (2019)
34. Xia, L., Sheng, B., Wu, W., Ma, L., Li, P.: Accurate gaze tracking from single camera using gabor corner detector. Multimed. Tools Appl. **75**(1), 221–239 (2014). https://doi.org/10.1007/s11042-014-2288-4
35. Salunkhe, P., Patil, A.R.: A device controlled using eye movement. In: Proceedings of 2016 International Conference on Electrical, Electronics, and Optimization Techniques, pp. 732–735 (2016)
36. Lupu, R.G., Bozomitu, R.G., Păsărică, A., Rotariu, C.: Eye tracking user interface for Internet access used in assistive technology. In: Proeceedings of 2017 E-Health and Bioengineering Conference, Sinaia, Romania (2017)
37. Tamura, Y., Takemura, K.: Estimating point-of-gaze using smooth pursuit eye movements without implicit and explicit user-calibration. In: ETRA 2020 Short Papers: ACM Symposium on Eye Tracking Research and Applications, pp. 1–4 (2020)
38. Drummond, J., Themessl-Huber, M.: The cyclical process of action research. Action Res. **5**(4), 430–448 (2007)
39. Chauhan, S., Agarwal, N., Kar, A.: Addressing big data challenges in smart cities: a systematic literature review. Info **18**(4), 73–90 (2016)
40. Morais, C., Sadok, D., Kelner, J.: An IoT sensor and scenario survey for data researchers. J. Braz. Comput. Soc. **25**(1), 1–17 (2019). https://doi.org/10.1186/s13173-019-0085-7
41. Thiébaud, E., Hilty, L., Schluep, M., Widmer, R., Faulstich, M.: Service lifetime, storage time, and disposal pathways of electronic equipment: a swiss case study. J. Ind. Ecol. **22**(1), 196–208 (2017)
42. Eldakar, M.A.M.: Who reads international Egyptian academic articles? An altmetrics analysis of Mendeley readership categories. Scientometrics **121**(1), 105–135 (2019). https://doi.org/10.1007/s11192-019-03189-7

43. Asemi, A., Heydari, M.: Correlation between the articles citations in web of science (WoS) and the readership rate in Mendeley and research gate (RG). J. Scientometr. Res. 7(3), 145–152 (2018)
44. Patak, A.A., Naim, H.A., Hidayat, R.: Taking mendeley as multimedia-based application in academic writing. Int. J. Adv. Sci. Eng. Inf. Technol. 6(4), 557–560 (2016)
45. Morimoto, C., Koons, D., Amir, A., Flickner, M.: Pupil detection and tracking using multiple light sources. Image Vis. Comput. 18(4), 331–335 (2000)
46. Bitbrain. https://www.bitbrain.com/blog/eye-tracking-devices. Accessed 28 May 2022
47. Ching, W.: Detection of delirium through eye-tracking methods. Thesis, Boston University (2018)
48. Ameen, M.A.H., Aldridge, C.M., Zhuang, Y., Yin, X.: Investigating the need for calibration to track eye. Research Square (2021)
49. Artemia, M., Liu, H.: A user study on user attention for an interactive content-based image search system. In: Proceedings of BIRDS 2021: Bridging the Gap between Information Science, Information Retrieval and Data Science (2021)
50. Fliorent, R., Cavanaugh, G., LLerena, C.: Measuring the engagement of children with autism spectrum disorder using eye-tracking data. NSUWorks (2020)
51. Saleema, M. R., Straus, A., Napolitano, R.: Interpretation of historic structure for non-invasive assessment using eye tracking. Int. Arch. Photogr. Remote Sens. Spat. Inf. Sci. 46, 653–660 (2021)
52. Mele, M. L., Millar, D., Rijnders, C. E.: Explicit and implicit measures in video quality assessment. In: Proceedings of the 14th International Joint Conference on Computer Vision, Imaging and Computer Graphics Theory and Applications (VISIGRAPP 2019), pp. 38–49 (2019)
53. Contero-López, P., Torrecilla-Moreno, C., Escribá-Pérez, C., Contero, M.: Understanding fashion brand awareness using eye-tracking: the mix-and-match approach. In: Markopoulos, E., Goonetilleke, R.S., Ho, A.G., Luximon, Y. (eds.) AHFE 2021. LNNS, vol. 276, pp. 432–440. Springer, Cham (2021). https://doi.org/10.1007/978-3-030-80094-9_51
54. Sulikowski, P., Zdziebko, T.: Deep learning-enhanced framework for performance evaluation of a recommending interface with varied recommendation position and intensity based on eye-tracking equipment data processing. Electron. (Switz.) 9(2), 266 (2020)
55. Katona, J., et al.: Recording eye-tracking parameters during a program source-code debugging example. In: Proceedings of 10th IEEE International Conference on Cognitive Infocommunications, pp. 335–338 (2019)
56. Katona, J., et al.: The examination task of source-code debugging using GP3 eye tracker. In: Proceedings of 10th IEEE International Conference on Cognitive Infocommunications, pp. 329–334 (2019)
57. Wu, W., et al.: Faster single model vigilance detection based on deep learning. IEEE Trans. Cogn. Dev. Syst. 13(3), 621–630 (2021)
58. Gomolka, Z., Twarog, B., Zeslawska, E., Kordos, D.: Registration and analysis of a pilot's attention using a mobile eyetracking system. In: Zamojski, W., Mazurkiewicz, J., Sugier, J., Walkowiak, T., Kacprzyk, J. (eds.) DepCoS-RELCOMEX 2019. AISC, vol. 987, pp. 215–224. Springer, Cham (2020). https://doi.org/10.1007/978-3-030-19501-4_21
59. Morita, K., et al.: Eye movement abnormalities and their association with cognitive impairments in schizophrenia. Schizophrenia Res. 209, 255–262 (2019)
60. Russell, L.L., et al.: Novel instructionless eye tracking tasks identify emotion recognition deficits in frontotemporal dementia. Alzheimer's Res. Ther. 12, 39 (2021)
61. Cutumisu, M., et al.: Eye tracking the feedback assigned to undergraduate students in a digital assessment game. Front. Psychol. 10, 1931 (2019)

62. Bender, L., Guerra, I., Ito, G., Vizcarra, I., Schianchi, A.: Mirada, tiempo y acción: visualizaciones alternativas de experimentos de seguimiento ocular con escenas dinámicas, In: Proceedings of Sexto Congreso Argentino de la Interacción-Persona Computador@, Telecomunicaciones, Informática e Información Científica (2017)

63. Hooge, I., Hessels, R., Nyström, M.: Do pupil-based binocular video eye trackers reliably measure vergence? Vision. Res. **156**, 1–9 (2019)

64. Costescu, C., et al.: Assessing visual attention in children using GP3 eye tracker. In: Proceedings of 10th IEEE International Conference on Cognitive Infocommunications, pp. 343–348 (2019)

65. Iskander, J., Hettiarachchi, I., Hanoun, S., Hossny, M., Nahavandi, S., Bhatti, A.: A classifier approach to multi-screen switching based on low cost eye-trackers. In: Proceedings of 2018 Annual IEEE International Systems Conference, pp. 1–6 (2018)

66. Mannaru, P., Balasingam, B., Pattipati, K., Sibley, C., Coyne, J.T.: Performance evaluation of the gazepoint GP3 eye tracking device based on pupil dilation. In: Schmorrow, D.D., Fidopiastis, C.M. (eds.) AC 2017. LNCS (LNAI), vol. 10284, pp. 166–175. Springer, Cham (2017). https://doi.org/10.1007/978-3-319-58628-1_14

67. Saisara, U., Boonbrahm, P., Chaiwiriya, A.: Strabismus screening by eye tracker and games. In: Proceedings of 2017 14th International Joint Conference on Computer Science and Software Engineering, pp. 1–5 (2017)

68. Kovari, A., et al.: Analysis of gaze fixations using an open-source software. In: Proceedings of 10th IEEE International Conference on Cognitive Infocommunications, pp. 325–328 (2019)

69. Yadav, D., Kohli, N., Kalsi, E., Vatsa, M., Singh, R., Noore, A.: Unraveling human perception of facial aging using eye gaze. In: Proceedings of 2018 IEEE/CVF Conference on Computer Vision and Pattern Recognition Workshops, pp. 2221–2227 (2018)

70. Seha, S., Papangelakis, G., Hatzinakos, D., Zandi, A.S., Comeau, F.J.: Improving eye movement biometrics using remote registration of eye blinking patterns. In: Proceedings of 2019 IEEE International Conference on Acoustics, Speech and Signal Processing, pp. 2562–2566 (2019)

71. Thampan, J., Mohammed, F., Tijin, M., Prabhu, P., Rince, K.M.: Eye based tracking and control system. Int. J. Innov. Sci. Mod. Eng. (IJISME) **4**(10), 13–17 (2017)

72. Mohanraj, I., Siddharth, S.: A framework for tracking system aiding disabilities. In: Proceedings of 2017 IEEE International Conference on Current Trends in Advanced Computing (ICCTAC), pp. 1–7 (2017)

73. WebGazer.js. https://webgazer.cs.brown.edu/. Accessed 25 Mar 2022

74. Höffner, S.: Gaze tracking using common webcams. Master's thesis, Osnabrück University (2018)

75. Liu, Y.: Real-time pupil localization using 3D camera. UC Davis Works (2016)

76. Huan, Y., Osman, M., Jong, J.: An innovation-driven approach to specific language impairment diagnosis. Malays. J. Med. Sci. **28**(2), 161–170 (2021)

77. Papoutsaki, A., Gokaslan, A., Tompkin, J., He, Y., Huang, J.: The eye of the typer: a benchmark and analysis of gaze behavior during typing. In: Proceedings of 2018 ACM Symposium on Eye Tracking Research & Applications, pp. 1–9 (2018)

78. Acharjee, J., Deb, S.: Identification of significant eye blink for tangible human computer interaction. In; Proceedings of the 2021 International Conference on Advance Computing and Innovative Technologies in Engineering (ICACITE), pp. 179–183 (2021)

79. Kathpal, K., Negi, S., Sharma, S.: iChat: interactive eyes for specially challenged people using OpenCV Python. In: Proceedings of the 9th International Conference on Reliability, Infocom Technologies and Optimization (Trends and Future Directions), pp. 1–5 (2021)

80. Urunkar, A.A., Shinde, A.D., Khot, A.: Drowsiness detection system using OpenCV and raspberry pi: an IoT application. In: Sanyal, G., Travieso-González, C.M., Awasthi, S., Pinto, C.M.A., Purushothama, B.R. (eds.) International Conference on Artificial Intelligence and Sustainable Engineering. LNEE, vol. 837, pp. 1–5. Springer, Singapore (2022). https://doi.org/10.1007/978-981-16-8546-0_1

81. Tuhkanen, S., Pekkanen, J., Wilkie, R., Lappi, O.: Visual anticipation of the future path: predictive gaze and steering. J. Vision **21**(8), 25 (2021)

82. Schweizer, T., Wyss, T., Gilgen-Ammann, R.: Eyeblink detection in the field: a proof of concept study of two mobile optical eye-trackers. Milit. Med. **187**, e404–e409 (2021)

83. Velisar, A., Shanidze, N.: Noise in the machine: sources of physical and computation error in eye tracking with pupil core wearable eye tracker: wearable eye tracker noise in natural motion experiments. In: Proceedings of ETRA 2021 Adjunct: ACM Symposium on Eye Tracking Research and Applications, pp. 1–3 (2021)

84. Patayon, U., Gallegos, J.M., Mack, P., Bacabis, R., Vicente, C.: Signaling and pacing: a comparative study on evidence based stimuli using an eye tracking device. Proc. Comput. Sci. **179**(4), 313–320 (2021)

85. Guimaraes, L., Schirlo, G., Gasparello, G., Bastos, S., Pithon, M., Tanaka, O.: Visual facial perception of postsurgical cleft lip scarring assessed by laypeople via eye-tracking. J. Orthodontic Sci. **10**(1) (2021)

86. Al-Lahham, A., Souza, P., Miyoshi, C., Ignácio, S., Meira, T., Tanaka, O.: An eye-tracking and visual analogue scale attractiveness evaluation of black space between the maxillary central incisors. Dent. Press J. Orthodont. **26**(1) (2021)

87. Banire, B., Al Thani, D., Qaraqe, M., Mansoor, B., Makki, M.: Impact of mainstream classroom setting on attention of children with autism spectrum disorder: an eye-tracking study. Univ. Access Inf. Soc. **20**(4), 785–795 (2020). https://doi.org/10.1007/s10209-020-00749-0

88. Pierdicca, R., Paolanti, M., Quattrini, R., Mameli, M., Frontoni, E.: A visual attentive model for discovering patterns in eye-tracking data—a proposal in cultural heritage. Sensors **20**(7), 2101 (2020)

89. Sendi, Y., Khan, N.: A new approach towards evaluating the performance of maritime officers by the utilization of mobile eye tracking system and facial electromyography. Int. J. Recent Adv. Multidisc. Res. **8**(3), 6700–6706 (2021)

90. Zardari, B.A., Hussain, Z., Arain, A.A., Rizvi, W.H., Vighio, M.S.: QUEST e-learning portal: applying heuristic evaluation, usability testing and eye tracking. Univ. Access Inf. Soc. **20**(3), 531–543 (2020). https://doi.org/10.1007/s10209-020-00774-z

91. Sogo, H.: Sgttoolbox: utility for controlling SimpleGazeTracker from Psychtoolbox. Behav. Res. Methods **49**(4), 1323–1332 (2016). https://doi.org/10.3758/s13428-016-0791-4

92. Navarro-Tuch, S., Gammack, J., Kang, D., Kim, S.: Axiomatic design of a man-machine interface for Alzheimer's patient care. In: IOP Conference Series: Materials Science and Engineering, p. 1174 (2021)

93. Li, J., Chowdhury, A., Fawaz, K., Kim, Y.: Kaleido: real-time privacy control for eye-tracking systems. In: Proceedings of 30ty USENIX Security Symposium, pp. 1793–1810 (2021)

94. Hanke, M., Mathôt, S., Ort, E., Peitek, N., Stadler, J., Wagner, A.: A practical guide to functional magnetic resonance imaging with simultaneous eye tracking for cognitive neuroimaging research. In: Pollmann, S. (ed.) Spatial Learning and Attention Guidance. NM, vol. 151, pp. 291–305. Springer, New York (2019). https://doi.org/10.1007/7657_2019_31

95. Papoutsaki, A., Laskey, J., Huang, J.: SearchGazer: webcam eye tracking for remote studies of web search. In: Proceedings of the 2017 Conference on Conference Human Information Interaction and Retrieval, pp. 17–26 (2017)

96. Kim, K., Son, K.: Eyeball tracking and object detection in smart glasses. In: Proceedings of the 2020 International Conference on Information and Communication Technology Convergence (ICTC), pp. 1799–1801 (2020)

97. Aqel, M., Alashqar, A., Badra, A.: Smart home automation system based on eye tracking for quadriplegic users. In: Proceedings of the 2020 International Conference on Assistive and Rehabilitation Technologies (iCareTech), pp. 76–81 (2020)

98. Shlyamova, E., Ezhova, K., Fedorenko, D.: The capabilities of developing eye tracking for AR systems on the base of a microcontroller Raspberry Pi. In: Proceedings of the VI Conference Optics, Photonics and Digital Technologies for Imaging Applications (2020)

99. Kwiatkowska, A., Sawicki, D.: Eye tracking as a method of controlling applications on mobile devices. In: Proceedings of the 15th International Joint Conference on e-Business and Telecommunications (ICETE 2018), pp. 373–380 (2018)

100. Wanluk, N., Visitsattapongse, S., Juhong, A., Pintavirooj, C.: Smart wheelchair based on eye tracking. In: Proceedings of the 9th Biomedical Engineering International Conference (BMEiCON), pp. 1–4 (2017)

101. Hausamann, P., Sinnott, C., MacNeilage, P.R.: Positional head-eye tracking outside the lab: an open-source solution. In: ACM Symposium on Eye Tracking Research and Applications (ETRA 2020 Short Papers), pp. 1–5 (2020)

102. Caspi, A., Barry, M.P., Patel, U.K., Salas, M.A.: Eye movements and the perceived location of phosphenes generated by intracranial primary visual cortex stimulation in the blind. Brain Stimul. 14(4), 851–860 (2021)

103. Gasparello, G.G., et al.: The influence of malocclusion on social aspects in adults: study via eye tracking technology and questionnaire. Prog. Orthod. 23(1), 1–9 (2022). https://doi.org/10.1186/s40510-022-00399-3

104. Tanaka, O., Farinazzo, R., Ceiti, C., Martins, T., Souza, E., Melo, M.: Laypeople's and dental students' perceptions of a diastema between central and lateral incisors: Evaluation using scanpaths and colour-coded maps. Orthod. Craniofac. Res. 23(4), 493–500 (2020)

105. Moreva, A., Kompaniets, V., Lyz, N.: Development and oculographic research of the website design concept for inclusive education. In: 019 Ural Symposium on Biomedical Engineering, Radioelectronics and Information Technology (USBEREIT), pp. 276–279 (2019)

106. Ujbányi, T.: Examination of eye-hand coordination using computer mouse and hand tracking cursor control. In: 2018 9th IEEE International Conference on Cognitive Infocommunications (CogInfoCom), pp. 353–354 (2018)

107. Sari, J.N., Nugroho, L.E., Insap Santosa, P., Ferdiana, R.: Modeling of consumer interest on E-commerce products using eye tracking methods. In: Ghazali, R., Deris, M.M., Nawi, N.M., Abawajy, J.H. (eds.) SCDM 2018. AISC, vol. 700, pp. 147–157. Springer, Cham (2018). https://doi.org/10.1007/978-3-319-72550-5_15

108. Du, N., Yang, X.J., Zhou, F.: Psychophysiological responses to takeover requests in conditionally automated driving. Accid. Anal. Prev. 148, 105804 (2020)

109. Baee, S., Pakdamanian, E., Ordonez, V., Kim, I., Feng, L.: EyeCar: modeling the visual attention allocation of drivers in semi-autonomous vehicles. Cornel University arciv (2019)

110. Burdzik, R., Celiński, I., Młyńczak, J.: Study of the microsleep in public transport drivers. In: Siergiejczyk, M., Krzykowska, K. (eds.) ISCT21 2019. AISC, vol. 1032, pp. 63–73. Springer, Cham (2020). https://doi.org/10.1007/978-3-030-27687-4_7

111. Popelka, S., Dolezalova, J., Beitlova, M.: New features of scangraph - a tool for revealing participants' strategy from eye-movement data. In: Proceedings of the 2018 ACM Symposium on Eye Tracking Research & Applications (ETRA 2018) (2018)

112. Karthick, S., Madhav, K., Jayavidhi, K.: A Comparative study of different eye tracking system. In: AIP Conference Proceedings, p. 2112 (2019)

113. Hassija, V., Chamola, V., Saxena, V., Jain, D., Goyal, P., Sikdar, B.: A survey on IoT security: application areas, security threats, and solution architectures. IEEE Access 7, 82721–82743 (2019)

114. Sundstedt, V., Navarro, D., Mautner, J.: Possibilities and challenges with eye tracking in video games and virtual reality applications. In: SIGGRAPH ASIA 2016 Courses (SA 2016) (2016)
115. Raptis, G.E., Katsini, C.: Analyzing scanpaths from a field dependence-independence perspective when playing a visual search game. In: ACM Symposium on Eye Tracking Research and Applications, pp. 1–7 (2021)
116. García-Baos, A., et al.: Novel interactive eye-tracking game for training attention in children with attention-deficit/hyperactivity disorder. Primary Care Companion CNS Disord. **21**, 26348 (2019)
117. Harper, L., et al.: The ESPU research committee: the impact of COVID-19 on research. J. Pediatr. Urol. **16**(5), 715–716 (2020)
118. Eyegaze Edge®. https://eyegaze.com/users/. Accessed 19 Apr 2022
119. Lapakko, D.: Communication is 93% nonverbal: an urban legend proliferates. Commun. Theater Assoc. Minnesota J. **34**(1), 2 (2007)

Group Decision-Making Involving Competence of Experts in Relation to Evaluation Criteria: Case Study for e-Commerce Platform Selection

Zornitsa Dimitrova[1] , Daniela Borissova[1,2(✉)] , Rossen Mikhov[1] , and Vasil Dimitrov[1]

[1] Institute of Information and Communication Technologies at the Bulgarian Academy of Sciences, 1113 Sofia, Bulgaria
{zornitsa.dimitrova,daniela.borissova,rossen.mikhov, vasil.dimitrov}@iict.bas.bg
[2] University of Library Studies and Information Technologies, 1784 Sofia, Bulgaria

Abstract. The ongoing digital transformation focuses on adapting companies to new digital technologies. In the digital age, the IT market proposes a huge amount of different software systems with a variety of capabilities. In this regard, the current article deals with the problem of evaluation and selection of the most appropriate e-commerce platform as a promising tool for improving business processes. For this purpose, groups of main evaluation criteria have been identified and an additional set of sub-criteria for each of them has been assigned. To form the final group decision, the decision makers' preferences are integrated in such a way to take into account each evaluation with different importance. This is done by formulation a new mathematical group decision model considering the differences in background experience, expertise, and qualifications of each decision maker. Instead of considering the weighted coefficient for the DM's expertise with equal importance towards all evaluated criteria, it is expressed by a vector representing the expertise of DM to each criterion. This allows for a more accurate assessment when forming the final group decision. In addition, a set of objective criteria have been defined, with filled-in normalized assessments of alternatives based on research of the options that are offered by the specific alternatives. The numerical testing demonstrates the applicability of the proposed mathematical model for group decision-making considering the decision makers' expertise expressed via a matrix of weighted coefficients over evaluation criteria.

Keywords: Group decision-making · Multi-criteria decision analysis · Digital transformation · e-Commerce platforms

1 Introduction

Today, digital capabilities are critical, as companies with an e-commerce manager are more likely to take digital exports faster than companies that rely on a traditional manager, regardless of the size of the company [1]. Some authors show that the processes

D. Simian and L. F. Stoica (Eds.): MDIS 2022, CCIS 1761, pp. 42–53, 2023.
https://doi.org/10.1007/978-3-031-27034-5_3

of international e-commerce development require significant initial investment and an appropriate e-commerce strategy to ensure optimal use of assets [2]. On the other hand, the adoption of e-commerce can enable a fully electronic transaction process for efficient information and capital flow management [3]. Using digital technologies contribute for creation or modification business processes according to the different business needs [4]. These processes are the core of digital transformation. In this regard, any attempt to improve the business models could contribute not only to companies' innovation but could be considered a major step for companies' competitiveness. For example, using a business intelligence framework for decision-making for effective management [5]. To cope with the variety of multi-factor features of business decisions the proposed flexible approach aiming to support the group decision making (GDM) using different strategies could be helpful [6].

In the digital age, every business is affected to a different degree from the contemporary trends related to e-business. This is due to the growing development of information and communication technology and lately, it is motivated by the pandemic Covid-19 situation. To improve and manage the processes related to e-commerce, different software products are developed. Such e-commerce platforms provide an opportunity for companies to access global markets. It should mention recent investigation were the results show the improved user attitudes toward e-commerce after the COVID-19 outbreak to avoid risks and social distance [7]. Authors found that existent winning opportunities for retail divisions when sharing data platform's sales [8]. The relation between the e-commerce platform and online selling through the e-commerce platform is challenging due to the different parameters of platforms. A significant factor in determination of the customer's value is the awareness of the potential of the web and the ability to use it [9]. That is why the use of a proper e-commerce platform is highly important as each of them are own specific parameters. It is shown that provided contracts by e-commerce platforms affect the operation and financing decisions of online retailers. This fact is based on the results that recommend a fixed commission rate of e-commerce platform that is not too large as the profits from online retail could be greatly reduced [10]. During the selection and evaluation process of a suitable e-commerce platform, the deployment and security issues need to be taken into account [11]. It is worth to mention also that e-commerce platforms ecosystems are related to the content management systems [12]. Along with this, it should consider also the availability of corresponding digital competencies regarding security when using such platforms [13].

For the selection of the most appropriate e-commerce platform, it is needed to identify some evaluation criteria. Once these criteria have been established, a group of competent experts is needed to carry out the evaluation and selection process. All of this makes such problems be considered as problems of multi-criteria analysis. To cope with such kind of problems there are a variety of classic models like SAW, WPM, AHP/ANP, TOPSIS, ELECTRE, PROMETHEE, etc. Some of them are known as outranking where pairwise comparison between alternatives is needed like ELECTRE, and PROMETHEE [14]. Other models rely on multi-attribute utility theory like SAW and WPM [15]. AHP generates trade-offs that are useful to the decision maker (DM), while TOPSIS determine the alternative that has the shortest geometric distance from the positive ideal solution [16–18]. In addition to these well-known models, new models for group decision-making are developing to cope with particular problems for fast evaluation [19]. A systematic review of method combinations for solving multi-criteria decision analysis problems is presented in [20]. The models based on the Multi-Attribute Utility Theory could be seen as promising as they rely on the evaluation of attributes performance and overall estimation is realized by aggregating them via the usage of a unique utility function [21].

Identification of the criteria for e-commerce platforms evaluation is a critical stage as the performance of these criteria will determine the best choice. The second critical stage is the determination of experts and the relation of their competency regarding the particular decision-making problem. These two essential stages are the subject of the current article where the main criteria are expressed by sub-criteria and a proper group decision-making model for evaluation and selection of e-commerce platform is formulated.

The rest of the article follows the structure: Sect. 2 provide a detailed description of the problem along with determined evaluation criteria, Sect. 3 describe the proposed mathematical group decision-making model for aggregation of individual preferences, Sect. 4 illustrates the numerical application of the proposed group decision-making model, Sect. 5 contains the relists analysis and discussion, while Sect. 6 provides some conclusions.

2 Problem Description

The management of an e-commerce business is a complex initiative and for its successful development it is necessary to apply knowledge and skills from different fields – sales, marketing, information technology, business organization, and management. The current problem concerns the evaluation and the selection of the most appropriate e-commerce platform as a promising tool for improving business processes. The particular case is related to a small company that have been identified a set of three suitable and well-established B2C platforms – WooComerce (https://woocommerce.com/), Shopify-Basic (https://www.shopify.com/), and BigComerce Essentials-Standard (https://www.bigcom merce.com/) to choose from. Considering the variety of activities that should be taken into account it is needed to divide the evaluation criteria into different groups. The authors have been identified four main groups of criteria have been for the specific problem. Each of these main groups is composed of a different number of sub-criteria, as shown in Table 1.

The first group includes general and objective criteria that are important for the operation of the business and the use of the e-commerce platform as a whole. It is important that the platform corresponds to the goals, capabilities and scale of the business in terms of price, staff, and plans for future development. The second group of criteria is related to commerce and the organization of the sales process. There are criteria such as inventory management (C6), which is important for the organisation of the sales, or analytics criterion (C9) that contributes to the improvement of the planning. The third group covers the technical criteria. Since the operational environment is online, the e-commerce platform has to meet requirements for speed, accessibility, information and payment security.

Table 1. Groups of criteria for e-commerce platforms evaluation.

Group 1: Common objective criteria	
C1	Pricing Model
C2	Payment Options variety
C3	Staff Accounts number
C4	Support Channels
C5	Free Templates
Group 2: Commerce criteria	
C6	Inventory Management
C7	Simplified Shopping Process
C8	Shipping Management
C9	Analytics
C10	Discount Management
Group 3: Technical criteria	
C11	Deployment
C12	Data and Payment Process Security
C13	Responsive Design
C14	Integrations and Apps
C15	Page Speed Performance
C16	SEO
Group 4: Marketing criteria	
C17	UX/UI Personalization
C18	Customer reviews Management
C19	Email/Multichannel Marketing

Maintenance, deployment methods and the ability to be detected by search engines are also important criteria in this group. The criteria included in group fourth are related to marketing. It is highly important for the e-commerce platform to be able to attract customers through its design and to be convenient and easy to use. Another key aspect is the potential for the business to be able to reach its target group of customers through various communication channels.

The group of experts that is going to choose the most preferred alternative is selected in such a way that it covers all the fields where expertise is needed. It is proposed that in the so formulated groups of criteria, different DMs expert weights be set in a manner that will reflect the key competencies and knowledge of the individual DM. The DM-1 is the Business Owner – he is supposed to have a vision for the development and the running of the business as well as the global goals that are pursued. The selected DM-2 is the Sales Manager. He should organize and manage sales, his expertise is in the field of supply and demand analysis as well as in pricing, so he receives the greatest relative weight in the criteria of Group 2. The responsibilities of the Software Engineer (DM-3) include the operation and technical maintenance of the platform that corresponds to the Group of criteria 3. The focus of the Marketing Manager (DM-4) is on the user experience and the capabilities for communication and interaction with customers. Another important aspect of his work is attracting new customers and he receives the highest weight in the criteria Group 4.

Along with these criteria groups, experts and alternatives, it is also important to have a proper utility function to aggregate the individual preferences of the DMs. It should be mention, that when forming the final group decision, the DMs' preferences should be integrated in such a way to take into account each preference with different importance. This is important as each DM has a different background experience, expertise, and qualification, and to be the final group decision more transparent this is a required condition.

3 Group Decision-Making Involving Competence of Experts in Relation to Evaluation Criteria

One of the well-known group decision-making model is the Simple Additive Weighting (SAW) known also as a weighted linear combination. Given number of alternatives are evaluated toward predefined evaluation criteria where each criterion is evaluated with an appropriate weighting factor expressing its importance [22, 23]. To get the group decision, the weighted average value for each alternative is determined by multiplying the weighting factor of each criterion by the evaluation of the alternative against this criterion A_i as follows [24]:

$$A_{SAW}^* = max \sum_{j=1}^{N} w_j a_{ij}, i = 1, 2, \ldots, M \tag{1}$$

$$\sum_{j=1}^{N} w_j = 1 \tag{2}$$

where the number of alternatives is denoted by index i, the number of criteria by N, w_j express the weight for criteria importance, and a_{ij} corresponds to the value of the i-th alternative in respect with j-th criterion.

In this formulation, all of the experts within group are taken with equal importance. Due to the fact that each expert has his own level of experience and knowledge and must be taken into account, a modification of SAW is proposed [25]:

$$A_{SAW}^{GDM} = max \sum_{k=1}^{Q} \lambda^k \sum_{j=1}^{N} w_j^k a_{ij}^k, i = 1, 2, \ldots, M \tag{3}$$

$$\sum_{j=1}^{N} w_j = 1 \tag{4}$$

$$\sum_{k=1}^{Q} \lambda^k = 1 \tag{5}$$

where λ^k is the weighted coefficient for the overall expertise of k-th DM from the formed group of Q experts ($k = 1, 2, .., Q$). The number of evaluation criteria is denoted by index j ($j = 1, 2, \ldots, N$) and alternatives are expressed index i ($i = 1, 2, \ldots, M$).

In contrast to the model (1)–(2) and model (3)–(5), in the current article the vector of weighted coefficients for the expertise of DMs $\{\lambda^1, \lambda^2, \ldots, \lambda^Q\}$ is represented by the following matrix $\Lambda = \{\lambda_j^k\}$:

$$
\Lambda = \begin{matrix} & DM^1 \ldots DM^k \ldots DM^Q \\ \begin{matrix} C_1 \\ \ldots \\ C_N \end{matrix} & \begin{bmatrix} \lambda_1^1 \ldots & \lambda_1^k \ldots & \lambda_1^Q \\ \ldots \ldots & \lambda_j^k & \ldots \\ \lambda_N^1 \ldots & \lambda_N^k \ldots & \lambda_N^Q \end{bmatrix} \end{matrix} \tag{6}
$$

The elements of this matrix are the coefficients λ_j^k that show the competency of particular DM with respect to evaluation criteria. Instead of considering the weighted coefficient for the DM's expertise with equal importance towards all evaluated criteria, it is expressed by a vector representing the expertise of DMs to each criterion. The use of this matrix allows for a more accurate to assess of the competence regarding the expertise of each DM to the specific evaluation criterion. When the set of evaluation criteria is quite large they could be divided into relevant subgroups. In this situation, the matrix (6) should contain the weighted coefficients for the importance of a particular DM with respect to the formed subgroups.

In such a way instead to consider individual contribution of DM to the evaluated criteria with equal importance, the coefficient that express the competency of DM (λ^k) is considered as a vector ($\lambda_1^k, \lambda_2^k, \ldots, \lambda_j^k, \ldots, \lambda_N^k,$) that represent expertise of DM to the particular criterion.

4   Case Study for Choosing an e-Commerce Platform

The numerical testing of the proposed modified SAW model was applied to the described problem of a selection of an e-commerce platform from given three alternatives, determined in advance four DMs, and four groups of criteria composed of selected overall 19 separate criteria. The range of evaluation scores is between 0 and 1, where a bigger value indicates better performance. The input data are shown in Table 2.

Table 2. Distributed expertise of DMs in accordance to the formulated groups of criteria.

Criteria Groups	DM1	DM2	DM3	DM4
Group 1 (Common objective criteria)	0.25	0.25	0.25	0.25
Group 2 (Commerce criteria)	0.15	0.50	0.15	0.20
Group 3 (Technical criteria)	0.14	0.14	0.57	0.15

(continued)

Table 2. (*continued*)

Criteria Groups	DM1	DM2	DM3	DM4
Group 4 (Marketing criteria)	0.25	0.25	0.10	0.40

For Group 1, all DMs have equal importance as common and objective criteria are considered. Group 2 reflecting the commerce criteria and therefore Sales Manager (DM-2) is with the highest expert weight of 0.50. The Software Engineer (DM-3), whose key competencies meet the criteria of Group 3 has the highest weight in it 0.57. The expertise of Marketing Manager (DM-4) with 0.40 is similarly reflected in Group 4.

The presented matrix in Table 3 shows the alternatives along with groups and evaluations toward criteria.

Table 3. Decision matrix – weights for the evaluation criteria and evaluations by DMs.

Criteria Groups	Criteria	Weights				A1 (WooComerce)				A2 (Shopify)				A3 (BigComerce)			
		DM1	DM2	DM3	DM4	DM1	DM2	DM3	DM4	DM1	DM2	DM3	DM4	DM1	DM2	DM3	DM4
Group 1	C1	0.10	0.03	0.02	0.01	0.667				0.667				0.667			
	C2	0.06	0.08	0.02	0.07	0.70				1.00				0.55			
	C3	0.05	0.02	0.05	0.01	0.10				0.20				1.00			
	C4	0.07	0.04	0.10	0.04	0.50				1.00				1.00			
	C5	0.02	0.05	0.03	0.07	0.25				0.75				1.00			
Group 2	C6	0.08	0.10	0.07	0.04	0.39	0.45	0.47	0.44	0.35	0.50	0.51	0.53	0.75	0.88	0.69	0.77
	C7	0.03	0.07	0.04	0.06	0.50	0.51	0.49	0.48	0.62	0.58	0.49	0.53	0.49	0.42	0.51	0.55
	C8	0.05	0.08	0.02	0.03	0.41	0.33	0.44	0.39	0.52	0.67	0.55	0.61	0.48	0.40	0.49	0.52
	C9	0.10	0.06	0.06	0.06	0.66	0.71	0.59	0.61	0.69	0.89	0.66	0.71	0.60	0.71	0.59	0.61
	C10	0.03	0.07	0.03	0.06	0.35	0.42	0.47	0.38	0.55	0.72	0.71	0.62	0.51	0.62	0.55	0.49
Group 3	C11	0.08	0.03	0.09	0.02	0.56	0.52	0.95	0.40	0.61	0.40	0.30	0.45	0.59	0.39	0.30	0.44
	C12	0.07	0.08	0.06	0.04	0.61	0.69	0.71	0.59	0.48	0.59	0.60	0.49	0.51	0.55	0.59	0.41
	C13	0.04	0.03	0.06	0.06	0.31	0.40	0.65	0.25	0.42	0.50	0.59	0.62	0.36	0.55	0.57	0.45
	C14	0.03	0.05	0.07	0.04	0.69	0.72	0.85	0.74	0.33	0.42	0.33	0.50	0.45	0.46	0.40	0.42
	C15	0.03	0.04	0.04	0.05	0.35	0.40	0.62	0.33	0.55	0.61	0.50	0.46	0.51	0.60	0.49	0.44
	C16	0.03	0.03	0.05	0.08	0.51	0.46	0.63	0.44	0.53	0.57	0.49	0.60	0.49	0.50	0.38	0.50
Group 4	C17	0.03	0.06	0.09	0.09	0.32	0.37	0.41	0.39	0.47	0.42	0.39	0.39	0.51	0.41	0.42	0.44
	C18	0.06	0.04	0.05	0.07	0.66	0.63	0.72	0.82	0.67	0.72	0.74	0.81	0.76	0.77	0.67	0.79
	C19	0.04	0.04	0.05	0.10	0.61	0.62	0.71	0.62	0.34	0.45	0.55	0.49	0.77	0.82	0.88	0.85

This matrix contains the weighting coefficients for the importance of the criteria that DMs have set among the 19 separate criteria. The sum of the weighting coefficients of each DM is in accordance with the given limit equal to 1. Complying with this condition, each DM can distribute the weighting coefficients in accordance with his point of view.

A special case is the evaluation of the alternatives according to the criteria C1-C5, which are part of Group 1 – Common and objective criteria. This allows the elicitation of objective evaluations for each of the alternatives and in this case to be assumed as common evaluation given by all of the DMs. For example, for the Pricing Model (C1), these calculations can be performed based on the subscription plans that are offered. The evaluation of Payment Options (C2) variety, Staff Accounts number (C3) and Free Templates (C5) can be performed according to the number of options offered. The evaluation of Support Channels (C4) can be performed according to the available channel options, such as Email Help Desk, 24/7 Live Operator, Chat, FAQs Forum, etc.

The evaluation of the alternatives according to the other criteria C6-C19 reflects the opinion of the individual DMs. The analysis of the results of the conducted experiment with the data from Table 2 and Table 3 using the proposed model (3)–(6) are discussed in the next section.

5   Results and Discussions

The final group decision result which gives the most preferred alternative to the e-commerce platform selection problem is illustrated in Fig. 1.

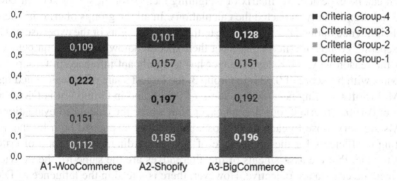

Fig. 1. The best choice of e-commerce platform in accordance to the influence of the groups of criteria.

The most preferred is A3-BigCommerce Essentials-Standard with an aggregate score of 0.668, followed by A2-Shopify-Basic with 0.640 and A1-WooCommerce with 0.594. Each of the alternatives has an advantage over the others in at least one of the groups of criteria. As it can be seen, WooCommerce is significantly more preferred by Criterion Group 3, receiving the highest score of 0.222, compared to 0.157 for Shopify and 0.151 for BigCommerce. However, it remains in the last place of the overall ranking as its result on the other groups of criteria is either the lowest or close to the lowest. Shopify gets the highest score in Criteria Group 2 (0.197), where it has a small lead over BigCommerce

(0.192), but at the same time, it has the lowest score in criteria Group 4 (0.101) where the difference with BigCommerce (0.128) is significant. BigCommerce is the favourite in criteria Group 1 with a score of 0.196 and Group 4 criteria with a score of 0.128. In the overall ranking, BigCommerce is the most preferred alternative because it has a balanced result and a good score on all groups of criteria. Although criteria Group 2 is not composed of the largest number of separate criteria, it has the greatest influence in the final result formed with 0.541. The smallest is the influence of criteria Group 4 with 0.339. The distribution of the calculated scores that DMs have given in accordance with the groups of criteria is given in Fig. 2.

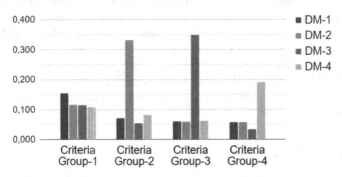

Fig. 2. DMs assessments per groups of criteria.

As it can be expected, the matrix of weighting factors for the importance of DM in Table 2 has significantly impacted the calculations. In all the groups of criteria, there is a visible influence of one DM. This is due to the higher weights of the expertise of DMs in the separate groups of criteria that meet their specific knowledge and competencies. This is the reason why, the Sales Manager has a significant influence in Group 2 and contributes with his score of 0.331, in Group 3, the largest share of the calculated result has DM-3 (Software Engineer) with a score of 0.349, and in Group 4 it is DM-4 with a score of 0.190. Criteria Group 1 is an exception, where the calculated evaluations of the DMs are very close to each other close. This can be explained both by the same weighting coefficients for the importance of DMs according to this group of criteria and by the fact that the assessments of the alternatives are the same for the individual DMs since the criteria are objective. However, there is a lead in the influence of DM-1 (0.154) and a lag of DM-4 (0.108). The reason for this is because DM-1 has given higher weight coefficients to the individual criteria of Group-1 compared to the other DMs and respectively DM-4 coefficients are lower. In summary, the DM with the largest share in the calculations is DM-2 with a score of 0.565, and the one with the smallest share is DM-1 with a score of 0.343.

The DMs assessment about different alternatives are illustrated in Fig. 3.

The highest scores of DM-1 (0.060) and DM-4 (0.072) were calculated for A3-BigCommerce (Fig. 3a). According to the DM-2, the C6 (Inventory Management) is the most important criterion and the alternative A-2 (Shopify) with overall score of 0.124 (Fig. 3b). The importance weight of C11-Deployment and the highest assessment of due to the open-source and flexibility in deployment make WooCommerce favourite with

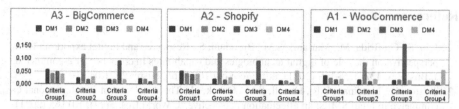

Fig. 3. DMs assessments per alternative.

result of 0.161 (Fig. 3c). The Software Engineer (DM-3) has a clear position in support of WooCommerce (A3) as can be seen in Fig. 3c.

The e-commerce platforms are the basis on which businesses build their online presence, and depending on the type of customers that the business focuses on, they can be targeted at either end-users like business to customers (B2C) or business to business (B2B). That is why the selection of the most appropriate e-commerce platform is of high importance for the business. Regardless of which e-commerce platform will be chosen, it should be compatible with the used devices and protocols, ensuring the proper data collection and processing. The use of appropriate software platform will create a prerequisite for creation of autonomous operations that bring processes together to tasks. It is worth noting that the use of such a model will contribute to better economic sustainability, as it is possible to take into account various factors tailored to the specific needs and requirements of e-commerce companies.

6 Conclusions

The current article deals with digital transformation and more precisely with adapting to new digital technologies related to e-commerce. The described problem concerns the identification and selection of the most appropriate e-commerce platform. For the goal, four main groups of criteria are determined and each of these groups is composed of sub-criteria. Along with the evaluation criteria, a group of competent experts determined in advance is involved to conduct the process of evaluation. The major contribution consists of the formulated mathematical model for aggregating DM evaluation scores in determining the final group decision. Instead of considering the weighted coefficient for the DM's expertise with equal importance towards all evaluated criteria, it is expressed by a vector representing the expertise of DM to each criterion. This allows the final group decision to be formed objectively, taking into account the level of competence of each DM with respect to the particular criterion. The proposed group evaluation and selection decision-making model has been applied to the selection of e-commerce platforms. The obtained results show the applicability of the described model when using defined groups of criteria and their sub-criteria. This well-structured problem can easily be coded as a web-based group decision-making tool. These activities are planned as future developments.

Acknowledgment. This work is supported by the Bulgarian National Science Fund by the project "Mathematical models, methods and algorithms for solving hard optimization problems to achieve high security in communications and better economic sustainability", KP-06-N52/7/19-11-2021.

References

1. Elia, S., Giuffrida, M., Mariani, M.M., Bresciani, S.: Resources and digital export: an RBV perspective on the role of digital technologies and capabilities in cross-border e-commerce. J. Bus. Res. **132**, 158–169 (2021). https://doi.org/10.1016/j.jbusres.2021.04.010
2. Tolstoy, D., Nordman, E.R., Hanell, S.M., Ozbek, N.: The development of international e-commerce in retail SMEs: an effectuation perspective. J. World Bus. **56**(3), 101165 (2021). https://doi.org/10.1016/j.jwb.2020.101165
3. Orji, I.J., Ojadi, F., Okwara, U.K.: The nexus between e-commerce adoption in a health pandemic and firm performance: the role of pandemic response strategies. J. Bus. Res. **145**, 616–635 (2022). https://doi.org/10.1016/j.jbusres.2022.03.034
4. Rachinger, M., Rauter, R., Muller, C., Vorraber, W., Schirgi, E.: Digitalization and its influence on business model innovation. J. Manuf. Technol. Manag. **30**(8), 1143–1160 (2019). https://doi.org/10.1108/JMTM-01-2018-0020
5. Borissova, D., Cvetkova, P., Garvanov, I., Garvanova, M.: A framework of business intelligence system for decision making in efficiency management. In: Saeed, K., Dvorský, J. (eds.) CISIM 2020. LNCS, vol. 12133, pp. 111–121. Springer, Cham (2020). https://doi.org/10.1007/978-3-030-47679-3_10
6. Borissova, D., Korsemov, D., Keremedchieva, N.: Generalized approach to support business group decision-making by using of different strategies. In: Saeed, K., Dvorský, J. (eds.) CISIM 2020. LNCS, vol. 12133, pp. 122–133. Springer, Cham (2020). https://doi.org/10.1007/978-3-030-47679-3_11
7. Kawasaki, T., Wakashima, H., Shibasaki, R.: The use of e-commerce and the COVID-19 outbreak: a panel data analysis in Japan. Transp. Policy **115**, 88–100 (2022). https://doi.org/10.1016/j.tranpol.2021.10.023
8. Niu, B., Dong, J., Dai, Z., Liu, Y.: Sales data sharing to improve product development efficiency in cross-border e-commerce. Electron. Commer. Res. Appl. **51**, 101112 (2022). https://doi.org/10.1016/j.elerap.2021.101112
9. Caputa, W., Krawczyk-Sokolowska, I., Pierscieniak, A.: The potential of web awareness as a determinant of dually defined customer value. Technol. Forecast. Soc. Chang. **163**, 120443 (2021). https://doi.org/10.1016/j.techfore.2020.120443
10. Chang, S., Li, A., Wang, X., Wang, X.: Joint optimization of e-commerce supply chain financing strategy and channel contract. Eur. J. Oper. Res. (2022). https://doi.org/10.1016/j.ejor.2022.03.013
11. Yoshinov, R., Kulikov, I., Zhukova, N.: Methods of composing hierarchical knowledge graphs of telecommunication networks. Probl. Eng. Cybern. Robot. **72**, 69–78 (2020). https://doi.org/10.7546/PECR.72.20.07
12. Engert, M., Evers, J., Hein, A., et al.: The engagement of complementors and the role of platform boundary resources in e-commerce platform ecosystems. Inf. Syst. Front. (2022). https://doi.org/10.1007/s10796-021-10236-3
13. Chehlarova, N., Tsochev, G., Kotseva, M., Miltchev, R.: Digital competencies of public administration employees related to cybersecurity. In: 12th National Conference with International Participation (ELECTRONICA), pp. 1–4 (2021). https://doi.org/10.1109/ELECTRONICA52725.2021.9513705
14. Corrente, S., Greco, S., Slowinski, R.: Multiple criteria hierarchy process with ELECTRE and PROMETHEE. Omega **41**(5), 820–846 (2013). https://doi.org/10.1016/j.omega.2012.10.009
15. Huang, Y.-S., Chang, W.-C., Li, W.-H., Lin, Z.-L.: Aggregation of utility-based individual preferences for group decision-making. Eur. J. Oper. Res. **229**(2), 462–469 (2013). https://doi.org/10.1016/j.ejor.2013.02.043

16. Saaty, T.L.: Fundamentals of the analytic hierarchy process. In: Schmoldt, D.L., Kangas, J., Mendoza, G.A., Pesonen, M. (eds.) The Analytic Hierarchy Process in Natural Resource and Environmental Decision Making. Managing Forest Ecosystems, vol. 3. Springer, Dordrecht (2001). https://doi.org/10.1007/978-94-015-9799-9_2

17. Hwang, C.L., Yoon, K.: Methods for multiple attribute decision making. In: Multiple Attribute Decision Making. LNEMS, vol. 186, pp. 58–191. Springer, Berlin, Heidelberg (1981). https://doi.org/10.1007/978-3-642-48318-9_3

18. Zyoud, S., Fuchs-Hanusch, D.: A bibliometric-based survey on AHP and TOPSIS techniques. Expert Syst. Appl. **78**, 158–181 (2017). https://doi.org/10.1016/j.eswa.2017.02.016

19. Borissova, D., Dimitrova, Z., Dimitrov, V.: How to support teams to be remote and productive: Group decision-making for distance collaboration software tools. Inf. Secur. **46**, 36–52 (2020). https://doi.org/10.11610/isij.4603

20. Marttunen, M., Lienert, J., Belton, V.: Structuring problems for multi-criteria decision analysis in practice: a literature review of method combinations. Eur. J. Oper. Res. **263**(1), 1–17 (2017). https://doi.org/10.1016/j.ejor.2017.04.041

21. Bystrzanowska, M., Tobiszewski, M.: How can analysts use multicriteria decision analysis? TrAC Trends Anal. Chem. **105**, 98–105 (2018). https://doi.org/10.1016/j.trac.2018.05.003

22. Afshari, A., Mojahed, M., Yusuff, R.M.: Simple additive weighting approach to personnel selection problem. Innov. Manag. Technol. **1**(5), 511–515 (2010)

23. Ginevicius, R., Podvezko, V.: Multicriteria evaluation of Lithuanian banks from the perspective of their reliability for clients. J. Bus. Econ. Manag. **9**(4), 257–267 (2008). https://doi.org/10.3846/1611-1699.2008.9.257-267

24. Triantaphyllou, E.: Multi-criteria decision making methods. In: Multi-criteria Decision Making Methods: A Comparative Study. Applied Optimization, vol. 44, p. 320. Springer, Boston, MA (2000). https://doi.org/10.1007/978-1-4757-3157-6_2

25. Korsemov, D., Borissova, D.: Modifications of simple additive weighting and weighted product models for group decision making. Adv. Model. Optim. **20**(1), 101–112 (2018)

Transparency and Traceability for AI-Based Defect Detection in PCB Production

Ahmad Rezaei[1]([✉]) [iD], Johannes Richter[1,2] [iD], Johannes Nau[1] [iD],
Detlef Streitferdt[1] [iD], and Michael Kirchhoff[1] [iD]

[1] Technische Universität Ilmenau, Helmholtzplatz 5, 98693 Ilmenau, Germany
{ahmad.rezaei,johannes.richter,johannes.nau,detlef.streitferdt,
michael.kirchhoff}@tu-ilmenau.de
[2] GÖPEL Electronic GmbH, Jena, Germany
j.richter@goepel.com

Abstract. Automatic Optical Inspection (AOI) is used to detect defects in PCB production and provide the end-user with a trustworthy PCB. AOI systems are enhanced by replacing the traditional heuristic algorithms with more advanced methods such as neural networks. However, they provide the operators with little or no information regarding the reasoning behind each decision.

This paper explores the research gaps in prior PCB defect detection methods and replaces these complex methods with CNN networks. Next, it investigates five different Cam-based explainer methods on eight selected CNN architectures to evaluate the performance of each explainer. In this paper, instead of synthetic datasets, two industrial datasets are utilized to have a realistic research scenario. The results evaluated by the proposed performance metric demonstrate that independent of the dataset, the CNN architectures are interpretable using the same explainer methods. Additionally, the Faster Score-Cam method performs better than other methods used in this paper.

Keywords: PCB defect detection · Automatic optical inspection · Cam explainer methods

1 Introduction

Current PCB production chains use several rounds of debugging and physical observations to find defective components or the root cause wherever possible. To support this procedure, emerging Industry 4.0 led to advancements in fault detection and classification approaches [4]. These advancements replaced the effort-intensive labor with data gathering sensors and cameras, high-speed data networks, and machine learning algorithms, which provided vast volumes of data, real-time access to the data, and a higher level of automation in performing the debugging experiments. Nowadays, defect detection is embedded into the production chains and is partly achieved by AOI systems with higher performance

D. Simian and L. F. Stoica (Eds.): MDIS 2022, CCIS 1761, pp. 54–72, 2023.
https://doi.org/10.1007/978-3-031-27034-5_4

and lower fault detection time. AOI is further improved by using AI instead of traditional heuristic algorithms. This is beneficial as it needs fewer parameter settings to configure; less time is spent on the programming phase; AI speeds up the fault detection procedure and delivers traceable results; and finally, AI enables the operator to define WHAT they want to inspect, not HOW.

The routine for manufacturing PCBs and utilizing AOI involves several steps. First, the client shares CAD data with the electronics manufacturing service (EMS). Next, EMS uses the provided CAD data and sets up the final printed circuit board assembly (PCBA) inspection program based on these data. Subsequently, in this setup, the inspection program uses position and dimension data for each electronic component and its sub-components(i.e., solder leads) to locate and crop images (taken by a high-resolution industrial camera). Upon these cropped images, the AOI system executes a classifier function resulting in a pass/fail for a component or sub-component. Previously, this classifier function was based on heuristics, but currently, it is replaced with AI-based approaches to reach more detailed and faster classification. Additionally, the initial images are currently cut into sub-images of electronic components down to solder pads using more promising AI-based techniques that better cover the components and defective areas. The paper only focuses on the prior approaches (defects classification for the PCB domain), and the latter is out of the scope of this paper.

To elaborate on the shortcomings and explore the research gaps, the common PCBA defects are first determined, and state-of-the-art defect detection approaches are explored. Then, the open issues in the publications are highlighted. Next, the paper emphasizes the development of a more practical defect debugging scenario, which maintains the PCBA diagnosis constraints.

Common PCBA defects leading to malfunction or decreased reliability are listed on the NPL Electronics Interconnection Group [1]. These defects mainly refer to board defects, component defects, component misplacement, soldering issues, foreign objects, or pseudo faults. Common defects used in the literature and this paper are represented in Table 1 using [21,25].

In the literature, the current state-of-the-art methods focus on object detection networks, which supersede conventional heuristic methods in terms of accuracy, time consumption, and precision. Authors in TDD-net [10] first proposed a PCB defect detection based on the Faster-RCNN architecture for PCB tiny defects. They further improved their research by using upcoming versions of object detectors along with modifications to make them useful for tiny object detection. For instance, the following models are used to fulfill speed, low false detection rate, and average detection accuracy; YOLOv2 [3], ensemble architecture of YOLOv2 and Faster-RCNN [18], improved YoloV3 [17], and improved versions of YOLOV4 [19,20]. However, these state-of-the-art methods lack the viability of being used in an industrial setup due to three reasons; firstly, the target position of all components and solder joints is given by the CAD data, and deviations from position and orientation should not be corrected but handled as a defect; secondly, available datasets (Peking university synthetic defect dataset in [10], PCB synthetic dataset in [15,32], and Deep PCB dataset in [32]) only focus on the board defects and not the components; finally, they do not con-

Table 1. Common defects in PCB manufacturing environments.

Defect name	Defective part	Brief description
Tombstone or billboard	Component	The component is disconnected from soldering pads or is standing on the side
Missing component	Component	The component is either not placed by the pick-and-place-arm, or it was removed during the process
Displaced component	Component	The component is either in the wrong orientation or displaced by the surface tension of the solder
Pseudo defects	Component/ soldering pan	Caused by mobile particles on the board; can lead to pseudo defect detection
Not soldered	Soldering pan	The soldering paste has been removed before the melting process
Not enough solder	Soldering pan	The solder paste is insufficient, and the joint is partly connected
Short circuit	Soldering pan/ board	Unwanted connection between board areas or joints leads to an unwanted electrical flow
Open circuit	Board	Demolished electrical connection in the middle of copper lines in the board
Spurious copper	Board	Damaged parts of the copper, which do not disconnect the electrical flow
Missing hole	Board	Not manufactured THT holes on the board

sider the volume of the image taken by an industrial camera (up to a Gigabyte per image), the useful PCB component data for localizing the components, and deterioration of detection quality in presence of trivial scale down and image cropping. Therefore, as long as the complete localization and classification process using CAD is not replaced with the AI systems, pure classifier architectures are a better drop-in replacement for current defect detection functions.

Sophisticated heuristic approaches do not give any insights into how the decisions on defects were made. Thus, a gap persists between the quality requirements of the EMS and the provided solution of the AOI system. The same applies to neural networks and makes them hard to understand for a (non-expert) user. The main motivation of this paper is the reasoning for defect classifiers to bridge this gap. Hence, AI does not decide everything alone in its black box. It is further embedded in processes involving human experts to provide reasoning on why and how a defect was identified.

Due to the intra-class similarity in various defects (happening in the same parts) and after reaching a high accuracy, models are inspected using explainer methods to avoid using biased reasoning. Several so-called explainer methods are used to provide the EMS with reasoning on classifications (elaborated more in the Sect. 2). To perform this procedure, PCB inspection training data is needed, which varies considerably based on the PCB and manufacturing line for different EMSs. Henceforth, each client/production line/PCB product requires a specific dataset to train the classifiers. This paper uses pre-trained networks to reduce the technical hurdle and limitations of collecting large datasets in this data-scarce environment. An additional problem is a potential change in manufacturers for the used components and, thus, the need to recollect and retrain classifiers. It seems more economical to stick with pre-trained architectures and small problem-centric datasets.

This paper is organized as follows. Section 2 provides insights on used CNN models and explainer methods. Section 3 defines the information about datasets, training setup and evaluation metrics. Section 4 proposes a single performance metric upon already defined evaluation metrics and explains the experiment procedure to evaluate various explainer methods using multiple datasets and models. The designed experiments evaluate various explainer methods across different models and datasets. The results for these experiments are discussed in Sect. 5 and in Sect. 7, a conclusion of the findings is given and the potential future directions of research are highlighted.

2   Preliminary Knowledge

This paper suggests using CNN classifier architectures for PCB defect diagnosis and investigates the reasoning behind each prediction using several explainer methods. This section provides more insights on selected CNN models for the experiments. The next section presents explainer methods to choose a set of these methods for the reasoning of CNN models during the experiments.

2.1   CNN Architecture Selection

To perform the classification task, eight pre-trained CNN architectures (publicly available in the Tensorflow framework [2]) are used. The next part and Table 2 briefly introduce the models used. VGG16 and VGG19 [27] are among early deep neural network classifiers, which use $3 * 3$ convolution instead of $7 * 7$ convolution to reduce parameter quantity. They have several convolutional and pooling layers for feature extraction and feed the activation maps into three fully-connected layers for classification. MobileNetV1 [14] and MobileNetV2 [23] have fewer parameters and perform faster inference in comparison with other models. Additionally, the proposed experiments include some models from ResNet family [13] that use residual connections to extract enhanced features from the input images. Finally, Big Transfer (BiT) CNN is used in this paper as this model is deemed to reach the large dataset regimes even with very few samples per class [16].

As discussed earlier, each PCB production line has its specific application domain (avionics or consumer electronics) dataset to train the AOI system. In this paper, two different industrial datasets (discussed in Sect. 3.1) are used for training the CNN networks. These datasets are classified by experienced industry operators, making them suitable for proceeding with a realistic scenario. The

Table 2. Information regarding the selected CNN models (top-1 accuracy on ImageNet [9] dataset is reported).

Model	#Parameters (M)	Speed (ms)	Top-1 Acc. (%)
VGGNet-16	134.2	49.4	72.2
VGGNet-19	139.6	51.2	–
MobileNet-v1	3.2	45.6	70.6
MobileNet-v2	2.2	48.5	72.0
ResNet-50	23.6	51.3	76.3
ResNet-101	42.6	59.3	77.2
ResNet-152	58.3	67.7	77.7
BiT	23.5	62.4	85.3

pre-trained CNN models are retrained using these PCB datasets to reach the highest potential prediction accuracy. Based on related work research and our own expertise the following section presents twelve explainer methods and selects the most prominent ones for evaluating the PCB defect classification.

2.2 Explainer Methods for CNNs

The High-Level Expert Group on AI presented set of guidelines and ethics to fulfill trust issues in AI applications [28]. The transparency and traceability of any decision made by AI models are one of the seven key requirements for integrating AI into real-life applications. The traceability (explanation on each decision) is vital for AOI systems in PCBA production lines as well; for instance, if a PCBA produced for health care sectors passes the AOI system that uses unreliable defect detectors, utilizing this PCB board may lead to life-threatening catastrophes in the presence of undetected defects on PCB.

Based on authors in [6], black box explainers are divided into two subcategories of global and local methods. Global methods deliver insights into the overall knowledge of the network. However, local methods explain each target image fed into the model. Furthermore, there are ante-hoc and post-hoc explainers. Ante-hoc methods apply changes to the input and go through model parameters to measure the importance of present features in the image for the final prediction result. Post-hoc methods entail the explanation by only considering the impact of model parameters on the outcome. Finally, explainers may be model-specific (MS) or model-agnostic (MA) if used for a specific model.

In this research scenario, a local model-agnostic explainer method, whether post-hoc or ante-hoc, is used to examine the reasoning behind classifying each PCB image. Table 3 provides information regarding the investigated explainer methods, such as two important factors execution time and discriminativity. Execution times in the table are measured in a setup with Tesla P40 GPU, except for the times with the literature reference. The discussion is continued by representing various methods and finalizing selected explainer methods.

Explainers represented in Table 3 follow one of the following approaches.

Table 3. Investigated explainer methods (the selected methods are highlighted).

Explainer's name	Properties	Execution time (seconds)	Discriminative
Lime [22]	local, Ante-hoc, MA	19.1	✓
Vanilla-Gradient(VG) [29]	local, Post-hoc, MA	-	✗
Guided-Backpropagation(GB) [29]	local, Post-hoc, MS	-	✗
Cam [34]	local, Post-hoc, MA	-	✗
Grad-Cam [24]	local, Post-hoc, MA	0.491	✓
Grad-Cam++ [7]	local, Post-hoc, MA	0.307	✓
Score-Cam [31]	local, Post-hoc, MA	2.220	✓
Faster Score-Cam	local, Post-hoc, MA	0.513	✓
Group-Cam [33]	local, Post-hoc, MA	1.030	✓
Cluster-Cam	local, Post-hoc, MA	1.340	✓
DDeconv [26]	local, Post-hoc, MA	18.75 [26]	✓

(MS = model-specific MA = model-agnostic)

1. Gradient-based: They analyze CNN model gradients and visualize them on a sensitivity map, i.e., Vanilla gradient and Guided-Backpropagation [29].
2. Super-pixel: These methods derive the explanation by classifying several proposed areas from the input. The explanation is in the form of Super-pixels to show the importance, i.e., Lime [22].
3. Class activation mapping (Cam)-based: The explanation is derived by considering feature activation maps in the CNN architecture. Cam [34] methods visualize the results as a heat map to show the relative importance in the explanation.

CNN model extract features related to all available classes, and then the classifier decides on one class based on the strength of the feature related to that class. However, using only gradient-based methods does not focus on specific features per class and highlights unnecessary features for the predicted class. This drawback (not achieving the discriminatory explanation for each class) hinders further usage of these methods in this paper.

Furthermore, Lime, Score-Cam [31], and DDeconv [26] methods are out of scope for this paper due to slower performance compared to other methods. In this regard, Lime does not provide the user with the relative importance of its explanation. In contrast, Cam methods use a heat map to show the relative importance of each explanation.

The Cam [34] method suggests using feature activation maps as an explanation for CNN predictions. This approach is the foundation of a series of Cam-based methods that were and are constantly improved to achieve better reasoning. First, the discrimination between classes is added by using gradients concerning the fully-connected layer as a weight for feature maps in Grad-Cam [24]. Next, Grad-Cam uses the first derivative to achieve the gradients, so Grad-Cam++ [7] improved this method by taking into account the second and third derivatives for calculating the gradients. Subsequently, authors in 2020 proposed Score-Cam [31] which computes the importance of each feature map by using it as a mask and comparing the probability drop on the target class. Problems such as vanishing gradients or noisy gradients make gradients not reliable for explaining CNNs. Hence, this method relies only on this importance as a weight for feature maps, but executing all feature maps as masks on the input is time-consuming. Henceforth, this paper uses Faster Score-Cam as a faster version of this method, which applies the same procedure over a few groups of feature maps. The areas outside the mask are set to zero, which may lead to an erroneous explanation. After the proposal of Score-Cam, the authors in 2021 have published the Group-Cam method [33] that benefits from both using gradients and applying masks on the input. Group-Cam involves feature map grouping for faster execution and uses the image's Gaussian Blur instead of zero values for areas outside the mask to enhance the explainer quality. Later in 2021, and following the same approach as Group-Cam, Spectral-Clustering Self-Matching Cam (SC-SM-Cam) method [11] was proposed that benefits from a spectral clustering algorithm for feature map grouping. After deriving the explanation, it leverages a post-processing technique (Self-matching Cam [12]), applicable to every Cam-based method, to improve the explanation localization drop during the upscaling procedure. This paper only focuses on the basic explainer methods, the SC-SM-Cam method is applied without the post-processing step (named as Cluster-Cam) as the post-processing step is an additional step that can be applied to other methods as well.

In conclusion, a series of five Cam methods are selected in this paper to provide reasoning for PCB defect classification (highlighted in the Table 3). In the next section, the paper elaborates on datasets, model settings, and evaluation metrics used.

3   Experiment Requirements and Metrics

First, the datasets and training setups are introduced to proceed with the explainer assessments. Then, information is provided on evaluation metrics used for comparing explainer methods.

3.1   Datasets and Training Setup

Using synthetic defect datasets due to privacy issues and scarcity of real-world data is common in the literature [10,15,32]. However, for having a realistic analysis, this paper benefits from two industrial datasets (represented in Table 4). The defect labels and bounding boxes are provided in the datasets for assessing the explainer methods.

For the preprocessing step, the data is transformed into (224, 224, 3) size and normalized between 0 and 1. Next, 80% are used for training and 20% of the datasets for evaluation purposes. To reduce the impact of imbalance present in the datasets, the SMOTE-TOMEK algorithm is used [5]. The data is oversampled with the synthetic minority over-sampling technique (SMOTE) algorithm [8], and then cleaned by removing similar data points with the TOMEK algorithm [30]. This results in 714 and 852 images for training the Dataset_v1 and Dataset_v2.

Table 4. Industrial PCB domain datasets used in this paper.

Dataset_v1		Dataset_v2	
Defect	#Images	Defect	#Images
Component missing	8	Component missing	178
Not soldered	50	Not soldered	71
Short circuit	52	Short circuit	34
Pseudo defect	127	Pseudo defect	52
Tombstone - Billboard	18	Tombstone	45
Component moved	10	Component moved	10
Too much solder	10		
Total	275	Total	390

All CNN models are trained based on the same hyper-parameter setting in Table 5, which achieve maximum accuracy after training. A learning rate of 0.01 is set for the VGGNet family and 0.001 for other models. The training results are evaluated using the accuracy and F1 score, of which the latter provides more valuable insights into how well the model has learned imbalanced classes.

Table 5. Train hyper-parameters.

Variable	Associated value
#Classes	7 or 6
batch	4
max_epochs	100
learning rate	0.01 and 0.001
optimizer	Adam
patience	3
min_delta	0.1%
image_size	(224, 224, 3)

3.2   Explainer Evaluation Metrics

The efficacy of the heat maps is evaluated through several qualitative and quantitative metrics. The prior is done via observing a subset of samples for a predefined goal, which is not scalable and prone to biased suggestions of each individual. The latter, however, is a measurable variable that interprets the importance of each heat map for the classification. The paper focuses on these scalable and reliable quantitative metrics, further developed in the next section as a performance metric for easier comparison across explainers.

Here, three important quantitative metrics are studied to evaluate the explainer methods. First, to compute the conservation metric, the heat map is used as a mask for the input image to measure the relevance of features within the heat map for top-class prediction probability (see Eq. 1). In other words, the input image area which is within the heat map area higher than a given threshold is preserved, and the conservation metric is computed using the masked input and the original input prediction probabilities from Eq. 1. The threshold value used for heat map areas mask is set to 0.5 in this paper.

The conservation value for image I ranges between $Conservation_{I_{cons.}^{max}} = (\frac{\frac{1}{N}-1}{\frac{1}{N}})$ to 1, where $Conservation_{I_{cons.}^{max}}$ contributes to the highest probability increase for N number of classes. For instance, conservation for the case of having $N = 6, 7$ maps to ranges $(-5, 1]$ and $(-6, 1]$. Negative values for the conservation mask are due to the increased probability for the preserved area. They do not reveal whether the explainer performs better than having a 0 probability drop for the preserved area. In more detail, a preserved area in such cases has more relevant features than the features the model focuses on, which means the model is performing poorly in focusing on the relevant feature in the feature maps. Thus, the negative values are neglected as they imply the potential for improvement on the model rather than the explainer performance. Furthermore, only the range $[0, 1]$ is considered for minimum and maximum probability drop.

$$Conservation_I = \frac{Pr(I_{original}) - Pr(I_{preserved})}{Pr(I_{original})} \quad (1)$$

Second, the occlusion test deduces how well the heat map covers the important features in the decision-making by observing the probability drop due to occluding heat map regions (0.5 threshold) from the original image (see Eq. 2). Upon the increase in the classification probability, the occlusion metric has the same positive and negative range as the conservation metric. Only the range $[0, 1]$ is included with the same set of arguments, contributing to the explainer's evaluation.

$$Occlusion_I = \frac{Pr(I_{original}) - Pr(I_{occluded})}{Pr(I_{original})} \quad (2)$$

Finally, the localization metric (formulated in Eq. 3) assesses how much of the heat map covers the defective areas within the provided bounding box ($Bbox$) for the image I. This metric ranges from 0 to 1, and values close to one imply the explainer is referring to the defective area. However, the values close to zero indicate that the explainer focuses on areas outside the defective area.

$$Localization_I = \frac{Area_{cam} \in Bbox}{Area_{cam}} \quad (3)$$

The following section highlights the need for a unified metric for explainer comparison, and a single performance metric is proposed for further experiments.

4 Proposed Explanation Score and Experiment Design

In assessing the underlying reasoning behind PCB defect classifications, the variability of the models and data in use should be considered. In this regard, the impact of different models on the same explainer methods is investigated, and a conclusion is drawn on the suitable Cam method used for PCB defect detection models. Next, a performance comparison between different explainer methods is conducted for the same CNN models across different datasets. This is useful to investigate the performance variability in the presence of different data. Finally, both of these experiments select suitable explainer methods, which achieve better ES results.

To conduct these three experiments and upon the forenamed evaluation metrics, a general performance metric was developed to compare explainers with a single unified metric. After that, the general procedure of experiments in Sect. 5 is described.

4.1 Explanation Score Performance Metric

To get to a unifying performance metric, comparing two models is not a simple task. Hence, to facilitate this comparison task, we develop a single metric, the Explanation Score (ES). It is a performance metric to calculate how accurate an explanation is regarding the localization and coverage of defect features in a PCB image. Figure 1 indicates how the described metrics are calculated and unified to form the ES metric. Out of the three metrics involved, occlusion and conservation analyze the feature relevance, and the localization metric analyzes the correct localization of defective parts. The defect localization should be prioritized over the others as a heat map can score high in contrast to the prior metrics by only considering a wrong set of features involved, which can be observed by measuring the localization metric. Furthermore, the priority of occlusion and conservation metrics is of the same level as their focus is on probability drop across heat-map masks.

First, the relevance of features (feature score (FS)) is averaged by considering the occlusion and transformed conservation metrics. The conservation is first inverted and then subtracted from 1. As a result of this transformation, both metrics measure the same relevance with different approaches on the same scale $[0, 1]$. For instance, zero on both metrics shows complete irrelevancy of features in the explanation, and one translates to full relevance of features within the explanation area. Next, the outcome of the feature relevance is multiplied by the localization metric, which acts as a weight for the prior metrics. The ES performance metric (shown in Eq. 4) has the same range $[0, 1]$ with 0 and 1 margins for worst and best cases.

$$ES_I = Localization_I * \frac{((1 - Conservation_I) + Occlusion_I)}{2} \quad (4)$$

The Feature Score diagram and the 3D ES diagram are depicted in Fig. 2 to show the relation between the three metrics involved. Sub-Figure 2a shows the

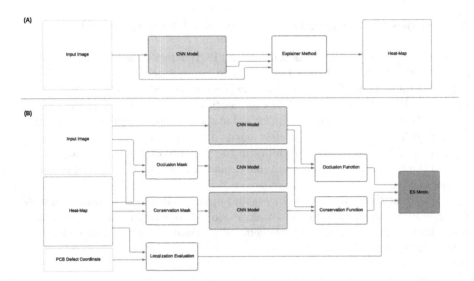

Fig. 1. (A) Heat map created by explainer methods. (B) Unifying quantitative metrics to form a single performance metric (ES).

reverse correlation between occlusion and conservation. Furthermore, Sub-Fig. 2b shows the priority of localization over the other metrics.

4.2   Experiment Design

The ES performance metric denotes how well an explainer method can localize and prioritize a classification result. The Algorithm 1 is a pseudo-code for the calculation and recording procedure of the ES across the dataset. This pseudo-code requires a list of trained models, cam methods, and a dataset and creates an evaluation data frame ($Eval\_df$) consisting of evaluation information. At first, parameters are initialized, and the code enters the first loop over the models in $Model\_list$. A second loop iterates on $Dataset$ and takes an image, label, and $Bbox$ information. Upon this data, the initial probability of top class ($Original\_score$) is computed, and the code enters a loop on available cam methods. The cam method is set up within this third loop, and the ES metric is calculated for each cam method in the loop. The ES metric and the class number are inserted into their respective location in the $Eval\_df$ data frame, and the counter i is incremented. The $Eval\_df$ data frame is used in the next Section to answer the research questions discussed in the paper.

5   Results and Discussion

5.1   Train Results

The selected models in Table 2 are trained using the settings in Table 5. The results shown in the Table 6 for both datasets across different CNN models rep-

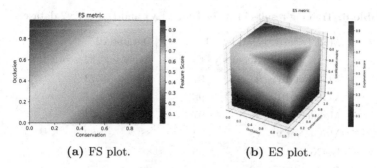

(a) FS plot.   (b) ES plot.

Fig. 2. 2D and 3D plots for FS and ES formula.

Algorithm 1. Pseudo-code for creating ES metric data frame.

Require: Model_list, Method_list, Dataset; **Output:** Eval_df;
 1: **Initialize** Evaluation_dataframe, i;
 2: **for** model_name in Model_list **do**
 3:     **Configure** Eval_model(model_name);
 4:     **for** data in Dataset **do**
 5:         Image, Label, Coordinates ← data;
 6:         Prediction_result = Eval_model.predict(Image);
 7:         Target_class = Prediction_result.argmax();
 8:         Original_pred = Prediction_result[Target_class];
 9:         **for** cam_method in Method_list **do**
10:             **Configure** args(cam_method);
11:             H_I = cam_method(Image, args);
12:             ES_I = Performance(Image, Coordinates, H_I, Original_score);
13:             Eval_df(model_name, cam_method, i) ← ES_I, Target_class;
14:             $i+=1$;
15:         **end for**
16:     **end for**
17: **end for**

---

resent that all models achieve an accuracy higher than 78 after being terminated under the training constraints. The presented results in Table 6 show that the Dataset_v1 is easier to learn as the F1-score has reached higher values (on average, a 19.25% improvement) than the second dataset in all the models.

5.2 Evaluation Across Models

In the first part of the experimental results, the paper focuses on comparing Cam methods across different models. ES performance metric per class label is calculated for the trained models with the first dataset (following the algorithm 1). The results are illustrated in Fig. 3. Explainer methods evaluated in this figure have different ES metrics on each defect class; for instance, Grad-Cam and Grad-Cam++ cannot retrieve the reasoning behind the "too much solder" class in ResNet50, whereas they perform better for other classes. Henceforth, a more fine-grained investigation is feasible to find a suitable explainer method with the higher ES metric for the target class. This is beyond the scope of this paper, it considers the average ES metric for defect labels.

Table 6. Training results in % for Dataset_v1 and Dataset_v2 datasets.

Model	Dataset_v1		Dataset_v2	
	Accuracy	F1-score	Accuracy	F1-score
VGGNet-16	95	97	91	73
VGGNet-19	93	96	88	71
MobileNet-v1	95	97	91	74
MobileNet-v2	78	84	88	82
ResNet-50	96	95	88	72
ResNet-101	89	89	90	73
ResNet-152	96	95	88	74
BiT	98	98	88	78

Overall, by taking the average ES per seven classes (represented in Table 7), the superiority of the Faster Score-Cam method in comparison with other methods is evident as it scores 0.023 ES metrics points above other Cam methods. Furthermore, the average ES score in this Table highlights three main factors. First, the maximum ES value in each row shows how successful the model is in learning the meaningful features for a class. Therefore, all trained models and the most noticeable VGGNet-19 are deemed to learn the wrong features leading to a low ES average score. Second, a pairwise comparison between Cluster-Cam and Group-Cam shows that they achieve almost the same results and do not outperform Grad-Cam and Grad-Cam++ in all models. Finally, explainer methods behave differently across various CNN models; for instance, Grad-Cam++ achieves the higher ES across VGGNet-16, whereas Group-Cam and Faster Score-Cam achieve the maximum ES for respectively ResNet-50 and all the remaining models. Hence, it is worth investigating if the explainers behave the same across the Dataset_v2 for the same models.

Table 7. Dataset_v1 average ES metric (top results are highlighted).

Model	Grad-Cam	Grad-Cam++	Faster Score-Cam	Group-Cam	Cluster-Cam
VGGNet-16	0.026	0.039	0.039	0.026	0.026
VGGNet-19	0.002	0.001	0.001	0.002	0.002
MobileNet-v1	0.057	0.065	0.076	0.051	0.061
MobileNet-v2	0.087	0.072	0.111	0.094	0.092
ResNet-50	0.129	0.163	0.195	0.209	0.203
ResNet-101	0.109	0.150	0.173	0.122	0.110
ResNet-152	0.058	0.081	0.157	0.084	0.058
BiT	0.136	0.136	0.164	0.141	0.127
Average	0.075	0.088	0.114	0.091	0.084

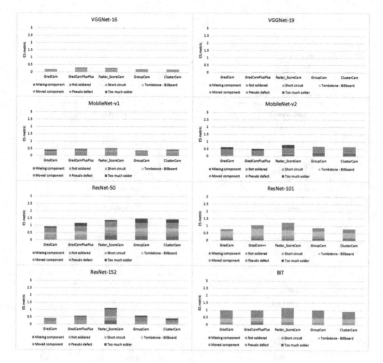

Fig. 3. ES score per class for Dataset_v1.

5.3 Evaluation Across Second Dataset

In this section, the second research question of whether the explainer methods generally perform the same across the dataset is answered by analyzing the ES metric for the Dataset_v2. Although the second dataset implied a lower F1-score and accuracy (in Table 6), the ES performance results shown in Fig. 4 demonstrate a meaningful improvement in comparison with the first dataset. In Fig. 3, the maximum of summed class ES metrics per explainer is close to 1.5, whereas in Fig. 4, it is close to 2.75 for less defect labels available in the second dataset.

The average ES metric per class (shown in Table 8) suggests the superiority in performance of Faster Score-Cam across different models. In conclusion, by considering the ranking of average ES metric per class reported in both Tables 7 and 8, the Faster Score-Cam outperforms the other methods, and the rest of the methods except Group-Cam have the same ranking. The Group-Cam with a negligible ES value is better than other methods that change the ranking.

5.4 Evaluation Across Models and Datasets

After comparing the explainer methods with each other and across different datasets, the paper investigates whether it is feasible to select a reasonable

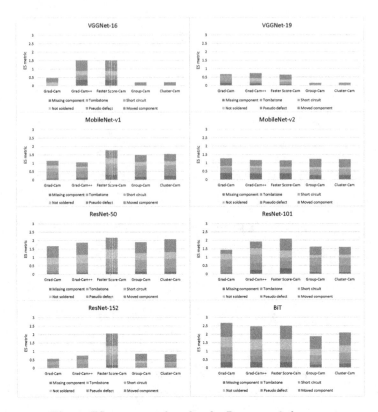

Fig. 4. ES score per class for the Dataset_v2 dataset.

explainer for a CNN model independent from the datasets in use. In Tables 7 and 8, four models (VGGNet-16, MobileNet-v1, ResNet-101, ResNet-152) have the same top performance explainer methods. Moreover, although in MobileNet-v2, ResNet-50, and BiT the top explainer method varies per dataset, they have similar explainer performance. This similarity is due to having the same second-ranked explainer method in one dataset with the second- or first-ranked method in the other dataset. Seven out of eight models observed similar performance across both datasets (VGGNet-19 poor performance on Dataset_v1 impels including this model).

A flowchart has been developed based on the above results (shown in Fig. 5), enabling users to select the most suitable Cam method for their PCB defect classification model. Cluster-Cam and Grad-Cam are rolled out of this flowchart as they do not achieve high values in both datasets.

6   Thread to Validity

This paper conducts research in the PCBA defect detection area, which is highly affected by the scarcity of industrial datasets due to confidentiality issues. This

Table 8. Dataset_v2 dataset ES metric (top results are highlighted).

Model	Grad-Cam	Grad-Cam++	Faster Score-Cam	Group-Cam	Cluster-Cam
VGGNet-16	0.077	0.251	0.251	0.033	0.033
VGGNet-19	0.112	0.121	0.105	0.024	0.025
MobileNet-v1	0.189	0.173	0.294	0.247	0.256
MobileNet-v2	0.209	0.192	0.189	0.204	0.200
ResNet-50	0.278	0.311	0.359	0.316	0.343
ResNet-101	0.238	0.318	0.346	0.268	0.264
ResNet-152	0.093	0.123	0.342	0.141	0.137
BiT	0.444	0.409	0.414	0.311	0.347
Average	0.205	0.237	0.287	0.193	0.200

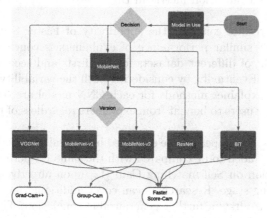

Fig. 5. A flowchart to select the suitable explainer method per CNN models.

issue is addressed in this paper by considering two private industrial datasets. Furthermore, the experiments are generalized across eight different CNN architectures, which leads to performing each evaluation metric on 16 different combinations of models and datasets and in a total of 520 experiments to cover five explainer methods and measure their corresponding ES metric per defect class.

The explainer results may not be reproducible due to the vanishing gradient problem discussed in the Sect. 2; or they may vary based on settings for the Gaussian Blur function used in Group-Cam and Cluster-Cam. The results of this paper can be extended by gathering more industrial datasets and improving the model's learning curve. Also, as the first paper discussing the classification explanation problem for PCB defects, this paper welcomes other researchers in this area to collaborate with the authors to pave the way for the robust use of CNN classifiers in the PCB defect industry.

7   Conclusion and Future Works

In this paper, the current AI approaches in AOI are discussed, and under an industrial scenario, CNN networks are chosen to perform defect classification. Next, the reliability issues when depending on the classification results are pointed out, and explainers are selected to provide reasoning on the prediction results. Moreover, eight CNN architectures and five prominent explainer methods are chosen based on electronics manufacturing constraints. Next, a unified performance metric called Explanation Score (ES) is proposed upon the current state of research, which facilitates the comparison of various explainer methods. ES metric utilizes occlusion and conservation metrics for feature quality evaluation and the localization metric in the assessment of focus on the defect areas.

The provided results conclude the superiority of Faster Score-Cam across both datasets, and similar performance of explainers is concluded with comparison in the light of different datasets for the first- and second-best ranking explainer methods. Eventually, by considering both the variability of models and datasets, suitable explainer methods for each CNN model are selected (Fig. 5), which enables other users to benefit from explainers regardless of the PCB defect dataset.

In future works, this paper can be extended by several factors. More explainer methods can be included to be compared with the same ES performance metric, e.g., an investigation on Self-matching Cam [12] upon already used explainers as a post-processing stage. Researchers can cover additional industrial datasets and CNN models to draw a further generalizable conclusion. The potential for improving CNN models with attention mechanisms can be investigated by measuring the negative values in the conservation metric.

Acknowledgments. Financial support for this study was provided by Thüringer Aufbaubank (TAB, 2021 FE 9036).

References

1. National physical laboratory industry defects database (2022). http://defectsdatabase.npl.co.uk/
2. Abadi, M., et al.: TensorFlow: large-scale machine learning on heterogeneous distributed systems. arXiv preprint arXiv:1603.04467 (2016)
3. Adibhatla, V.A., Chih, H.C., Hsu, C.C., Cheng, J., Abbod, M.F., Shieh, J.S.: Defect detection in printed circuit boards using you-only-look-once convolutional neural networks. Electronics **9**(9), 1547 (2020)
4. Angelopoulos, A., et al.: Tackling faults in the industry 40 era-a survey of machine-learning solutions and key aspects. Sensors **20**(1), 109 (2019)
5. Batista, G.E., Bazzan, A.L., Monard, M.C., et al.: Balancing training data for automated annotation of keywords: a case study. In: WOB, pp. 10–18 (2003)
6. Buhrmester, V., Münch, D., Arens, M.: Analysis of explainers of black box deep neural networks for computer vision: a survey. Mach. Learn. Knowl. Extract. **3**(4), 966–989 (2021)

7. Chattopadhay, A., Sarkar, A., Howlader, P., Balasubramanian, V.N.: Grad-CAM++: generalized gradient-based visual explanations for deep convolutional networks. In: 2018 IEEE winter conference on applications of computer vision (WACV), pp. 839–847. IEEE (2018)
8. Chawla, N.V., Bowyer, K.W., Hall, L.O., Kegelmeyer, W.P.: SMOTE: synthetic minority over-sampling technique. J. Artif. Intell. Res. **16**, 321–357 (2002)
9. Deng, J., Dong, W., Socher, R., Li, L.J., Li, K., Fei-Fei, L.: ImageNet: a large-scale hierarchical image database. In: 2009 IEEE Conference on Computer Vision and Pattern Recognition, pp. 248–255. IEEE (2009)
10. Ding, R., Dai, L., Li, G., Liu, H.: TDD-net: a tiny defect detection network for printed circuit boards. CAAI Trans. Intell. Technol. **4**(2), 110–116 (2019)
11. Feng, Z., Ji, H., Stanković, L., Fan, J., Zhu, M.: SC-SM CAM: An efficient visual interpretation of CNN for SAR images target recognition. Remote Sens. **13**(20), 4139 (2021)
12. Feng, Z., Zhu, M., Stanković, L., Ji, H.: Self-matching CAM: a novel accurate visual explanation of CNNs for SAR image interpretation. Remote Sens. **13**(9), 1772 (2021)
13. He, K., Zhang, X., Ren, S., Sun, J.: Deep residual learning for image recognition. In: Proceedings of the IEEE Conference on Computer Vision and Pattern Recognition, pp. 770–778 (2016)
14. Howard, A.G., et al.: MobileNets: efficient convolutional neural networks for mobile vision applications. arXiv preprint arXiv:1704.04861 (2017)
15. Huang, W., Wei, P.: A PCB dataset for defects detection and classification. arXiv preprint arXiv:1901.08204 (2019)
16. Kolesnikov, A., et al.: Big transfer (BiT): general visual representation learning. In: Vedaldi, A., Bischof, H., Brox, T., Frahm, J.-M. (eds.) ECCV 2020. LNCS, vol. 12350, pp. 491–507. Springer, Cham (2020). https://doi.org/10.1007/978-3-030-58558-7_29
17. Lan, Z., Hong, Y., Li, Y.: An improved YOLOv3 method for PCB surface defect detection. In: 2021 IEEE International Conference on Power Electronics, Computer Applications (ICPECA), pp. 1009–1015. IEEE (2021)
18. Li, Y.T., Kuo, P., Guo, J.I.: Automatic industry PCB board dip process defect detection with deep ensemble method. In: 2020 IEEE 29th International Symposium on Industrial Electronics (ISIE), pp. 453–459. IEEE (2020)
19. Liao, X., Lv, S., Li, D., Luo, Y., Zhu, Z., Jiang, C.: YOLOv4-MN3 for PCB surface defect detection. Appl. Sci. **11**(24), 11701 (2021)
20. Liu, G., Wen, H.: Printed circuit board defect detection based on MobileNet-YOLO-fast. J. Electron. Imaging **30**(4), 043004 (2021)
21. Nau, J., Richter, J., Streitferdt, D., Kirchhoff, M.: Simulating the printed circuit board assembly process for image generation. In: 2020 IEEE 44th Annual Computers, Software, and Applications Conference (COMPSAC), pp. 245–254. IEEE (2020)
22. Ribeiro, M.T., Singh, S., Guestrin, C.: "Why should i trust you?" Explaining the predictions of any classifier. In: Proceedings of the 22nd ACM SIGKDD International Conference on Knowledge Discovery and Data Mining, pp. 1135–1144 (2016)
23. Sandler, M., Howard, A., Zhu, M., Zhmoginov, A., Chen, L.C.: MobileNetv 2: inverted residuals and linear bottlenecks. In: Proceedings of the IEEE conference on computer vision and pattern recognition, pp. 4510–4520 (2018)
24. Selvaraju, R.R., Cogswell, M., Das, A., Vedantam, R., Parikh, D., Batra, D.: Grad-CAM: visual explanations from deep networks via gradient-based localization. Int. J. Comput. Vision **128**(2), 336–359 (2020)

25. Shi, W., Lu, Z., Wu, W., Liu, H.: Single-shot detector with enriched semantics for PCB tiny defect detection. J. Eng. **2020**(13), 366–372 (2020)

26. Si, N., Zhang, W., Qu, D., Chang, H., Zhao, D.: Fine-grained visual explanations for the convolutional neural network via class discriminative deconvolution. Multimed. Tools Appl. **81**(2), 2733–2756 (2022)

27. Simonyan, K., Zisserman, A.: Very deep convolutional networks for large-scale image recognition. arXiv preprint arXiv:1409.1556 (2014)

28. Smuha, N.A.: The EU approach to ethics guidelines for trustworthy artificial intelligence. Comput. Law Rev. Int. **20**(4), 97–106 (2019)

29. Springenberg, J.T., Dosovitskiy, A., Brox, T., Riedmiller, M.: Striving for simplicity: the all convolutional net. arXiv preprint arXiv:1412.6806 (2014)

30. Tomek, I.: Two modifications of CNN. IEEE Trans. Syst. Man Cybern. **6**, 769–772 (1976)

31. Wang, H., et al.: Score-CAM: score-weighted visual explanations for convolutional neural networks. In: 2020 IEEE/CVF Conference on Computer Vision and Pattern Recognition Workshops (CVPRW), pp. 111–119. IEEE (2020)

32. Wu, X., Ge, Y., Zhang, Q., Zhang, D.: PCB defect detection using deep learning methods. In: 2021 IEEE 24th International Conference on Computer Supported Cooperative Work in Design (CSCWD), pp. 873–876. IEEE (2021)

33. Zhang, Q., Rao, L., Yang, Y.: Group-CAM: group score-weighted visual explanations for deep convolutional networks. arXiv preprint arXiv:2103.13859 (2021)

34. Zhou, B., Khosla, A., Lapedriza, A., Oliva, A., Torralba, A.: Learning deep features for discriminative localization. In: 2016 IEEE Conference on Computer Vision and Pattern Recognition (CVPR), pp. 2921–2929. IEEE (2016)

Tasks Management Using Modern Devices

Livia Sangeorzan[1] , Nicoleta Enache-David[1]([⊠]) , Claudia-Georgeta Carstea[2] ,
and Ana-Casandra Cutulab[1]

[1] Transilvania University, Brasov, Bdul Eroilor 29, 500036 Brașov, Romania
nicoletadavid@gmail.com
[2] Henri Coanda Airforce Academy, Brasov, St. Mihai Viteazul 160, 500187 Brașov, Romania
claudia.carstea@afahc.ro

Abstract. Nowadays the people need tools that will help in their work and businesses. For this purpose, mobile applications have been developed for almost all
domains, in order to have a quick access to information. The architecture and
the design of a mobile application is very important. When looking for the best
mobile application architecture, the developers must choose between MVC, MVP
and MVVM models. As aprove of concept we implemented a mobile application for task management that uses the MVC model, the most popular in the
mobile app development. Generally, task management tools help the people to
work efficiently and to organize their tasks. The aim of this paper is to analyze
the advantages of using Java ME, XML, Firebase and its services in mobile applications development. The conclusion is that these technologies are suitable for
mobile task management applications implementation.

Keywords: Application · Tasks Management · Activities

1 Introduction

During the COVID-19 pandemic the number of mobile applications for task management
has increased. In the lockdown due to the COVID-19 pandemic, the use of mobile devices
has become more and more common and they are used by millions of people every single
day, for a variety of purposes such as time management or entertainment. But as the
number of mobile device users rises, so does the demand for applications to suit users'
needs.

One of the major problems that people can struggle with is lack of motivation, and this
can have a great impact on their personal or professional lives. With lack of motivation
comes lowered productivity and accumulating feelings of stress can only serve to worsen
the situation. There are different ways in which people can combat lack of motivation
and organize their time better in order to accomplish more throughout the day. One of
these methods is the creation of task lists, with tasks that can be checked off in order to
see daily progress and stay motivated to complete activities. It is exactly the purpose of
the developed application.

A similar application is Google Tasks, where the users can save and edit the tasks
from any device. It is also integrated with Gmail and Google Calendar [8]. Similar

D. Simian and L. F. Stoica (Eds.): MDIS 2022, CCIS 1761, pp. 73–86, 2023.
https://doi.org/10.1007/978-3-031-27034-5_5

applications have important advantages like the following: it helps the users to create tasks directly from emails, can define start dates, start time and deadlines. These tasks can be seen as events of the Google Calendar. It also can send the users reminders for their tasks.

Another similar application is ToDoIst, link: https://todoist.com/. This tool offers the opportunity to organize the tasks by setting priority settings and sharing the workload.

Another application for task management is Microsoft ToDo, link: https://todo.mic rosoft.com/tasks/en-us. This is a desktop application and a Microsoft ToDo mobile application that can be used to solve work tasks. Its main feature is that the users can share the workload online.

The developed application has the possibility for users to learn new information about different topics, improve memory and have fun, all through the help of quizzes presented in the Quiz Game section of the app. The main advantage of the application is that it can be used in tasks management, helping the people to stay organized and to have fun during their daily activities.

As a general notion, a task is a unit of activity or a part of a project required to be done [3, 10]. Due to the huge technology development, many tasks management application have been created. Generally, task management tools help the people to work efficiently, stay organized and meet deadlines.

There are two types of tasks management applications: for personal and for professional use. In case of personal use applications, these contain to-do list features. But for professional use applications, they are able to track progress on work, display the status of work, prioritize tasks and create planning timelines [11].

Mobile app architecture is obviously one of the fundamental factors that define future success and development strategy for the mobile application.

The scope of this article is to present the advantage of using the technologies like Java ME, XML, Firebase and its services (Firebase Firestore and Cloud Storage).

The article is organized as follows. In Sect. 2 we analyze the current state regarding the use of Android platform for mobile applications development. Section 3 describes the main results and the design and implementation of our developed application. In Sect. 4 we give some future development ideas, like multiplayer mode implementation. Conclusion is presented in Sect. 4.

2 Android Platform for Mobile Applications Development

Android is an open-source operating system and free development platform for mobile devices and phones, based on the Linux kernel. The system was developed by the Open Handset Alliance, of which the main commercial sponsor is Google and the original company was acquired by Google in 2005 [2]. There are a variety of libraries provided that simplifies the programming process for applications.

The latest version of Android is Android 12, which offers a number of changes to performance, design and usability including the "One Handed Mode". This means that developers can add innovative functions to apps and test them with the Android SDK (Software Development Kit) emulator.

Android Studio is the official Integrated Development Environment (IDE) for developing Android applications and is based on IntelliJ IDEA. This is the tool used for the development of Android applications and it offers a number of developer tools and features, such as an emulator for virtual devices (AVD - Android Virtual Device) and GitHub integration.

There are four different types of components for Android applications that serve specific purposes: activities, services, broadcast receivers and content providers [4].

An activity represents a user interface and user interaction with the application. For example, an application can have the following activities: to show a list of hotels, to add a hotel, to book a hotel room. The activities work together to create a user experience but they are independent each other.

A service is a component of an application that runs in background and performs different long running operations. For example, while a user runs another application, a service can download an image, without interrupting the user's current activity.

A broadcast receiver is a component that permits the system to send events outside of the normal user flow in an app, so that the application can react to system-wide broadcast announcements. The system can also send broadcasts to apps that are not currently running. An application can send a notification in order to inform the user of an event that is going to happen in near future.

A content provider can organize a shared data that is stored in a database, file system, web, or another permanent location. The application must have access to this location. Other apps can use the content provider to query or change the data, if the content provider allows it.

3   Main Results

We developed an Android application for task management to prove the Android API 26 (also known as Android 8.0 Oreo) capabilities. The application, named Evolve Now, was designed to contain different layers like the following: Presentation Layer, Business Layer, Data Layer.

Evolve Now was developed as an Android application. Android is a hugely used mobile operating system, with around 2.5 billion users worldwide [1]. It is also open-source, provides a variety of useful libraries for development, and is free, which also lowers development costs.

The current application was developed for Android API 26 (also known as Android 8.0 Oreo), which will run on approximately 88.2% of mobile devices.

The Evolve Now application offers users the possibility to create tasks, set their importance and due date and track them on a list. Here tasks can be anything, from a doctor's appointment or a daily activity like doing the laundry, to a work-related task. Once a task is completed, it can be checked off and the user receives points which are displayed on the user profile. These points connect to another part of the app which is meant to provide entertainment: a quiz game.

Quiz categories can be unlocked with points from completed tasks and offer the user an incentive to complete more tasks or complete them quicker. The purpose of the application is to help users be more motivated and organize their activities and time better, while also providing a means of entertainment in the same application.

3.1 Technologies and Programming Languages

The programming languages and technologies used for the implementation of the task management usually are the most as follows: Java ME, XML, Firebase and its services (Firebase Firestore and Cloud Storage).

Java Micro Edition (also known as Java ME) is a version of the Java programming language used for the development of portable and secure applications for mobile and embedded devices [6].

XML (Extensible Markup Language) [7] is a language used for the creation of clearly structured documents. This language was used for the creation of files for layouts, drawings, colors, themes and many more within the Evolve Now Android project.

Firebase Authentication adds user authentication to projects and was used in the developed application. This service is easy to use and can be configured to work with different authentication methods, such as username/password or access with other accounts such as Google and Facebook. The service has been integrated with Evolve Now so that users can create accounts with email and password, confirm their email and sign in, or even reset their password if needed.

Firebase Firestore is a storage service offered by the Google Firebase platform and was used as a database for the application. Firestore's structure consists of collections, each containing documents which have fields in the form of key-value pairs [9].

Firebase Cloud Storage is a storage service that is used by the developers to store different media files such as videos, photos, audio on a cloud platform, in a so-called bucket, so that they can be used in various applications. With the help of the service, media files can be uploaded and down loaded regardless of the network quality. If there is a problem with the network connection and the process stops, it will simply continue afterwards with better network quality and that even saves band width [9].

The following flowchart illustrated in Fig. 1 shows some of the interactions between the application and different Firebase Services:

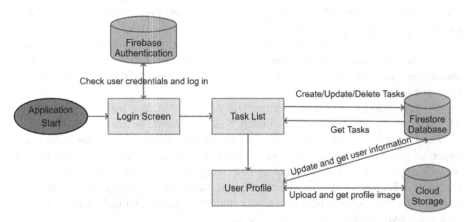

Fig. 1. Use of firebase services in Evolve Now.

After starting the application, if the user is not logged in, they are taken to the Login Screen, where they must register or log in. This authentication process uses the Firebase

Authentication service, which provides methods for registering and authenticating users, as well as resetting passwords. After logging in, the user has access to the application and is taken to the Task List, where they can start adding and completing tasks.

Tasks are stored in the Firestore Database and each operation performed by the user (creating, editing, deleting and completing tasks) uses methods provided by Firebase to apply the necessary changes to the Firestore Database. Every time the list is reloaded or a change is made, the database is queried in order to get the task list and display it.

In addition to creating or deleting tasks, the user can also complete tasks. Whenever a task is completed, the newly gained points are added to the total points the user has and these can then be viewed in the User Profile, along with other pieces of information like the username, which can be changed and total number of tasks completed by the user. The User Profile screen can be reached from the application's navigation menu. Here the user can also upload a profile picture, which is then stored in the Cloud Storage offered by Firebase and loaded into the User Profile when accessing the screen or after changing the picture.

3.2 Mobile Applications Architecture

The architecture of a mobile application usually consists of the following layers: Presentation Layer, Business Layer and Data layer. Presentation Layer contains User Interface components and the other components that process them. Business Layer is made by workflows, different business entities and other components. Data layer consists of data utilities and different data access parts.

There are six mobile applications architecture principles: sustainability, **maintainability and** manageability, reusability, testability, security and performance.

Sustainability means the possibility that a mobile application would deal with environmental challenges. These changes could be some technology updates which would involve the functionality of servers and databases. A sustainable mobile application could handle the updates and minimize their effects. Actually, a sustainable mobile application features will offer the support to the developers to update it with a minimum cost.

Maintainability and Manageability are qualities that define how easily and efficiently applications can be updated, fixed, improved, monitored, and optimized.

According with [10] there are four important factors that defines if the software can be modified and updated. These modifications could be changes in the software in order to obtain a good functionality.

The factors that affect the maintainability of software are the following: analyzability which is a measure of the possibility to find failures in software; changeability that is the possibility to change or modify the software products; stability that is the capability to find different bad effects from modifying the software product; testability which is the possibility to test and finally validate the software. In the paper [10] the authors define some metrics that each factor depends on.

Reusability of code components allows the application developers to produce the code for new features faster. Any well-developed architecture includes some reusable code that assures a rapid development of new versions.

In the paper [10] the authors presents the software reuse in the Android mobile applications market, taking into count two perspectives: class reuse and reuse by inheritance.

The conclusion is that in practice, most of the base classes that are reused by developers are from the Android API or a third party API.

Testability principle ensures the consistency and sustainability of an application under various conditions and modes. Testability is a characteristic of a mobile application to support different testing processes.

Security is one of the most important issues for building a successful mobile application, because it prevents the data from being stolen.

Performance means that a mobile application should be web responsive and should perform quickly. The users expect their applications to be fast and very good for their businesses.

3.3   Common Architecture for Mobile Application

When developing a mobile application, the designer must choose between one of the MVC, MVP and MVVM models.

The MVC (Mode—View-Controller) model means to have three components. The class or file of the application must be categorized into the following layers: Model, View and Controller.

In the Model component the data is stored; the model can handle different businesses rules from the real world and the relationship with the database. The View layer is User Interface (UI) and it is a component that holds the things that the users can see on the screen; it is the interface for the date stored in de Model layer. The Controller layer manages the relationship between the View and the Model; it contains the application logic.

MVP (Model-View-Presenter) provides an easier structure of the code. It provides also modularity, testability and a clear codebase. It consists of the following components: Model, View and Presenter.

The Model layer is used for storing data and it has similar responsibilities as the Model layer from MVC. The View layer is UI (User Interface) that gives the data visualization and keep the history of users actions in order to notify the Presenter layer. Finally, the Presenter component uses the UI logic to decide the things to be displayed.

MVVM (Model-View-ViewModel) is the model used specially in industry software architecture. Its principle is to separate the data presentation layer (View or UI) from the logic part of the application. The Model layer and ViewModel get, save and store data. The View layer informs the ViewModel about the actions made by the user. This layer does not contain any code of application logic. The ViewModel layer is a link between the Model and the View.

MVC is considered one of the most popular in the mobile web application development. The MVC model helps to create platform applications development and consists of plain logic and content. These parts are standardized User Interface application design.

In Fig. 2 the MVC model for Evolve Now application is illustrated.

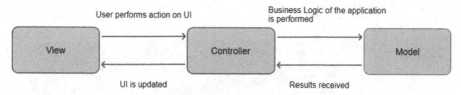

Fig. 2. MVC model for Evolve Now application.

3.4 Application Features

The architecture of the Evolve Now application is illustrated in Fig. 3. Within the application there are three main layers: the Presentation Layer, the Business Layer and the Data Layer.

The Presentation Layer contains the UI components and processes of the application, which define how the application will look when used. The Business Layer contains all of the components related to the logic of the application, the code which defines how the application will work and the Data Layer encompasses data-related components, such as utilities and helpers.

In the case of Evolve Now, the UI components from the Presentation Layer are those that define the layouts for the screens and certain graphical elements such as icons, modals and buttons, the Business Layer is made of all code that defines application logic, such as Activities, and the Data Layer contains the application helpers. The user interacts with the Client Application and the Client Application uses Firebase Databases and Services from the Support Infrastructure to store data and perform operations on that data.

The main features of Evolve Now are its task list, the in-app calendar and the quiz game. In Fig. 4 there are presented the main screen of the application, the applications' features and My Tasks screen.

When starting the application, the user is greeted by a login screen. In order to access the app, the user must first register with an email and a password and confirm the email address by accessing the confirmation link in their inbox. The next step is to return to the application and after logging in for the first time, a tutorial or on boarding screen will appear, with five slides containing information about the app's features and the rules for gaining points and unlocking quiz categories.

In the "My Tasks" screen of the application, every user can see their own list of tasks, create new tasks and edit or delete existing ones. Every created task has a level of importance established by every user. Task importance levels are displayed with the help of colors: green is for low, orange for medium and red for high.

There are 3 available importance levels: Low, Medium or High. Depending on the importance level attributed to a task, the user will receive a different amount of points upon task completion. These points are defined in the application "rules", which can be found on the second slide of the tutorial screen. For low importance tasks the user will receive 10 so-called xPoints, for medium importance 15 xPoints and for high importance 20 xPoints. For example, "10 xPoints" represent a number of 10 points.

The "Calendar" screen contains a custom calendar view where the current month is displayed. The user can scroll through the months using the arrows at the top of the

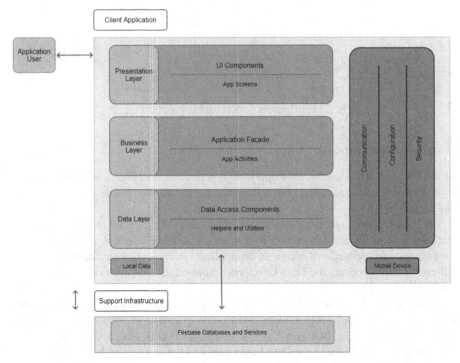

Fig. 3. The architecture of the Evolve Now application. (Color figure online)

screen and view upcoming due dates. If there are any tasks due on a certain date, a text will appear on that date which reads "x tasks due" where x is the number of tasks due on that day. The current date is always marked on the calendar.

Navigating to the "Quiz Game" part of the application takes the user to the list of available quiz question categories for the game. The quiz categories are the following: Art & Culture, General, Music, Science, Sports, Language, Geography and Computer Science.

Here a user can choose a category from the list and tap on "Play" to start the game. Some of the categories will be locked and the user will need to meet certain conditions in order to unlock them like reaching a certain number of xPoints or a certain total number of tasks completed. Whenever a user unlocks a new category, this information is also saved in the database and can later be used to generate statistics regarding the usage of the quiz game and the categories which were unlocked the most by users.

The questions from the quiz will appear in a random order, so each time the user starts a new quiz, they can be different. Information about the current quiz is displayed at the top of the screen and underneath it there is information about which question the user is on out of the total number of questions, the current score and progress marked as correct or wrong answers and a timer which is ticking down. The question text is displayed within a box and each question has one correct answer out of 4 possible answers.

In Fig. 5 there is a scheme that presents the steps of quiz unlocking and playing the quiz.

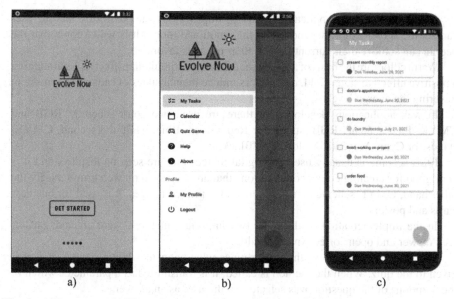

Fig. 4. a) First screen; b) The main screen of the application; c) My Tasks screen. (Color figure online)

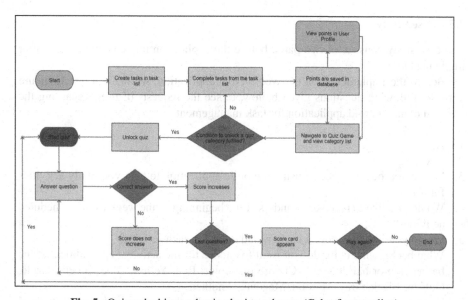

Fig. 5. Quiz unlocking and quiz playing scheme. (Color figure online)

The goal of the game is to choose correct answers for as many questions as possible in order to accumulate points.

A task management application could be improved by the possibility that the user can set the screen color, depending on the emotional mood. It is well known that the people can achieve their current tasks if their mood is favorable.

Warm colors are the basis of a good mood, having a stimulating effect and maintaining a positive affective state. Cold, dark colors induce mental states of sadness, depression, lowering the energy level.

In web applications development, there are used two color schemes: RGB and CMYK. First scheme, RGB, stands for Red, Green, Blue, while the second, CMYK, stands for Cyan, Magenta, Yellow, Key/Black.

RGB scheme is usually used for digital images that are seen on displays and it is very good for creating colors combinations that are displayed on web browsers. By the other hand, CMYK scheme is recommended for printed images. It is used for business cards and posters.

In the implementation of the quiz, two important methods were created: onCorrectAnswer and openCorrectAnswerDialog.

The onCorrectAnswer method determines what happens when a correct answer is given to the quiz. When the method is called, it is transmitted as a parameter which of the 4 options of the question was selected by the user as an answer.

The openCorrectAnswerDialog method opens a window with information about the correct answer given by the user.

3.5 Case Study

The case study consists of two parts: before the application implementation and after implementation.

Before the implementation of Evolve Now application, we made a questionnaire with the following questions given bellow, to see the requests of users regarding the creation of an Android application for task management.

1. Age
2. Sex
3. Would you be interested in using a mobile application to plan your daily activities for a week?
4. Would you like to receive reminders at the beginning of the week for the scheduled activities?
5. Would you like to choose a background color for each activity?
6. What background color do you want to choose for the activity "appointment at the barbershop or hairdresser"? Choose one color: Red, Yellow, Blue, Green, Purple, Golden, Black, Gray, Maroon, Magenta, Another.
7. What background color do you want to choose for the activity "scheduling medical investigations"? Choose one color: Red, Yellow, Blue, Green, Purple, Golden, Black, Gray, Maroon, Magenta, Another.
8. What background color do you want to choose for the activity "participating in a party"? Choose one color: Red, Yellow, Blue, Green, Purple, Golden, Black, Gray, Maroon, Magenta, Another.
9. What background color do you want to choose for the "do laundry" activity?

The questionnaire was tested on 54 participants with varied in a much wider age range (18–65). From 54 participants, 15 were women (27.78%) and 39 were men (72.22%). The answers obtained led to the conclusion that regardless of gender and age, all respondents are interested in this type of application.

In the second part of this case study, the application was tested by 21 participants having the ages between 21 and 58. There were 10 women and 11 men. The participants, after running the application, answered the questions from Fig. 6.

SECTION 1

Age ☐

SEX
F ☐ M ☐

SECTION 2

Application installation:
Easy ☐
Moderate ☐
Difficult ☐

Navigation into application:
Easy ☐
Moderate ☐
Difficult ☐

The information is clear:
Agreement ☐
Disagreement ☐

Graphical interface is:
Attractive ☐
Useful ☐
Difficult ☐

SECTION 3

Quiz utility:
Very useful ☐
Useful ☐
Useless ☐

The importance of being able to choose the quiz field:
Very useful ☐
Useful ☐
Useless ☐

SECTION 4

Application scope:

Business Management ☐
Business Marketing ☐
Personal ☐

Fig. 6. Questionnaire of the application Evolve Now.

Application installation – from a total of 21 participants, 33.33% answered "easy", 28.57% "moderate" and 38.10% "difficult".

Navigation into application – from a total of 21 participants, 38.10% answered "easy", 33.33% "moderate" and 28.57% "difficult". This means that the most of participants were satisfied using the application.

The information is clear – 47.62% answered "agreement" and 52.38% "disagreement".

Graphical interface – 28.62% answered "easy", 57.14% answered "moderate" and 14.29% answered "difficult". This result suggests we could improve the graphical user interface of the application.

Application utility – 42.86% answered "very useful", 28.57% answered "useful" and 28.57% answered "useless". In contrast to the graphical interface, this result shows that the participants are delighted by the features offered by the application.

The importance of being able to choose the quiz field – 28.57% answered "very useful", 42.86% answered "useful" and 28.57% answered "useless". The conclusion is that the application' utility is very high.

Application scope – 19.05% answered "Business management", 52.38% answered "Business marketing" and 28.57% answered "Personal".

The conclusion of the study is that the application is of interest and can be improved using colors for each proposed activity.

Table 1 shows some properties of some of these colors with some general characteristics [5].

Table 1. General characteristics of colors.

Nr.crt	Color name	General characteristics
1	GREEN	It is the symbol of new beginning, hope, protection, and life
2	BROWN	It is the holiday symbol
3	RED	It is the symbol of power, joy, dynamism
4	VIOLET	It is the symbol for dignity, mysticism, and protection
5	BLUE	It is the symbol for truth, consistency, and seriousness

It is no coincidence that 24.1% of the subjects chose the color green for questions 6 and 9. If we analyze the meaning of this color, we notice that it suggests harmony, meaning order (hairdressing), cleanliness and hope (health).

4 Future Development

There are several ideas which could be implemented in the future in order to further develop and improve Evolve Now, including the following: adding a multiplayer mode which allows user to connect, create their own questions or quizzes and send these to other players; adding more quiz categories and questions in future content updates; adding an achievement system for certain goals within the application (such as completing a quiz with a perfect score) and expanding the user profile to display these achievements as trophies.

Also, as a future development one could add options to sort, filter and search for tasks in order for the users to have quick access to desired information. Adding options to customize the application: fonts, colors, font size, a "night mode" theme or adding an option for choosing the color scheme that the user likes depending on the emotional mood, could be also good ideas for application's improvement.

User statistics regarding productivity (how many tasks a user has completed in a certain amount of time) and motivation (how quickly users complete tasks) and usage of the quiz game (which categories have been unlocked the most times, quizzes played).

Over last years, the AI (Artificial Intelligence) is integrated in more and more domains, consumer applications, and business applications. The tasks management applications could benefit also from the AI implementation, especially for the optimization process.

Also, the AI is very useful in project management since it has the ability to forecast different scenarios and project tasks. Due to continuous technology development, in future it will be able to handle very complex project management tasks.

Using the emotional state of each user, highlighted by the color chosen for a task, we can set a certain background for the tasks. The colors chosen by users for a certain task can be stored in a database and analyzed with AI algorithms. This could lead to a better quality of the Evolve Now application.

5 Conclusion

In this paper we analyzed the advantages of using Java ME, XML, Firebase and its services in mobile applications development. We wanted to prove that Android is a suitable platform for developing mobile applications.

To prove the concept, we have implemented a tasks management mobile application that uses these technologies.

The idea for the application presented was the desire to provide a solution for lack of motivation and stress relief in the modern world, especially in the context of the COVID-19 pandemic.

The app's task list and the calendar offer a clear and organized view of new and completed tasks, along with their importance as established by the user, while also providing the user with a means to stay productive and motivated through the accumulation of points for each task completed. This then connects to the quiz game, which provides entertainment and an opportunity to learn new information from different fields.

Together, all of these features make up Evolve Now, an Android application meant to offer users a pleasant experience that combines productivity and fun.

The AI has won a good perspective and its goal is to help the people in their work and businesses. Applications such Evolve Now could improve the people's work in future. The conclusion is that these technologies are suitable for mobile task management applications implementation.

References

1. Curry, D: Android Statistics, Business of Apps (2022)
2. Callaham, J: The history of android: the evolution of the biggest mobile OS in the world, Android Authority (2021)
3. Drucker, P.F.: Management: Tasks, Responsibilities, Practices. Harper & Row, New York (1974)

4. Griffiths, D., Griffiths, D.: Head First Android Development: A Brain-Friendly Guide. O'Reilly Media, Inc., Sebastopol (2015)
5. Heller, E.: Wie Farbenwirken, Farbpsychologie – Farbsymbolik – Kreative Farbgestaltung. Verlag GmbH, Reinbekbei Hamburg (1999)
6. Kanjilal, J.: How Java is used in android app development (2016)
7. Kumar, A.: Mastering Firebase for Android Development: Build real-time, scalable, and cloud-enabled Android apps with Firebase, 1st Edition, Packt Publishing (2018)
8. Ray, E.T.: Learning XML. O'Reilly Media Inc., Sebastopol (2001)
9. Moroney, L.: The Definitive Guide to Firebase. Build Android Apps on Google's MobilePlatform, Apress, New York (2017)
10. Ruiz, I.M.J, Nagappan, M., Adams, B., Hassan, A.E.: Understanding reuse in the android market. In: 20th IEEE International Conference on Program Comprehension (ICPC), pp. 113–122, IEEE, Passau, Germany (2012)
11. Rumane, A.R.: Quality Tools for Managing Construction Projects. Taylor & Francis Inc., Milton Park (2013)

Machine Learning

A Method for Target Localization by Multistatic Radars

Kiril Alexiev$^{(\boxtimes)}$ ⓘ and Nevena Slavcheva

Institute of Information and Communication Technologies, Bulgarian Academy of Science,
1113 Sofia, Bulgaria
`alexiev@bas.bg`

Abstract. A method for target localization by multistatic radar system is proposed. Two cases are considered – 2D case and 3D case. 2D case considers targets located near ground level (such as low flying drones) and for which the z-coordinate can be ignored. Localization of targets is carried out by finding the intersection points of ellipses. The problem is defined also in 3D space where the targets are located at a higher altitude that can't be neglected. A solution for finding intersection of ellipsoids is proposed. Target detection is realized by solving nonlinear systems of equations by using homotopy. After target detection target tracking should be realized with big variety of modern methods applicable in different scenarios/cases. Special attention should be paid on data association problem due to unknown labels of detected targets.

Keywords: Multistatic radars · MIMO radar systems · Target detection

1 Introduction

The multistatic radar system (MSRS) consists of a number of transmitters and receivers, situated at different points in space, which carry out collaborate work for aerial target detection [1]. The localization of targets is calculated based on the time delays of emitted by transmitters signals reflected from the targets and received by the receivers. In spite of similarity in definition with classical radar systems there are several significant differences in determining the location of a target with classic radar and MSRS. In classical radar, the transmitter and receiver are located at the same place, and the target's position can be determined from the direction in which the signal was emitted, respectively from the direction from which the signal was received (these are azimuth angle and elevation angle of antenna). The distance to the target is determined by the travel time of the reflected from the target signal.

In the presence of a multistatic radar system several approaches are practiced to determine the location of the target, depending on the architecture of the sensor network. As initial conditions for such a system, we will assume that the target is moving in the area of operation of the network of sensors whose locations are known in advance. In the specialized literature these sensors are often called anchors. The simplest localization occurs when the target receives the broadcast messages of the anchors. The messages

D. Simian and L. F. Stoica (Eds.): MDIS 2022, CCIS 1761, pp. 89–103, 2023.
https://doi.org/10.1007/978-3-031-27034-5_6

contain exact time of broadcasting. Knowing anchor coordinates and the travel time of the message the target calculates its location somewhere on the circle, drown with center located at the anchor coordinates and radius calculated as a product of travel time and velocity of electromagnetic waves in the atmosphere.This type of localization belongs to the class of Time of Arrival (TOA) target localization systems [16]. It should be noted, that at least 3 transmitters are needed for exact position determination of a target in 2D space (if the transmitters are two, there could be two crossings of the circles) and 4 for 3D space. A major disadvantage of this approach is that the clocks of all transmitters and the receiver must be kept synchronized. This is not infeasible for the transmitters and is realized with the help of extremely precise atomic clocks (for example, in global navigation satellite systems), but for the receiving station it could be a serious technical limitation. It is overcome by the involvement of an additional transmitter, with the help of which the clock offset of the receiving station can be calculated.

The next Time Difference Of Arrivals (TDOA) localization scheme uses the difference in time of arrival of the signal emitted by the target at a pair of receivers [16]. This results in a simpler scheme of operation and lack of requirements for clock synchronization. The difference in signal travel times for two receivers defines a hyperbola on which the emitter should lie (in the 2-D case). The location is uniquely determined by the intersection of at least two hyperbolas, so at least 3 receivers are required for 2D case.

The last and the most complex target localization scheme involves multiple transmitters and receivers. They could be radars with spatially distributed receivers and transmitters, or several broadcasting transmitters and several also spatially distributed receivers. The transmitters emit signals, which are reflected by a target and received from the receivers. In some literary sources this localization scheme is called Time Sum Of Arrival (TSOA) [17] because it uses the travel time of signal received at the receiver, which is obtained from the sum of two travel times: 1) the travel time of the signal from the transmitter to the target and 2) the travel time of the reflected from the target signal to the receiver.This abbreviation is not very correct due to the fact that the method actually reuses the difference of the travel times of the signal from one transmitter and two receivers or two transmitters and one receiver. In this article, only the scheme described above will be considered, although there are many similar ones using the Doppler shift [18–21], and angle of arrival [22, 23].

Solving the problem of localization of the reflecting object can be easily reduced to the TDOA scheme. Let's assume we have one transmitter and several receivers. The waves emitted by the transmitter are reflected from the target and hit the receivers. The path taken by the electromagnetic waves from the transmitter to the target is the same and the differences in signal travel times in the respective receivers are determined only by the distance of the target to the receivers - a situation in which the TDOA scheme is implemented. Two approaches using the TDOA scheme have been used to solve the above problem: the spherical interpolation (SI) method [9, 14] and the spherical intersection (SX) method [9, 12]. Without going into the analysis of these methods, we will note only that at least 4 receivers in the 3D space are necessary in the presence of one transmitter for target localization.

In [14] a justification of the localization method SX is given. The authors point out that one of the main problems of TDOA localization is related to the numerical solution of the hyperbolic equations. Even a small change in the parameters of one of the hyperboloids can move the intersection point significantly. The hyperboloids are not spatially bounded! Therefore, the proposed methods SX and SI transform the problem of hyperboloids intersection to sphere intersection/interpolation. They are spatially limited and a change in their parameters cannot degrade the solution process so brutally.

In this article a slightly different solution is proposed. We define the localization problem in MSRS as the problem of finding the intersection of ellipses or ellipsoids depending on the dimensionality of the space. These figures are also spatially limited. Due to different sites of transmitter and receiver, the geometric location of a target is determined by the time delay of reflected by target signal (transmitter - target – receiver) in regard to directly received signal (transmitter - receiver) and the places of corresponding transmitter and receiver. It is an ellipse in the 2D case and an ellipsoid in the 3D case with transmitter and receiver in the foci. When multiple transmitters and receivers are available the location of the target can be found as crossing point of a group of ellipses or ellipsoids defined by different transmitter-receiver pairs. The intersections of the ellipses/ellipsoids determine the exact position of the targets. At least three ellipses (which can be set by one transmitter and three receivers or one receiver and three transmitters) in the 2D case determine the exact target position and at least four ellipsoids in the 3D case.

In some cases, the nonlinear system of equations is difficult to be solved. To improve the convergence of the process we apply homotopy. Another feature of the proposed solution is that the minimum number of sensors is used - 2 to form two ellipses in the 2D space and 3 to form three ellipsoids in the 3D space.

Solving nonlinear systems of equations allows estimation of the exact position of the target. The detection of target coordinates is only the first step in the information technology of MSRS data processing. A stable target tracking has to be organized also, based on classic Kalman filter and its improved versions. Some of these algorithms are featured in the present article providing possibility for tracking separate or group of targets, in presence of clutter and/or non-Gaussian noise, tracking of highly maneuverable targets, targets in complex air environment and etc.

The paper is organized as follows. The next section discuses classical MSRS target localization problem as time differences of arrival problem. The proposed approach for target localization using homotopies is described in Sect. 3. How to apply the method is described in Sect. 4. The application of target tracking algorithms with MSRS target detectors has its own specifics and the most suitable of them were shortly commented in Sect. 5. Section 6 summarizes the received results and outlines the most important conclusions.

2 Target Localization Problem

In the specialized literature on the subject, the problem of finding the position of a target by a MSRS is usually considered as a problem of solving the task of detecting the location of a transmitter by the differences of the signal arrival times at several receivers [1, 10]. Switching from one task to another is trivial and will be shown below. Let the system under consideration consist of several transmitters and one receiver. This initial assumption does not limit the formulation of the problem in any way, and it can easily be reformulated for multiple emitters and multiple receivers. We will denote the transmitters by $Tr_1, Tr_2, \ldots, Tr_n$, and the receiver by R. If the target is denoted by T, and we know the signal emission time from the corresponding transmitter, then upon simultaneous reflection of the signal from the target, we will receive reflected signal in the receiver with a time delay, respectively t_{Tr_i-T-R}, where i denotes the source of the signal (transmitter Tr_i). It is clear that $t_{Tr_i-T-R} = t_{Tr_i-T} + t_{T-R}$. If denote the time difference of delays of signals, emitted from transmitters i and j by $t_{i,j}$, we can express its value as: $t_{i,j} = t_{Tr_i-T-R} - t_{Tr_j-T-R} = t_{Tr_i-T} + t_{T-R} - t_{Tr_j-T} - t_{T-R} = t_{Tr_i-T} - t_{Tr_j-T}$, i.e. the problem could be considered as a problem of estimation the position of an emitter if the time difference of delays are known. In such a formulation, the problem has been well known since the sixties of the last century and is called hyperbolic positioning, where a position is calculated by measuring time differences of arrival (TDOA). There is plentiful number of articles on this topic [10–14]. Many algorithms were derived for solving TDOA systems but the most prominent between them are spherical interpolation (SI) and spherical intersection (SX). Both algorithms rely on closed-form equations. Two important remarks will be made here. The first one is that the precise estimation of the target – receiver range has a fundamental influence on the overall estimation accuracy [1]. The second one concerns the number of necessary sensors – they should be at least with 1 more than the minimal number of four (the minimal number is three transmitters and one receiver, which determine three ellipsoids for target position detection). Moreover, having more sensors will enable us to calculate the target's position more robustly.

3　Homotopies for Target Localization

Providing reflected signals from one and the same target from multiple sensors is problematic in a number of cases. Therefore, in the present paper, a solution based on directly solving the classic issue of crossing ellipses (in 2D space) or ellipsoids (in 3D space) is proposed. The task requires iterative solving of a system of nonlinear equations. The classical approaches for solving such type of systems do not work due to poor convergence. To find the solution with increased reliability, homotopy is used - a standard approach for solving systems of nonlinear equations. A detailed description of the method for 2D case is given bellow.

The input information consists of sensor coordinates. The speed of electromagnetic waves in the vacuum is also known $- c = 299792458 \text{ m/s}^2$ and the constant for refraction in the air $n = 1.0003$. Let us denote by $dt_{i,j}$ the time difference between time of flight of emitted by transmitter Tr_i signal, reflected by target and received by receiver j and time of flight of emitted by transmitter i signal and received directly by receiver j (Fig. 1).

$$dt_{i,j} = t_{Tr_i-T-R_j} - t_{Tr_i-R_j} \tag{1}$$

$$D_{i,j} = \sqrt{\left(x_{Tr_i} - x_{R_j}\right)^2 + \left(y_{Tr_i} - y_{R_j}\right)^2} \tag{2}$$

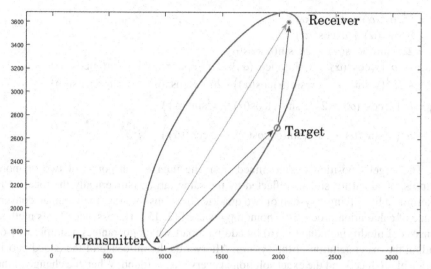

Fig. 1. Problem setting with one receiver, one transmitter and one target.

$$D_{i-T-j} = dt_{i,j} * \frac{c}{n} + D_{i,j},$$ (3)

Here $D_{i,j}$ is the distance between transmitter i and receiver j; D_{i-T-j} is full travelled distance - transmitter Tr_i, target and receiver R_j. The ellipse parameters are: major ellipse semiaxis $a_{i,j} = \frac{D_{i-T-j}}{2}$; minor ellipse semiaxis $b_{i,j} = \sqrt{a_{i,j}^2 - \left(\frac{D_{i,j}}{2}\right)^2}$ and the angle of ellipse rotation $\alpha_{i,j} = arctg\frac{y_{Tr_i} - y_{R_j}}{x_{Tr_i} - x_{R_j}}$.

The ellipse's parameters are used to determine the coefficients of the equation for the analytical description in the space. The ellipse at the axes origin:

$$\frac{X^2}{a_{i,j}^2} + \frac{Y^2}{b_{i,j}^2} = 1$$ (4)

should be rotated and translated/shifted by s_x and s_y:

$$\begin{bmatrix} x \\ y \end{bmatrix} = \begin{bmatrix} \cos(\alpha_{i,j}) & -\sin(\alpha_{i,j}) \\ \sin(\alpha_{i,j}) & \cos(\alpha_{i,j}) \end{bmatrix}^{-1} \begin{bmatrix} X - s_x \\ Y - s_y \end{bmatrix},$$ (5)

where $s_x = \frac{x_{Tr_i} + x_{R_j}}{2}$ and $s_y = \frac{y_{Tr_i} + y_{R_j}}{2}$.

The generalized representation of ellipse is through the following quadratic equation:

$$a_1 x^2 + a_2 y^2 + a_3 xy + a_4 x + a_5 y + a_6 = 0$$ (6)

To simplify mathematical expressions for a considered ellipse we will omit the indexes i and j pointing out corresponding transmitter and receiver for this ellipse. The coefficients are calculated as follow:

$a_1 = b^2\cos^2(\alpha) + a^2\sin^2(\alpha)$

$a_2 = b^2\sin^2(\alpha) + a^2\cos^2(\alpha)$

$a_3 = 2a^2\sin(\alpha)\cos(\alpha) - 2a^2\sin(\alpha)\cos(\alpha)$

$a_4 = -2b^2\left(s_x\cos^2(\alpha) + s_y\sin(\alpha)\cos(\alpha)\right) - 2a^2\left(s_x\sin^2(\alpha) + s_y\sin(\alpha)\cos(\alpha)\right)$

$a_5 = -2b^2\left(s_y\sin^2(\alpha) + s_x\sin(\alpha)\cos(\alpha)\right) - 2a^2\left(s_y\cos^2(\alpha) + s_x\sin(\alpha)\cos(\alpha)\right)$

$a_6 = b^2\left(s_x^2\cos^2(\alpha) + 2s_x s_y\sin(\alpha)\cos(\alpha) + s_y^2\sin^2(\alpha)\right)$

$\qquad + a^2\left(s_x^2\sin^2(\alpha) + 2s_x s_y\sin(\alpha)\cos(\alpha) + s_y^2\cos^2(\alpha)\right) - a^2 b^2$

The target's position is determined from the intersection points of two or more ellipses, obtained for signals reflected by the same target. Numerically the location is determined by solving a system of two quadric equations. In order to guarantee convergence of calculation process the homotopy is used [2, 15]. The essence of this method consists of modifying equation (6) by adding a part which guarantees a simple way of finding the exact solution of the system. Afterwards the influence of the added part is iteratively reduced and the exact solution at every step is found. When the change of the added part becomes negligible it is assumed that the solution to the system is found [2]. If we denote the equations of the two ellipses, which define the location of one target, with $f_1(x, y)$ and $f_2(x, y)$:

$$\begin{cases} f_1(x, y) = a_{11}x^2 + a_{12}y^2 + a_{13}xy + a_{14}x + a_{15}y + a_{16} \\ f_2(x, y) = a_{21}x^2 + a_{22}y^2 + a_{23}xy + a_{24}x + a_{25}y + a_{26} \end{cases} \tag{7}$$

Respectively the system of two homotopic equations will look like:

$$H : \begin{cases} H_1(t, x, y) = (1 - t)\left(p_1^2 x^2 - q_1^2\right) + t f_1(x, y) \\ H_2(t, x, y) = (1 - t)\left(p_2^2 y^2 - q_2^2\right) + t f_2(x, y) \end{cases} \tag{8}$$

In (8) t is a parameter while p_i and q_i are randomly generated complex numbers. The parameter t changes from 0 to 1 with sufficiently small step. For $t = 0$ the equations have the following form:

$$\begin{cases} H_1(0, x, y) = p_1^2 x^2 - q_1^2 \\ H_2(0, x, y) = p_2^2 y^2 - q_2^2 \end{cases} \tag{9}$$

This system has four possible solutions:

$$x_1 = \frac{q_1}{p_1}; y_1 = \frac{q_2}{p_2}$$

$$x_1 = -\frac{q_1}{p_1}; y_1 = \frac{q_2}{p_2}$$

$$x_1 = \frac{q_1}{p_1}; y_1 = -\frac{q_2}{p_2}$$

$$x_1 = -\frac{q_1}{p_1}; y_1 = -\frac{q_2}{p_2}$$

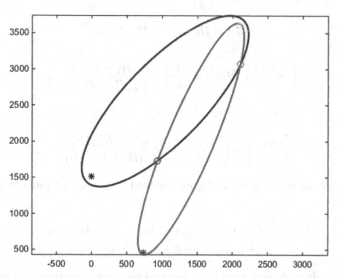

Fig. 2. Determination of the intersection points of two ellipses.

For each of them an iterative search for solutions is applied with t increasing from 0 to 1. Two solutions in the real domain are depicted on Fig. 2 using the marker "○"; the transmitter is marked with "△" and the two receivers is marked with "*" and "*":

The nonlinear polynomial system of two or more Eqs. (8) is solved by using Newton's method. The Newton's method characterizes with start at a known vector and the new vector is computed through solving a linear system for the update vector. The considered system is denoted by H (homotopic Eqs. (8)). If the update vector is $\Delta z = \begin{bmatrix} \Delta x \\ \Delta y \end{bmatrix}$, then the linear system looks like:

$$\frac{dH(z_0)}{dz}\Delta z = -H(z_0) \tag{10}$$

Here $z_0 = \begin{bmatrix} x_0 \\ y_0 \end{bmatrix}$ is the starting vector.

The next vector is calculated by:

$$z_1 = z_0 + \Delta z \tag{11}$$

Applying the Eqs. (10) and (11) to (8) we receive:

$$H(z) = H(x, y) = \begin{bmatrix} H_1(x, y) \\ H_2(x, y) \end{bmatrix} \tag{12}$$

$$\frac{dH(z)}{dz} = \begin{bmatrix} \frac{\partial H_1(x,y)}{\partial x} & \frac{\partial H_1(x,y)}{\partial y} \\ \frac{\partial H_2(x,y)}{\partial x} & \frac{\partial H_2(x,y)}{\partial y} \end{bmatrix} = \begin{bmatrix} H_{1,1} & H_{1,2} \\ H_{2,1} & H_{2,2} \end{bmatrix} \tag{13}$$

$$\left(\frac{dH(z)}{dz}\right)^{-1} = \frac{1}{H_{1,1}H_{2,2} - H_{1,2}H_{2,1}} \begin{bmatrix} H_{2,2} & -H_{1,2} \\ -H_{2,1} & H_{1,1} \end{bmatrix} \tag{14}$$

$$det = H_{1,1}H_{2,2} - H_{1,2}H_{2,1} \tag{15}$$

$$\left(\frac{dH(z)}{dz}\right)^{-1} H(z) = \frac{1}{det}\begin{bmatrix} H_{2,2}H_1 - H_{1,2}H_2 \\ -H_{2,1}H_1 + H_{1,1}H_2 \end{bmatrix} \tag{16}$$

Finally:

$$\begin{bmatrix} x_{k+1} \\ y_{k+1} \end{bmatrix} = \begin{bmatrix} x_k - \frac{1}{det}(H_{2,2}H_1 - H_{1,2}H_2) \\ y_k - \frac{1}{det}(-H_{2,1}H_1 + H_{1,1}H_2) \end{bmatrix} \tag{17}$$

In 3D case the equation of an ellipsoid defined by transmitter i and receiver j is a little longer:

$$f_{ij}(x, y, z) = a_{1ij}x^2 + a_{2ij}y^2 + a_{3ij}z^2 + a_{4ij}xy + a_{5ij}xz$$
$$+ a_{6ij}yz + a_{7ij}x + a_{8ij}y + a_{9ij}z + a_{10ij} = 0 \tag{18}$$

Further, to simplify mathematical expressions for a considered ellipsoid we will omit the indexes i and j. The coefficients in the above equation are calculated as follow:

$$a_1 = b^2c^2m_{11}^2 + a^2c^2m_{21}^2 + a^2b^2m_{31}^2$$

$$a_2 = b^2c^2m_{12}^2 + a^2c^2m_{22}^2 + a^2b^2m_{32}^2$$

$$a_3 = b^2c^2m_{13}^2 + a^2c^2m_{23}^2 + a^2b^2m_{33}^2$$

$$a_4 = b^2c^2 2m_{11}m_{12} + a^2c^2 2m_{21}m_{22} + a^2b^2 2m_{31}m_{32}$$

$$a_5 = b^2c^2 2m_{11}m_{13} + a^2c^2 2m_{21}m_{23} + a^2b^2 2m_{31}m_{33}$$

$$a_6 = b^2c^2 2m_{12}m_{13} + a^2c^2 2m_{22}m_{23} + a^2b^2 2m_{32}m_{33}$$

$$a_7 = -b^2c^2\left(2m_{11}^2 s_x + 2m_{11}m_{12}s_y + 2m_{11}m_{13}s_z\right) - a^2c^2\left(2m_{21}^2 s_x + 2m_{21}m_{22}s_y + 2m_{21}m_{23}s_z\right)$$
$$- a^2b^2(2m_{31}^2 s_x + 2m_{31}m_{32}s_y + 2m_{31}m_{33}s_z)$$

$$a_8 = -b^2c^2\left(2m_{12}^2 s_y + 2m_{11}m_{12}s_x + 2m_{12}m_{13}s_z\right) - a^2c^2\left(2m_{22}^2 s_y + 2m_{21}m_{22}s_x + 2m_{22}m_{23}s_z\right)$$
$$- a^2b^2(2m_{32}^2 s_y + 2m_{31}m_{32}s_x + 2m_{32}m_{33}s_z)$$

$$a_9 = -b^2c^2\left(2m_{13}^2 s_z + 2m_{11}m_{13}s_x + 2m_{12}m_{13}s_y\right) - a^2c^2\left(2m_{23}^2 s_z + 2m_{21}m_{23}s_x + 2m_{22}m_{23}s_y\right)$$
$$- a^2b^2(2m_{33}^2 s_z + 2m_{31}m_{33}s_x + 2m_{32}m_{33}s_y)$$

$$a_{10} = b^2c^2(m_{11}^2 s_x^2 + m_{12}^2 s_y^2 + m_{13}^2 s_z^2 + 2m_{11}m_{12}s_x s_y + 2m_{11}m_{13}s_x s_z + 2m_{12}m_{13}s_y s_z)$$
$$+ a^2c^2(m_{21}^2 s_x^2 + m_{22}^2 s_y^2 + m_{23}^2 s_z^2 + 2m_{21}m_{22}s_x s_y + 2m_{21}m_{23}s_x s_z + 2m_{22}m_{23}s_y s_z)$$
$$+ a^2b^2(m_{31}^2 s_x^2 + m_{32}^2 s_y^2 + m_{33}^2 s_z^2 + 2m_{31}m_{32}s_x s_y + 2m_{31}m_{33}s_x s_z + 2m_{32}m_{33}s_y s_z)$$

$$- a^2 b^2 c^2$$

where a, b and c are the semiaxes of the corresponding ellipsoid (their calculation is identical to the semiaxes of the ellipse and c is equal to b), s_x, s_y and s_z are calculated analogously to the 2D case.

In the equations above there are elements of rotation matrix m_{kl}, $k = \overline{1,3}$, $l = \overline{1,3}$, calculated as follows. Let denote again (as in 2D case) the angle of ellipsoid rotation by $\alpha_{i,j}$. The angle of rotation is determined by the angle between ellipsoid major axis and the axis OX of the reference coordinate system. The ellipsoid major axes is determined by transmitter i and receiver j, placed on the foci of the ellipsoid. Additionally we calculate the cross product of these two vectors denoted here by C_r. In the definition of rotation matrix the coordinates of its normalized form will be used: $[x, y, z]$.

Let us denote $\tau = 1 - \cos(\alpha_{i,j})$. The rotation matrix is:

$$m = \begin{bmatrix} \tau x^2 + \cos(\alpha_{i,j}) & \tau xy - \sin(\alpha_{i,j})z & \tau xz + \sin(\alpha_{i,j})y \\ \tau xy + \sin(\alpha_{i,j})z & \tau y^2 + \cos(\alpha_{i,j}) & \tau yz - \sin(\alpha_{i,j})x \\ \tau xz - \sin(\alpha_{i,j})y & \tau yz + \sin(\alpha_{i,j})x & \tau z^2 + \cos(\alpha_{i,j}) \end{bmatrix}$$

The system of three homotopic equations will look like:

$$H : \begin{cases} H_1(t, x, y, z) = (1 - t)\left(p_1^2 x^2 - q_1^2\right) + t f_1(x, y, z) \\ H_2(t, x, y, z) = (1 - t)\left(p_2^2 y^2 - q_2^2\right) + t f_2(x, y, z) \\ H_3(t, x, y, z) = (1 - t)\left(p_3^2 x^2 - q_3^2\right) + t f_3(x, y, z) \end{cases} \tag{19}$$

For t $= 0$ the equations have the following form:

$$\begin{cases} H_1(t, x, y, z) = p_1^2 x^2 - q_1^2 \\ H_2(t, x, y, z) = p_2^2 y^2 - q_2^2 \\ H_3(t, x, y, z) = p_3^2 x^2 - q_3^2 \end{cases} \tag{20}$$

This system has eight possible initial solutions:

$$x_1 = \frac{q_1}{p_1}; y_1 = \frac{q_2}{p_2}; z_1 = \frac{q_3}{p_3}$$

$$x_2 = -\frac{q_1}{p_1}; y_2 = \frac{q_2}{p_2}; z_2 = \frac{q_3}{p_3}$$

$$x_3 = \frac{q_1}{p_1}; y_3 = -\frac{q_2}{p_2}; z_3 = \frac{q_3}{p_3}$$

$$x_4 = \frac{q_1}{p_1}; y_4 = \frac{q_2}{p_2}; z_4 = -\frac{q_3}{p_3}$$

$$x_5 = -\frac{q_1}{p_1}; y_5 = -\frac{q_2}{p_2}; z_5 = \frac{q_3}{p_3}$$

$$x_6 = -\frac{q_1}{p_1}; y_6 = \frac{q_2}{p_2}; z_6 = -\frac{q_3}{p_3}$$

$$x_7 = \frac{q_1}{p_1}; y_7 = -\frac{q_2}{p_2}; z_7 = -\frac{q_3}{p_3}$$

$$x_8 = -\frac{q_1}{p_1}; \; y_8 = -\frac{q_2}{p_2}; \; z_8 = -\frac{q_3}{p_3}$$

$$\Delta H = \begin{bmatrix} \frac{\partial H_1}{\partial x}(x,y,z,t) & \frac{\partial H_1}{\partial y}(x,y,z,t) & \frac{\partial H_1}{\partial z}(x,y,z,t) \\ \frac{\partial H_2}{\partial x}(x,y,z,t) & \frac{\partial H_2}{\partial y}(x,y,z,t) & \frac{\partial H_2}{\partial z}(x,y,z,t) \\ \frac{\partial H_3}{\partial x}(x,y,z,t) & \frac{\partial H_3}{\partial y}(x,y,z,t) & \frac{\partial H_3}{\partial z}(x,y,z,t) \end{bmatrix}$$

$$= \begin{bmatrix} (1-t)2x_1p_1^2 + t\frac{\partial f_1}{\partial x}(x,y,z) & t\frac{\partial f_1}{\partial y}(x,y,z) & t\frac{\partial f_1}{\partial z}(x,y,z) \\ t\frac{\partial f_2}{\partial x}(x,y,z) & (1-t)2y_2p_2^2 + t\frac{\partial f_2}{\partial y}(x,y,z) & t\frac{\partial f_2}{\partial z}(x,y,z) \\ t\frac{\partial f_3}{\partial x}(x,y,z) & t\frac{\partial f_3}{\partial y}(x,y,z) & (1-t)2z_3p_3^2\frac{\partial f_3}{\partial z}(x,y,z) \end{bmatrix}$$

$$\frac{dH}{dt} = \begin{bmatrix} -p_1^2x^2 + q_1^2 + f_1 \\ -p_2^2y^2 + q_2^2 + f_2 \\ -p_3^2z^2 + q_3^2 + f_3 \end{bmatrix} \tag{21}$$

$$DZ = \Delta H^{-1}H \tag{22}$$

The system of equations for localizing the target consists of three quadric equations in the form of (18). For each of them the corresponding homotopic equation is composed (19) and eight initial possible solutions are studied. The intersection point between the three ellipsoids is a real solution of the system of equations of type (19). An intersection point of three ellipsoids is shown on Fig. 3.

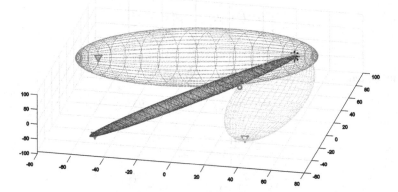

Fig. 3. Determination of the intersection point of ellipsoids.

4 Discussion on the Method

Solving systems of nonlinear equations is not a trivial problem. It is always accompanied by the possibility of two types of errors. The first of them is related to the impossibility of finding a solution at all. It is happened when Newton's method is divergent. If we recall from the theory, we must be close enough to the sought solution for the method to be convergent. This cannot be guaranteed when solving real tasks of the considered type. In the homotopy method, various options for space normalization, stepwise adaptive algorithms, and others modifications have been discussed in order to overcome this weakness.

The algorithm fails also when it falls into a local extremum and find a suboptimal solution. The standard way to avoid/resolve the problem is to restart algorithm with different starting point.

These disadvantages, however, are inherent to some extent even to other approaches that have found widespread use. The proposed algorithm has its place because of the specificity of the task. Typically, a network of sensors is deployed to guard or monitor an area. On Fig. 4 there are five sensors randomly placed in space. Typically, intruders enter the controlled zone from a direction. In order to be located, it is necessary to detect them by a certain number of pairs of sensors. It is obvious that the achievement of target detection by a larger number of sensor pairs will be possible only in cases of sufficient proximity of the target to the sensors. I.e. this will be reached significantly later compared to the proposed algorithm, which required a minimum number of pairs to detect the target. On the figure the zone which is covered by group of 4 receivers is shown in red. The proposed here algorithm requires simultaneous detection by three receivers in order to locate the target. The target will be detected in a significantly bigger area (displayed in blue). The application of this algorithm does not negate the use of methods established in the field. The new approach complements them and works primarily in cases where other algorithms cannot collect enough data to locate the target. The optimal solution is to run both algorithms in parallel: the new one will be applicable for distant or low observable targets and classical one will work in the vicinity.

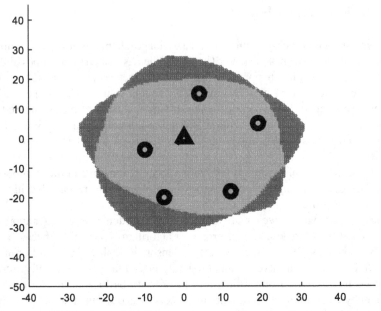

Fig. 4. A scenario with a target approaching controlled zone.

5 Target Tracking

There exists a wide range of algorithms for separate and group target tracking. They provide reliable target tracking in various situations. Here some of the most popular target tracking algorithms are described but with point of view of unlabeled input data (data association problem).

5.1 Kalman Filter

Kalman filter [3] is the base for almost all tracking algorithms. In terms of linear Markovian dynamic systems with Gaussian noise the Kalman filter is an optimal estimator. When one of the mentioned conditions isn't met it is necessary to use a modification of the filter. Some of them realized and described below.

5.2 Extended Kalman Filter (EKF)

EKF [4, 5] is a version of Kalman filter which can be applied to estimate nonlinear systems. The nonlinear system is approximated by Tailor series and instead of transition matrix its Jacobian is used for linear approximation or some of the other derivatives of higher order if more accurate approximation is required. Although nowadays EFK is considered as a good but non-optimal decision for nonlinear system estimation, in a number of cases when the system is strongly nonlinear the received results are incorrect.

5.3 Probabilistic Data Association (PDA) and Joint PDA (JPDA)

Next group of developed algorithms are PDA and JPDA [4, 5]. They are designed to solve association problem when more than one measurement are found in the gate. The first method solves the problem when a trajectory is present with many false alarms, while the second one finds the most plausible solution when targets are crossing or located sufficiently near each other (Fig. 5).

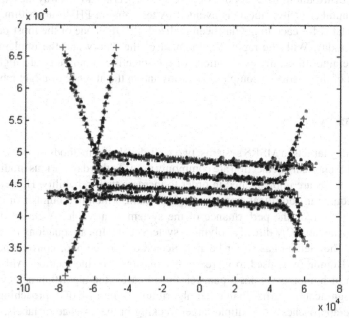

Fig. 5. Association of measurements and clustering of closely located targets with the help of JPDA.

5.4 Interacting Multiple Models (IMM)

The different versions of Interacting Multiple Models (IMM) [6] are developed for tracking maneuverable targets and objects with different dynamic states which require specific models. Estimators use the different models in parallel and at every step they evaluate the quality of each model.It is one of the most advanced target tracking algorithms available today. With more models and uncertainty about the origin of the measurements, however, the complexity of the algorithm grows exponentially with the number of models.

5.5 Sequential Monte Carlo Methods

The Sequential Monte Carlo Methods (also known as Particle filters) [7] are applied as an alternative to EKF for statistical modeling of distribution during filtering.The algorithm allows stable targets tracking in the presence of large non-linearities in behavior and non-Gaussian noise. The complexity of the algorithm is high, and with a relatively small number of targets, it fails to cope in real time.

5.6 Bayesian Multitarget Tracking

Bayesian Multitarget Tracking in the last few years has been also known as Probability Hypothesis Density filters (PHDF) [8]. In these filters the state vector includes individual probability distributions for every one of observed targets and in this way the exponential growth of number of hypothesis is avoided. A test version PHDF algorithm with the creation of label for each target has been realized. PHDF is one of the most promising algorithms today. With the rapid development of the theory and the implementation of real time algorithms, the expectations of the scientific community are high for the application of this theory to complex scenarios and in the absence of target labels.

6 Summary

When locating targets in MSRS systems, two well-developed methods of spherical interpolation and spherical intersection are commonly used. To this day, various modifications of these methods are implemented in order to increase their reliability. However, these methods require an additional amount of sensor pairs, which in a number of cases can significantly degrade the performance of the system as a whole. A classical method for target localization by directly solving a system of nonlinear equations is applied in the article. The method has so far been neglected due to the poor convergence of the algorithm. Homotopy is used to increase the robustness of the method. Although the application of homotopy is not a panacea for all cases, the algorithm achieves target detection significantly earlier than other algorithms in the case of approaching targets. Some basic approaches for multiple target tracking in the absence of labels, as is the case with MSRS detection systems, are also shortly reviewed.

Acknowledgement. This work was supported by the NSP SD program, which has received funding from the Ministry of Education and Science of the Republic of Bulgaria under the grant agreement no. Д01-74/19.05.2022.

References

1. Malanowski, M.: Signal Processing for Passive Bistatic Radar. Artech House, Boston (2019)
2. Morgan, A.: Solving Polynomial Systems Using Continuation for Engineering and Scientific Problems. SIAM (2009)
3. Kálmán, R.E.: A new approach to linear filtering and prediction problems. J. Basic Eng. **82**(1), 35–45 (1960)
4. Bar-Shalom, Y., Fortmann, T.: Tracking and Data Association. Academic Press, San Diego (1988)
5. Bar-Shalom, Y., Li, X.-R.: Multitarget-Multisensor Tracking: Principles and Techniques. YBS Publishing, Storrs (1995)

6. Mazor, E., Averbuch, A., Bar-Shalom, Y., Dayan, J.: Interacting multiple model methods in target tracking: a survey. IEEE Trans. AES **34**(1), 103–123 (1998)
7. Angelova, D., Mihaylova, L.: Joint target tracking and classification with particle filtering and mixture Kalman filtering using kinematic radar information. Digit. Signal Process. **16**(2), 180–204 (2006)
8. Vo, B.T., Vo, B.N., Cantoni, A.: Analytic implementations of the cardinalized probability hypothesis density filter. IEEE Trans. SP **55**(7, Part 2), 3553–3567 (2007)
9. Malanowski, M., Kulpa, K.: Two methods for target localization in multistatic passive radar. IEEE Trans. AES **48**(1), 572–580 (2012)
10. Friedlander, B.: A passive localization algorithm and its accuracy analysis. IEEE J. Oceanic Eng. **OE-12**(1), 234–245 (1987)
11. Mellen, G., Pachter, M., Raquet, J.: Closed-form solution for determining emitter location using time difference of arrival measurements. IEEE Trans. AES **39**(3), 1056–1058 (2003)
12. Schau, H., Robinson, A.: Passive source localization employing intersecting spherical surfaces from time-of-arrival differences. IEEE Trans. Acoust. Speech Signal Process. **35**(8), 1223–1225 (1987)
13. Schmidt, R.: A new approach to geometry of range difference location. IEEE Trans. AES **8**(6), 821–835 (1972)
14. Smith, J., Abel, J.: The spherical interpolation method of source localization. IEEE J. Oceanic Eng. **12**(1), 246–252 (1987)
15. Sommese, A., Verschelde, J., Wampler, C.: Homotopies for intersecting solution components of polynomial systems. The Abstract and manuscript in pdf format. SIAM J. Numer. Anal. **42**(4), 1552–1571 (2004)
16. Bensky, A.: Wireless Positioning Technologies and Applications, 2nd edn. (GNSS Technology and Applications). Artech House Publishers (2016)
17. Zekavat, R., Buehrer, M. (eds.): Handbook of Position Location: Theory, Practice, and Advances, 2nd edn. Wiley-IEEE Press (2019)
18. Du, Y., Wei, P.: An explicit solution for target localization in noncoherent distributed MIMO radar systems. IEEE Signal Process. Lett. **21**(9), 1093–1097 (2014)
19. Yang, H., Chun, J.: An improved algebraic solution for moving target localization in noncoherent MIMO radar systems. IEEE Trans. SP **64**(1), 258–270 (2016)
20. Amiri, R., Behnia, F., Sadr, M.: Efficient positioning in MIMO radars with widely separated antennas. IEEE Commun. Lett. **21**(7), 1569–1572 (2017)
21. Amiri, R., Behnia, F., Sadr, M.: Positioning in MIMO radars based on constrained least squares estimation. IEEE Commun. Lett. **21**(10), 2222–2225 (2017)
22. Amiri, R., Zamani, H., Behnia, F., Marvasti, F.: Sparsity-aware target localization using TDOA/AOA measurements in distributed MIMO radars. ICT Express **2**(1), 23–27 (2016)
23. Amiri, R., Behnia, F., Zamani, H.: Efficient 3-D positioning using timedelay and AOA measurements in MIMO radar systems. IEEE Commun. Lett. **21**(12), 2614–2617 (2017)

Intrusion Detection by XGBoost Model Tuned by Improved Social Network Search Algorithm

Nebojsa Bacanin[✉][iD], Aleksandar Petrovic[iD], Milos Antonijevic[iD],
Miodrag Zivkovic[iD], Marko Sarac[iD], Eva Tuba[iD], and Ivana Strumberger[iD]

Singidunum University, Danijelova 32, 11000 Belgrade, Serbia
{nbacanin,aleksandar.petrovic,mantonijevic,mzivkovic,
msarac,istrumberger}@singidunum.ac.rs, etuba@ieee.org

Abstract. The industry 4.0 flourished recently due to the advances in a number of contemporary fields, alike artificial intelligence and internet of things. It significantly improved the industrial process and factory production, by relying on the communication between devices, production machines and equipment. The biggest concern in this process is security, as each of the network-connected components is vulnerable to the malicious attacks. Intrusion detection is therefore a key aspect and the largest challenge in the industry 4.0 domain. To address this issue, a novel framework, based on the XGBoost machine learning model, tuned with a modified variant of social network search algorithm, is proposed. The introduced framework and algorithm have been evaluated on a challenging UNSW-NB 15 benchmark intrusion detection dataset, and the experimental findings were put into comparison against the outcomes of other high performing metaheuristics for the same problem. For comparison purposes, alongside the original version of social network search, harris hawk optimization algorithm, firefly algorithm, bat algorithm, and artificial bee colony, were also adopted for XGBoost tuning and validated against the same internet of things security benchmark dataset. Experimental findings proved that the best performing developed XGBoost model is the one which is tuned by introduced modified social network search algorithm, outscoring others in most of performance indicators which were employed for evaluation purposes.

Keywords: Intrusion detection · Swarm intelligence · XGBoost · Optimization · Social network search algorithm

1 Introduction

Internet of Things (IoT) advancements have led to a concept called Industrial Revolution 4.0. The idea is to apply IoT to the industrial process of factories, resulting in smart factories. The benefits of the IoT technology allow for communication between devices, machines, and equipment used in the manufacturing

D. Simian and L. F. Stoica (Eds.): MDIS 2022, CCIS 1761, pp. 104–121, 2023.
https://doi.org/10.1007/978-3-031-27034-5_7

process. This allows for serving of more customized client requests and increases the rate of productivity alongside quality.

The main concern with the application of this technology to the process lines is the security. Every IoT component in the process is vulnerable to attacks as it has network communication capabilities. Furthermore, the failure of devices also has to be considered as a more serious threat in comparison to the traditional methods. To mitigate these scenarios, real-time detection systems are implemented for failures and intrusions [37].

Artificial intelligence (AI) [20] and Machine Learning (ML) [21] field improvements contribute to the development of the security systems. Additionally, the computational resource availability has increased further enhancing the probability of a reliable intrusion detection system. Furthermore, the algorithms of AI and ML nature are being improved, which results in constant improvement in the field. These algorithms have performed well for this case as it belongs to a group of NP-hard problems, for which these algorithms have the most use.

The issues with these solutions is that they have high computational resource requirements and usually work with vast and complex datasets. While considering the complexity of the applied algorithm itself, abundant control parameters further complicate these problems. This is referred to as the problem of hyperparameter optimization, that is an NP-hard problem. The selection of these parameters has to be performed manually and they influence the successfulness of the process. For this reason ML solutions are optimized by outside algorithms that perform hyper-parameter optimization. Due to the no free lunch theorem [43], there is not a single solution that works for all of the problems. New options have to be explored and the old solutions are vigorously improved. eXtreme gradient boosting (XGBoost) [18] is considered as one of the best performers.

More stable performance can be achieved by ensemble learning in comparison to the single model. For the classification problems, the random forest model fares exceptionally well. The XGBoost brings further improvements upon the tree structure logic introducing regularization terms and second-order derivatives. The applications of this solution are seen in the fields of security [52], the industry [27], finance [22], as well as business [32].

Nonetheless, XGBoost is not without deficiencies. Considering the relatively large number of un-trainable parameters (hyper-parameters), optimizing (tuning) the XGBoost for specific problem is challenging. This issue was addressed by using the previously mentioned optimization metaheurisitcs. The group of algorithms that has proven well for this case is the swarm intelligence (SI) algorithms [11]. The SI is stochastic and population based approach, which works well with the ML techniques. Additionally, the swarm-based algorithms are popular group of methods for NP-hard problem solving [2]. The algorithm that authors have utilized for the purpose of XGBoost tuning is the social network search

(SNS) algorithm [39]. However, since the basic SNS exhibits some cons, in the proposed research, a modified version is devised and validated

The most notable contributions of this research are listed:

- Development of a modified variant of the SNS algorithm, designed to address the known issues of the original implementation.
- Application of the implemented algorithm in the framework to tune the parameters of the XGBoost machine learning model.
- Evaluation of the proposed hybrid XGBoost and metaheuristics framework on the well-known intrusion detection benchmark dataset - UNSW-NB 15.
- Comparison of the experimental outcomes to the results of other cutting-edge approaches.

The rest of the paper is organized in the following manner: Sect. 2 provides insight on the technologies required for this research, Sect. 3 presents the original algorithm and the applied improvements, Sect. 4 explains the experimental process and provides results, and finally Sect. 5 summarizes the research and gives final remarks.

2   Preliminaries and Related Works

This section first introduces basic background information related to methods used in the proposed research.

2.1   Overview of XGBoost

The XGBoost algorithm utilizes tree structures while introducing regularization terms and second-order derivatives. These modifications provide better performance compared to other similar algorithms, which is why XGBoost is commonly used for solution of problems of higher complexity. In order to improve objective function optimization, an additive training method is exploited by XGBoost. This way, every optimization process step is established based on outcome of the previous one. The t-th objective function of XGBoost model equation expression:

$$F_o{}^i = \sum_{k=1}^n l\left(y_k, \hat{y}_k^{i-1} + f_i\left(x_k\right)\right) + R(f_i) + const, \tag{1}$$

in which the $F_o{}^i$ denotes i-th objective function, l is the term loss for the $t - th$ round, $const$ represents the constants term, and R is the regularization term which be calculated based on Eq. (2)

$$R(f_i) = \gamma T_i + \tfrac{\lambda}{2} \sum_{j=1}^T w_j^2, \tag{2}$$

where γ and λ are customization parameters, while w_j denotes solution of weights and T_i represents tree structure evaluated in i-th round. It should be noted that the larger the γ and λ values are, the simpler is the tree structure.

The issue of over fitting may be addressed with the second-order Taylor expansion application to Eq. (1):

$$obj^{(t)} = \sum_{i=1}^{n}[l(y_i, \hat{y}_i^{t-1}) + g_i f_t(x_i) + \frac{1}{2}h_i f_t^2(x_i)] + \Omega(f_t) + const \tag{3}$$

First and second model derivatives, denoted as g and h, respectively are given as:

$$g_j = \partial_{\hat{y}_k^{i-1}} l\left(y_j, \hat{y}_k^{i-1}\right) \tag{4}$$

$$h_j = \partial_{\hat{y}_k^{i-1}}^2 l\left(y_j, \hat{y}_k^{i-1}\right) \tag{5}$$

The solution obtainable by combination of Eq. (2), Eq. (4) and Eq. (5) into Eq. (3) and forming the following equations:

$$w_j^* = -\frac{\sum_{t \in I_j} g_t}{\sum_{t \in I_j} h_t + \lambda} \tag{6}$$

$$F_o^* = -\frac{1}{2}\sum_{j=1}^{T} \frac{\left(\sum_{i \in I_j} g_i\right)^2}{\sum_{i \in I_j} h_i + \lambda} + \gamma T, \tag{7}$$

where F_o^* denotes the loss function score, the expression $\left(\sum_{i \in I_j} g_i\right)^2$ denote that the all instances are mapped to leaf j, and w_j^* represents the solution of weights.

2.2   Swarm Intelligence Applications in Machine Learning

The group of swarm intelligence algorithms is considered metaheuristic due to the inspiration for creation of such algorithms comes from nature. Animals that exhibit swarm-like behavior are used for inspiration as this translates well to algorithms. This group of algorithms continues to give excellent results with the NP-hard problems and therefore is considered a go-to optimizer for such problems. The best results are achieved with the use of hybridization. During this process, algorithms are combined with the purpose of eliminating each other deficiencies. The motivation for hybridization is the lack of performance in one of the two defining phases of swarm algorithms. Algorithm usually performs well in either exploration or exploitation, which both have different purposes. Exploration focuses on broadening the search, while the exploitation focuses on accuracy. The trade-off is required to be achieved between these phases for the finding of the global optima, which is the end goal.

Notable algorithms from the swarm family are FA [45], particle swarm optimization (PSO) [34], BA [47], ABC [26], elephant herding optimization (EHO) [41], and whale optimization algorithm (WOA) [33].

Newer high performing algorithms are grasshopper optimization algorithm (GOA) [30], monarch butterfly optimization (MBO) [42], salp swarm algorithm (SSA) [29], and HHO [23].

The application of these algorithms on the real world problem with NP-hard complexity is vast and some examples are given in further text. The swarm intelligence algorithms can be applied to wireless sensors networks (WSNs) [7,49], cryptocurrency trends estimations [36], artificial neural network optimization [3–6,8,13,17,28], cloud-edge computing and task scheduling [9,16], computer-conducted MRI classification and sickness determination [10,12,15,25,38], and finally the COVID-19 global epidemic-associated applications [1,14,31,48,50,51].

3 Proposed Method

3.1 Basic Social Network Search Algorithm

The SNS [39] is metaheuristic algorithm that uses human social behaviors to achieve more popularity. Interaction between users on various social networks represents the base of the SNS algorithm. This is done through the implementation of the user's moods, that include imitation, conversation, disputation and innovation, which guide the behavior of simulated users and with it manages the behavior of the algorithm. These moods represent simplified representations of real-world interaction between social network users. In developing new solutions, SNS uses one of the previously mentioned moods.

Imitation models represent one of the main characteristics that mimic real-world social media. Prevailing behavior on social network includes following other people and getting information about their posts and actions. Furthermore, if the published post expresses some challenging topics, users will aim to imitate or disprove presented views. The equations which describe this model are expressed in:

$$X_{i\,new} = X_j + rand(-1,1) \times R$$
$$R = rand(0,1) \times r \tag{8}$$
$$r = X_j - X_i$$

where X_i is the j-th users view vector selected at random with $i \neq j$. Random value selection is described with $rand(-1,1)$ and $rand(0,1)$, which are arbitrary vectors in intervals of $[-1,1]$ and $[0,1]$ respectively. The radius of shock R represents the influence of the j-th user, and multiple rs affect the magnitude. The r is calculated according to differences in opinions of j-th users. The last shock's influence is computed by multiplication of the random vector value in range $[-1,1]$ with positive or negative component values, depending if the shared opinions match or not.

Conversation mood represents communication between individual users via a private chat. Mirroring the real-world social behavior, simulated users gain insight into differences in opinions regarding various topics by exchanging information privately. This mechanism is formulated with the following equation:

$$X_{i\,new} = X_k + R$$
$$R = rand(0,1) \times D \tag{9}$$
$$D = sign(f_i - f_j) \times (X_j - X_i),$$

in which X_k is the randomly chosen vector of the subject of conversation, R is the chats effect that are opinion difference based and represent users' changes in perspective, shown with X_k. The difference of views that users have is represented by D. Randomly chosen vector $rand(0,1)$ between $[0,1]$ is a vector of a random chat view of a user. Additionally, chat vectors of randomly selected users i and j are represented by X_i and X_j, respectively. Note for this case $j \neq i \neq k$. The $sign(f_i - f_j)$ determines in which direction X_k will move via comparison, while the f_i and f_j denote fitness of solutions j and j, respectively.

Disputation mood mimics the users behavior when they elaborate and defend their views on certain topics among themselves. Additionally, users can create chat groups where they can be influenced by the different or opposite opinions of the others. The basic disputation mood model implies that a random number of users form a discussion group in which the opinions are calculated using following formula:

$$X_{i\,new} = X_i + rand(0,1) \times (M - AF \times X_i)$$
$$M = \frac{\sum_t^{N_r} X_t}{Nr} \tag{10}$$
$$AF = 1 + round(rand)$$

with X_i representing the i-th users view vector, $rand(0,1)$ is an arbitrary vector in the interval $[0,1]$ and M being the mean value of opinions. The admission factor AF is denoted by a random value of 1 or 2 and serves as an indication of insistence to their own opinion when discussing it with other people. The input is rounded to the nearest integer using the $round()$ function and $rand$ represents a random value in the interval $[0,1]$. The sum of people in a discussion group is depicted by N_r and is a arbitrary value between 1 and N_{user}, where N_{user} represents the overall users number of the network.

The process of users being able to understand the particular topic better because of their knowledge or expertise, or they discover original concepts on certain shared content is simulated in SNS by innovation mood. Sometimes a specific topic may have various features and by changing the perception about some of those features, the broad perception of the topic understanding might change resulting in a novel view. This approach can mathematically be described as:

$$X_{i\,new}^d = t \times x_j^d + (1-t) \times n_{new}^d$$
$$n_{new}^d = lb_d + rand_1 \times (ub_d - lb_d) \tag{11}$$
$$t = rand_2$$

in which the d-th variable chosen at random is shown as d in the range $[1, D]$, and the amount of available variables is shown by D. The $rand_1$ and $rand_2$ represent additional random values for the range $[0, 1]$. The minimum and maximum d-th values of n_{new}^d are given as ub_d and lb_d, respectfully. X_j^d represents the current idea for the d-th idea, and the user j selected at random to satisfy $j \neq i$. For the case of an user altering their opinion a new idea is generated which becomes n_{new}^d. Formed as an interpolation of the current idea, a new view x_{new}^d is created on the d-th dimension.

The following equation models the shift in dimension for the x_{new}^d, as it is considerable as a newly shareable view:

$$X_{i\,new} = [x_1, x_2, x_3, ..., x_{i\,new}^d, ..., x_D] \tag{12}$$

and the $x_{i\,new}^d$ shows a novel insight on an issue with the point of view d-th that is replaced with the current one x_i^d. Number of users, maximum iterations, and limits are required parameters for the setup of initial network. The next equation creates each initial view:

$$X_0 = lb + rand(0, 1) \times (ub - lb) \tag{13}$$

where for each user an initial view vector is shown as X_0, while a value chosen at random from the range $[0, 1]$ is shown as $rand(0, 1)$. The lower and upper bounds are given as ub and lb, respectively. The following equation is applied for the problems of maximization view limiting:

$$X_i = \begin{cases} X_i, & f(X_i) < f(X_{i\,new}) \\ X_{i\,new}, & f(X_{i\,new}) \geq f(X_i) \end{cases} \tag{14}$$

3.2 Improved SNS Approach

According to practical simulations with original SNS metaheuristics on standard bound-constrained benchmarks, it was observed that in some cases (runs), the search process converges towards sub-optimum regions of the search space as a consequence of low population diversity. In other words, if initial, random solutions are far from optimum, exploitation leads to premature convergence and final results are not satisfying.

Method showed in this paper addresses above mentioned issues by incorporating novel initialization scheme and strategy for maintaining population diversity throughout the whole algorithm's run.

3.2.1 Novel Initialization Scheme
The proposed method applies regular initialization strategy shown by the following equation:

$$X_{i,j} = lb_j + \psi \cdot (ub_j - lb_j), \tag{15}$$

for which the j-th component of i-th individual is shown as $X_{i,j}$, while the lower and upper bounds of the j parameter are lb_j and ub_j, respectively. Pseudo-random number provided by the normal distribution for range $[0, 1]$ is ψ. Even so, for application of quasi-reflection-based learning (QRL) on the Eq. (15) results in coverage of a wider search area, as seen in [35]. This results in a quasi-reflexive-opposite element (X_j^{qr}) for each individual parameter j (X_j).

$$X_j^{qr} = \text{rnd}\left(\frac{lb_j + ub_j}{2}, x_j\right), \tag{16}$$

which utilizes the rnd function for selection of pseudo-random value in the range $\left[\dfrac{lb_j + ub_j}{2}, x_j\right]$. It also should be noted that in case if $x_j < \dfrac{lb_j + ub_j}{2}$, the above expression becomes $\text{rnd}(x_j, \frac{lb_j+ub_j}{2})$. When considering the QRL, the given initialization scheme starts by initializing only $NP/2$, as it does not introduce overhead to the algorithm considering the $FFEs$. The applied scheme is given in Algorithm 1.

Algorithm 1. Initialization scheme based on QRL procedure pseudo-code

Step 1: Generate population P^{init} of $NP/2$ individuals by applying the Eq. (15)
Step 2: Generate QRL population P^{qr} from the P^{init} by using Eq. (16)
Step 3: Produce starting population P by merging P^{init} and P^{qr} $(P \cup P^{qr})$
Step 4: Obtain fitness for each individual in P
Step 5: Sort all solutions in P in respect of fitness

3.2.2 Mechanism for Maintaining Population Diversity

The population diversification is a means to monitor converging/diverging of the search procedure of the algorithm, described by [19]. The work presented utilizes the $L1$ population diversity metric [19] and it obtains diversity from individuals and the dimensionality components. The mentioned research suggests the use of $L1$ norm's dimensionality component for evaluation of the algorithm's search process.

Considering m as the amount of units in the population and n as the dimensionality problem, the $L1$ norm has following form as shown in Eqs. (17–19):

$$\overline{x_j} = \frac{1}{m} \sum_{i=1}^{m} x_{ij} \tag{17}$$

$$D_j^p = \frac{1}{m} \sum_{i=1}^{m} \left| x_{ij} - \overline{x}_j \right| \tag{18}$$

$$D^p = \frac{1}{n} \sum_{j=1}^{n} D_j^p, \tag{19}$$

for which the mean solutions positions in each dimension vector is shown as $\bar{x}$, $L1$ norm as vector of diversity of solutions' positions D_j^p, and for the entire population the diversity value is D^p, as a scalar value.

For the initial rounds execution duration of the algorithm in which the common initialization equation is applied (15), the diversity should be of high quality. On the other hand, the quality is predicted to drop in later iterations. To tackle this problem, the improved algorithm applies $L1$ norm for the dynamic diversity threshold (D_t) parameter.

Population diversity mechanism is performed as follows: firstly the D_t (D_{t0}) is obtained; during every iteration the $D^P < D_t$ condition is considered, while the D^P stands for the diversity of the current population; if the outcome of the condition is true, which suggests bad diversity, analogously produced random solutions replace nrs of the worst individuals. Auxiliary control parameter is represented by nrs (number of replaced solutions). Taking empirical simulations and theoretical analysis into consideration the resulting expression for obtaining D_{t0} can be used in following manner:

$$D_{t0} = \sum_{j=1}^{NP} \frac{(ub_j - lb_j)}{2 \cdot NP} \qquad (20)$$

The Eq. (20) relies on the prediction that the majority of the components will be generated at the mean of lower and upper boundaries of the parameters, seen in Eq. (15) and Algorithm 1. Presumingly that the process is functioning as planed and the algorithm is reaching towards the optimal region, the value of D_t is supposed to decline from the start value of $D_t = D_{t0}$ as suggested in the following equation:

$$D_{t+1} = D_t - D_t \cdot \frac{t}{T}, \qquad (21)$$

in which the current and subsequent iterations are denoted t and $t + 1$, respectively, and the maximum number of rounds per single run is T. To conclude, the previously described method will not be activated regardless of the D^P as D_t is decreasing dynamically.

3.2.3   Inner Workings and Complexity of the Proposed Algorithm

Considering the applied changes to the algorithm the new solution was proposed to be named diversity oriented SNS (DOSNS). When considering $FFEs$ the SNS algorithm enhanced version is not computationally more complex. Firstly, the improved initialization strategy does not require additional $FFEs$. The eligibility of the population diversity mechanism is not validated as it replaces the nrs worst solutions with new solutions whether the newly generated ones are better or worse. The final formula for calculating the complexity of the DOSNS in terms of $FFEs$ is given: $O(DOSNS) = O(NP) + O(T \cdot NP)$.

The pseudocode of the improved approach, DOSNS, is given in Algorithm 2. It is noted that the loop that goes through all solutions is omitted for the clarity reasons.

Algorithm 2. Pseudocode for the DOSNS algorithm

Set number of solutions (users), T, lb, ub, $t = 0$
Initialize starting population P of NP solutions according to Algorithm 1
Determine values of D_{t0} and D_t
Evaluate each user (solution) according to objective function
while $t \leq T$ **do**
    $Mood = rand(1, 4)$
    if $(Mood == 1)$ **then**
        Create new views based on Eq. (8)
    else if $(Mood == 2)$ **then**
        Create new views based on Eq. (9)
    else if $(Mood == 3)$ **then**
        Create new views based on Eq. (10)
    else if $(Mood == 4)$ **then**
        Create new views based on Eq. (11)
    end if
    Limit new views According to Eq. (14)
    Evaluate new view based on the objective function
    if (New view better then current view) **then**
        Keep old view, don't share new view
    else
        Replace old view with new view and share it
    end if
    Calculate D^P
    if $(D^P < D_t)$ **then**
        Replace worst nrs with solutions created as in (15)
    end if
    Asses the population
    Find the current best
    Update D_t by expression (21)
end while
Return Optimal Solution
Save and visualize overall and detailed optimization statistics

---

4 Experimental Reports and Discussion

This section first introduces properties of network intrusion dataset used in experiments with performance metrics, followed by comparative analysis with other cutting-edge approaches.

4.1 Datasets and Metrics

The UNSW-NB 15 dataset has been exploited for experimental purposes, available on https://www.kaggle.com/datasets/mrwellsdavid/unsw-nb15. The dataset was created by the use of IXIA PerfectStrom toolset for collection of raw traffic packages. This has been performed by the Cyber Range Laboratory of the Australian Center for Cyber Security (ACCS). The dataset was intended to be used as a set of regular network communication alongside cyber-attack patterns synthetic in nature. The dataset can be used both for problems of multi-class as well as binary classification. The use of the dataset in this research is for binary classification. The train and test split is applied to the set by default and for the purpose of this work they were combined. The string values were converted to integers and the instance that do not provide their state were removed as part of the preprocessing. The numerical values for the protocol state follow: FIN - 0, INT - 1, CON - 2, ACC - 3, REQ - 4, RST - 5. The class distribution and the heatmap of the variable correlation in UNSW-NB 15 dataset are shown in Fig. 1.

From the presented bar plot with classes distribution it can be seen that the utilized dataset is unbalanced. However, it should be noted that the handling unbalanced data is not conducted in this study.

To establish proper testing coefficients the authors use standard metrics for this type of problem. True positives (TP), true negatives (TN), false positives (FP), and false negatives (FN) are the base ground for further evaluative calculations. With their use a formula for the calculation of accuracy can be derived:

$$ACC = (TP + TN) \,/\, (TP + FP + TN + FN) \tag{22}$$

The other calculated metrics are precision, recall, and F-measure and are calculated by the formulas (23)–(25).

$$Precision = TP \,/\, (TP + FP) \tag{23}$$

$$Recall(sensitivity) = TP \,/\, (TP + FN) \tag{24}$$

$$F - measure = (2 \cdot Precision \cdot Recall)/(Precision + Recall) \tag{25}$$

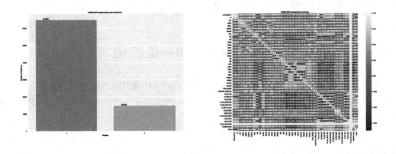

Fig. 1. Class distribution and heatmap of the correlation between the variables of the binary UNSW-NB 15 dataset

The test cases were executed for 10 solution over 8 iterations. The XGBoost parameters were setup by example of [24] in the same ranges.

4.2   Experimental Findings and Comparative Analysis

The performances of the introduced DOSNS metaheuristics were analyzed on the task of XGBoost tuning for the intrusion detection dataset provided above. The outcomes of the DOSNS algorithm have been verified against five other state of the art algorithms, that were evaluated in the same experimental setup and simulation conditions. These metaheuristics that were used as the reference included the basic SNS [40], HHO [23], FA [44], BA [46] and ABC [26]. These metaheuristics were independently implemented for the sake of this manuscript, and used the proposed control parameter setup as shown in their respective publications.

All above mentioned metaheuristics were executed in 30 independent runs with 8 iterations, where each approach developed XGBoost model by tuning six following hyper-parameters within pre-defined boundaries:

- learning rate (η), $[0.1, 0.9]$, continuous,
- min_child_weight, $[0, 10]$, continuous,
- subsample, $[0.01, 1]$, continuous,
- collsample_bytree, $[0.01, 1]$, continuous,
- max_depth, $[3, 10]$, integer and
- γ, $[0, 0.5]$, continuous.

Therefore, the solution's vector length is given by $L = 6$. It's noted that the parameters' boundaries were determined empirically.

Table 1 provides the experimental outcomes with respect to the achieved error rate over 30 runs. The best result in every category has been marked in the table. The introduced XGBoost-DOSNS algorithm achieved the superior best value in comparison to all other algorithms, while the XGBoost-HHO approach achieved slightly better values for worst, mean and median metrics.

Table 2 shows the detailed simulation metric on the UNSW-NB 15 dataset, where it can be noted that the proposed XGBoost-DOSNS model achieved the best accuracy of 99.6878%, in front of the XGBoost-HHO and XGBoost-SNS that achieved 99.6815%.

Table 1. Overall metrics of the observed metaheuristics on UNSW-NB 15 dataset

Method	XGBoost-DOSNS	XGBoost-SNS	XGBoost-HHO	XGBoost-FA	XGBoost-BA	XGBoost-ABC
Best	**0.003122**	0.003185	0.003185	0.003247	0.003247	0.003435
Worst	0.003934	0.004247	**0.003747**	0.004059	0.003809	0.004309
Mean	0.003616	0.003647	**0.003491**	0.003585	0.003566	0.003784
Median	0.003747	0.003528	**0.003497**	0.003622	0.003560	0.003747
Std	0.000267	0.000349	0.000186	0.000241	**0.000166**	0.000271
Var	0.000000	0.000000	0.000000	0.000000	0.000000	0.000000

Table 2 provides detailed metrics for the best generated solution (XGboost model) for every metaheuristics.

Table 2. Detailed metrics of the observed metaheuristics on USNW-NB 15 dataset

	XGBoost-DOSNS	XGBoost-SNS	XGBoost-HHO	XGBoost-FA	XGBoost-BA	XGBoost-ABC
Accuracy (%)	**99.6878**	99.6815	99.6815	99.6753	99.6753	99.6565
Precision 0	0.991339	**0.992321**	0.991009	0.991989	0.991989	0.991653
Precision 1	0.998155	0.997849	0.998155	0.997849	0.997849	0.997695
M.Avg. Precision	**0.996878**	0.996813	0.996816	0.996751	0.996751	0.996563
Recall 0	0.992	0.990667	0.992	0.990667	0.990667	0.99
Recall 1	0.998002	**0.998233**	0.997925	0.998156	0.998156	0.998079
M.Avg. Recall	**0.996878**	0.996815	0.996815	0.996753	0.996753	0.996565
F1 Score 0	**0.991669**	0.991493	0.991504	0.991328	0.991328	0.990826
F1 Score 1	**0.998079**	0.998041	0.99804	0.998002	0.998002	0.997887
M.Avg. F1 Score	**0.996878**	0.996814	0.996816	0.996752	0.996752	0.996564

To visualize performance of the proposed algorithm, Figs. 2, 3 and 4 show precision-recall (ROC) and receiver under operating characteristics (ROC) area under the curves (AUC), confusion matrix and one versus rest (OvR) ROC curves for the best generated solution.

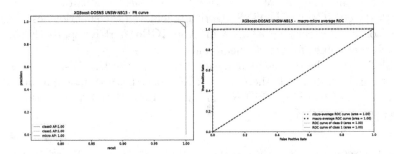

Fig. 2. PR and ROC plots of the proposed XGBoost-DOSNS approach on the UNSW-NB 15 dataset

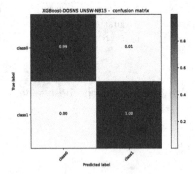

Fig. 3. The XGBoost-DOSNS confusion matrix for UNSW-NB 15 dataset

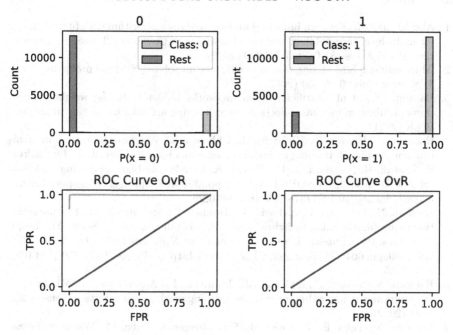

Fig. 4. The XGBoost-DOSNS one versus rest (OvR) ROC

5   Conclusion

This paper introduced a novel version of the SNS metaheuristics, that was devised in such way to tackle the common problems noted on the original SNS implementation. The novel algorithm was given the name DOSNS (diversity oriented SNS), and it was later utilized to tune the XGBoost model. The proposed XGBoost-DOSNS was evaluated on the intrusion detection benchmark dataset UNSW-NB 15, and the results were compared to 5 cutting-edge metaheuristic algorithms, utilized in the same setup on the same dataset. The experimental findings show that the suggested DOSNS algorithm obtained superior level of performances on this particular dataset, indicating possible application of this method in real-world intrusion detection systems. The future challenges in this domain should focus on the validation of the proposed hybrid model on additional intrusion detection datasets, with a goal to establish higher confidence in the results prior to the employment in the real-world scenario.

References

1. Abd Elaziz, M., et al.: An improved marine predators algorithm with fuzzy entropy for multi-level thresholding: real world example of Covid-19 CT image segmentation. IEEE Access **8**, 125306–125330 (2020)
2. Abdulrahman, S.M.: Using swarm intelligence for solving NP-hard problems. Acad. J. Nawroz Univ. **6**, 46–50 (2017)
3. Bacanin, N., et al.: Artificial neural networks hidden unit and weight connection optimization by quasi-refection-based learning artificial bee colony algorithm. IEEE (2021)
4. Bacanin, N., Bezdan, T., Zivkovic, M., Chhabra, A.: Weight optimization in artificial neural network training by improved monarch butterfly algorithm. In: Shakya, S., Bestak, R., Palanisamy, R., Kamel, K.A. (eds.) Mobile Computing and Sustainable Informatics. LNDECT, vol. 68, pp. 397–409. Springer, Singapore (2022). https://doi.org/10.1007/978-981-16-1866-6_29
5. Bacanin, N., Petrovic, A., Zivkovic, M., Bezdan, T., Antonijevic, M.: Feature selection in machine learning by hybrid sine cosine metaheuristics. In: Singh, M., Tyagi, V., Gupta, P.K., Flusser, J., Ören, T., Sonawane, V.R. (eds.) ICACDS 2021. CCIS, vol. 1440, pp. 604–616. Springer, Cham (2021). https://doi.org/10.1007/978-3-030-81462-5_53
6. Bacanin, N., Stoean, C., Zivkovic, M., Jovanovic, D., Antonijevic, M., Mladenovic, D.: Multi-swarm algorithm for extreme learning machine optimization. Sensors **22**, 4204 (2022)
7. Bacanin, N., Tuba, E., Zivkovic, M., Strumberger, I., Tuba, M.: Whale optimization algorithm with exploratory move for wireless sensor networks localization. In: Abraham, A., Shandilya, S.K., Garcia-Hernandez, L., Varela, M.L. (eds.) HIS 2019. AISC, vol. 1179, pp. 328–338. Springer, Cham (2021). https://doi.org/10.1007/978-3-030-49336-3_33
8. Bacanin, N., Zivkovic, M., Bezdan, T., Cvetnic, D., Gajic, L.: Dimensionality reduction using hybrid brainstorm optimization algorithm. In: Saraswat, M., Roy, S., Chowdhury, C., Gandomi, A.H. (eds.) Proceedings of International Conference on Data Science and Applications. LNNS, vol. 287, pp. 679–692. Springer, Singapore (2022). https://doi.org/10.1007/978-981-16-5348-3_54
9. Bacanin, N., Zivkovic, M., Bezdan, T., Venkatachalam, K., Abouhawwash, M.: Modified firefly algorithm for workflow scheduling in cloud-edge environment. Neural Comput. Appl. **34**, 9043–9068 (2022)
10. Basha, J., et al.: Chaotic Harris Hawks optimization with quasi-reflection-based learning: an application to enhance CNN design. Sensors **21**, 6654 (2021)
11. Beni, G.: Swarm intelligence. In: Sotomayor, M., Pérez-Castrillo, D., Castiglione, F. (eds.) Complex Social and Behavioral Systems. ECSSS, pp. 791–818. Springer, New York (2020). https://doi.org/10.1007/978-1-0716-0368-0_530
12. Bezdan, T., Milosevic, S., Venkatachalam, K., Zivkovic, M., Bacanin, N., Strumberger, I.: Optimizing convolutional neural network by hybridized elephant herding optimization algorithm for magnetic resonance image classification of glioma brain tumor grade. In: 2021 Zooming Innovation in Consumer Technologies Conference (ZINC), pp. 171–176 (2021)
13. Bezdan, T., et al.: Hybrid fruit-fly optimization algorithm with k-means for text document clustering. Mathematics **9**, 1929 (2021)
14. Bezdan, T., Zivkovic, M., Bacanin, N., Chhabra, A., Suresh, M.: Feature selection by hybrid brain storm optimization algorithm for COVID-19 classification. Mary Ann Liebert Inc., publishers 140 Huguenot Street, 3rd Floor New ... (2022)

15. Bezdan, T., Zivkovic, M., Tuba, E., Strumberger, I., Bacanin, N., Tuba, M.: Glioma brain tumor grade classification from MRI using convolutional neural networks designed by modified FA. In: Kahraman, C., Cevik Onar, S., Oztaysi, B., Sari, I.U., Cebi, S., Tolga, A.C. (eds.) INFUS 2020. AISC, vol. 1197, pp. 955–963. Springer, Cham (2021). https://doi.org/10.1007/978-3-030-51156-2_111

16. Bezdan, T., Zivkovic, M., Tuba, E., Strumberger, I., Bacanin, N., Tuba, M.: Multi-objective task scheduling in cloud computing environment by hybridized bat algorithm. In: International Conference on Intelligent and Fuzzy Systems, pp. 718–725 (2020)

17. Bukumira, M., Antonijevic, M., Jovanovic, D., Zivkovic, M., Mladenovic, D., Kunjadic, G.: Carrot grading system using computer vision feature parameters and a cascaded graph convolutional neural network, vol. 31, p. 061815. SPIE (2022)

18. Chen, T., et al.: XGBoost: extreme gradient boosting, vol. 1, pp. 1–4 (2015)

19. Cheng, S., Shi, Y.: Diversity control in particle swarm optimization. In: 2011 IEEE Symposium on Swarm Intelligence, pp. 1–9 (2011)

20. Dick, S.: Artificial intelligence. PubPub (2019)

21. El Naqa, I., Murphy, M.J.: What is machine learning? In: El Naqa, I., Li, R., Murphy, M.J. (eds.) Machine Learning in Radiation Oncology, pp. 3–11. Springer, Cham (2015). https://doi.org/10.1007/978-3-319-18305-3_1

22. Gumus, M., Kiran, M.S.: Crude oil price forecasting using XGBoost. In: 2017 International Conference on Computer Science and Engineering (UBMK), pp. 1100–1103 (2017)

23. Heidari, A.A., Mirjalili, S., Faris, H., Aljarah, I., Mafarja, M., Chen, H.: Harris hawks optimization: algorithm and applications. Future Gener. Comput. Syst. **97**, 849–872 (2019)

24. Jovanovic, D., Antonijevic, M., Stankovic, M., Zivkovic, M., Tanaskovic, M., Bacanin, N.: Tuning machine learning models using a group search firefly algorithm for credit card fraud detection. Mathematics **10**, 2272 (2022)

25. Jovanovic, L., Zivkovic, M., Antonijevic, M., Jovanovic, D., Ivanovic, M., Jassim, H.S.: An emperor penguin optimizer application for medical diagnostics. In: 2022 IEEE Zooming Innovation in Consumer Technologies Conference (ZINC), pp. 191–196 (2022)

26. Karaboga, D.: Artificial bee colony algorithm. Scholarpedia **5**, 6915 (2010)

27. Kiangala, S.K., Wang, Z.: An effective adaptive customization framework for small manufacturing plants using extreme gradient boosting-XGBoost and random forest ensemble learning algorithms in an industry 4.0 environment, vol. 4, p. 100024. Elsevier (2021)

28. Latha, R.S., Saravana Balaji, B., Bacanin, N., Strumberger, I., Zivkovic, M., Kabiljo, M.: Feature selection using grey wolf optimization with random differential grouping. Comput. Syst. Sci. Eng. **43**, pp. 317–332 (2022). https://doi.org/10.32604/csse.2022.020487, http://www.techscience.com/csse/v43n1/47062

29. Mirjalili, S., Gandomi, A.H., Mirjalili, S.Z., Saremi, S., Faris, H., Mirjalili, S.M.: Salp swarm algorithm: a bio-inspired optimizer for engineering design problems. Adv. Eng. Softw. **114**, 163–191 (2017)

30. Mirjalili, S.Z., Mirjalili, S., Saremi, S., Faris, H., Aljarah, I.: Grasshopper optimization algorithm for multi-objective optimization problems. Appl. Intell. **48**, 805–820 (2018)

31. Mohammed, S., Alkinani, F., Hassan, Y.: Automatic computer aided diagnostic for COVID-19 based on chest X-ray image and particle swarm intelligence. Int. J. Intell. Eng. Syst. **13**, 63–73 (2020)

32. Muslim, M.A., Dasril, Y.: Company bankruptcy prediction framework based on the most influential features using XGBoost and stacking ensemble learning, vol. 11 (2021)

33. Pham, Q.V., Mirjalili, S., Kumar, N., Alazab, M., Hwang, W.J.: Whale optimization algorithm with applications to resource allocation in wireless networks. IEEE Trans. Veh. Technol. **69**, 4285–4297 (2020)

34. Poli, R., Kennedy, J., Blackwell, T.: Particle swarm optimization: an overview. Swarm Intell. **1**, 33–57 (2007)

35. Rahnamayan, S., Tizhoosh, H.R., Salama, M.M.A.: Quasi-oppositional differential evolution. In: 2007 IEEE Congress on Evolutionary Computation, pp. 2229–2236 (2007). https://doi.org/10.1109/CEC.2007.4424748

36. Salb, M., Zivkovic, M., Bacanin, N., Chhabra, A., Suresh, M.: Support vector machine performance improvements for cryptocurrency value forecasting by enhanced sine cosine algorithm. In: Bansal, J.C., Engelbrecht, A., Shukla, P.K. (eds.) Computer Vision and Robotics. Algorithms for Intelligent Systems, pp. 527–536. Springer, Singapore (2022). https://doi.org/10.1007/978-981-16-8225-4_40

37. Stone-Gross, B., et al.: Your botnet is my botnet: analysis of a botnet takeover. In: Proceedings of the 16th ACM Conference on Computer and Communications Security, pp. 635–647 (2009)

38. Tair, M., Bacanin, N., Zivkovic, M., Venkatachalam, K.: A chaotic oppositional whale optimisation algorithm with firefly search for medical diagnostics. Comput. Mater. Contin. **72**, pp. 959–982 (2022). https://doi.org/10.32604/cmc.2022.024989, http://www.techscience.com/cmc/v72n1/46919

39. Talatahari, S., Bayzidi, H., Saraee, M.: Social network search for global optimization (2021). https://doi.org/10.1109/ACCESS.2021.3091495

40. Talatahari, S., Bayzidi, H., Saraee, M.: Social network search for global optimization. IEEE Access **9**, 92815–92863 (2021). https://doi.org/10.1109/ACCESS.2021.3091495

41. Wang, G.G., Deb, S., Coelho, L.D.S.: Elephant herding optimization. In: 2015 3rd International Symposium on Computational and Business Intelligence (ISCBI), pp. 1–5 (2015)

42. Wang, G.G., Deb, S., Cui, Z.: Monarch butterfly optimization. Neural Comput. Appl. **31**, 1995–2014 (2019)

43. Wolpert, D.H., Macready, W.G.: No free lunch theorems for optimization. IEEE Trans. Evol. Comput. **1**, 67–82 (1997)

44. Yang, X.-S.: Firefly algorithms for multimodal optimization. In: Watanabe, O., Zeugmann, T. (eds.) SAGA 2009. LNCS, vol. 5792, pp. 169–178. Springer, Heidelberg (2009). https://doi.org/10.1007/978-3-642-04944-6_14

45. Yang, X.S.: Firefly algorithm, stochastic test functions and design optimisation (2010)

46. Yang, X.S.: Bat algorithm for multi-objective optimisation. Int. J. Bio-Inspir. Comput. **3**, 267–274 (2011)

47. Yang, X.S.: Bat algorithm: literature review and applications (2013)

48. Zivkovic, M., et al.: Hybrid genetic algorithm and machine learning method for COVID-19 cases prediction. In: Shakya, S., Balas, V.E., Haoxiang, W., Baig, Z. (eds.) Proceedings of International Conference on Sustainable Expert Systems. LNNS, vol. 176, pp. 169–184. Springer, Singapore (2021). https://doi.org/10.1007/978-981-33-4355-9_14

49. Zivkovic, M., Bacanin, N., Tuba, E., Strumberger, I., Bezdan, T., Tuba, M.: Wireless sensor networks life time optimization based on the improved firefly algo-

rithm. In: 2020 International Wireless Communications and Mobile Computing (IWCMC), pp. 1176–1181 (2020)

50. Zivkovic, M., et al.: COVID-19 cases prediction by using hybrid machine learning and beetle antennae search approach. Sustain. Cities Soc. **66**, 102669 (2021)

51. Zivkovic, M., Jovanovic, L., Ivanovic, M., Krdzic, A., Bacanin, N., Strumberger, I.: Feature selection using modified sine cosine algorithm with COVID-19 dataset. In: Suma, V., Fernando, X., Du, K.-L., Wang, H. (eds.) Evolutionary Computing and Mobile Sustainable Networks. LNDECT, vol. 116, pp. 15–31. Springer, Singapore (2022). https://doi.org/10.1007/978-981-16-9605-3_2

52. Zivkovic, M., Stoean, C., Chhabra, A., Budimirovic, N., Petrovic, A., Bacanin, N.: Novel improved salp swarm algorithm: an application for feature selection. Sensors **22**, 1711 (2022)

Bridging the Resource Gap in Cross-Lingual Embedding Space

Kowshik Bhowmik[1,2(✉)] and Anca Ralescu[2]

[1] The College of Wooster, Wooster, OH 44691, USA
kbhowmik@wooster.edu
[2] University of Cincinnati, Cincinnati, OH 45220, USA
ralescal@ucmail.uc.edu

Abstract. The mapping based methods for inducing a cross-lingual embedding space involves learning a linear mapping from the individual monolingual embedding spaces to a shared semantic space where English is often chosen as the hub language. Such methods are based on the orthogonal assumption. Resource limitations and typological distance from English often result in a deviation from this assumption and subsequently poor performance for the low-resource languages. In this research, we will present a method for identifying optimal bridge languages to achieve better mapping for the low-resource languages in the cross-lingual embedding space. We also report Bilingual Induction Task (BLI) performances for the shared semantic space achieved using different cross-lingual signals.

Keywords: Cross-lingual word embeddings · Low-resource languages · Bilingual Lexicon Induction (BLI)

1 Introduction

Cross-lingual word embeddings are cross-lingual representation of words in a joint embedding space [1]. Mikolov et al. observed that upon the application of an appropriate linear transformation, words and their translations showed similar geometric arrangements in their respecting monolingual embedding spaces [2]. This observation led to the transformation of a vector space of the source language to that of the target language by learning a linear projection with a transformation matrix. The transformation matrix is learned using the most frequent words in the source embedding space and their respective translations in the target language embedding space. Stochastic Gradient Descent is used by minimizing the squared Euclidean distance between the seed source words, now transformed using the transformation matrix, and their translation word embeddings in the target language vector space. The basic regression method was later improved by constraining the transformation matrix to be orthogonal [3,4]. The solution under the orthogonal constraint can be computed using Singular Value Decomposition (SVD).

The mapping methods of inducing a cross-lingual embedding space by aligning monolingual vector spaces assume that the independently trained vector spaces are approximately isomorphic. This assumption - popularly referred to in the literature as the isomorphic assumption- does not always hold. As a result, the algorithms based on this assumption perform poorly for non-isomorphic spaces, or fail completely in some cases. Typological distance among languages have been cited as a factor contributing to non-isomorphism between vector spaces [5,6]. Vulić et al. argue that non-isomorphism also stems from degenerate vector spaces [13]. They show that the size of monolingual resource used to train the embeddings along with the duration of training contribute to the non-isomorphism as well.

A common setting for Natural Language Processing is the co-occurance of both data limitations and compute resource constraints. The recent advancements in language models have often resulted from additional parameters, making the size of the models grow exponentially. While these models have improved the quality of several downstream tasks, the increased size has driven up the training cost as well as the latency and memory footprint during inference. According to Ahia et al. low-resource languages are used in parts of the world where lack of computational resources is as severe a constraint as the lack of digital data on which to train the models [7]. They use the cost of data in those regions as a proxy for the cost of access to technology [8].

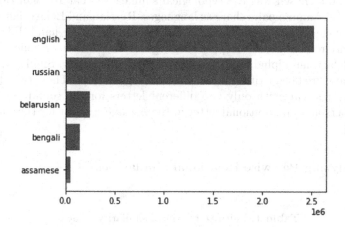

Fig. 1. Size of monolingual embedding spaces.

This research proposes to work around these limitations to achieve better representation for low-resource languages in a cross-lingual embedding space. The goal is to improve the Bilingual Lexicon Induction (BLI) task performance between distant, low-resource languages to English by utilizing the available resources of related higher-resource languages. Instead of directly imposing our existing knowledge of linguistic similarities, this paper proposed a simple method that leverages the degree of isomorphism between vector spaces as a measure of linguistic similarities to cluster related languages together. First, degrees of isomorphism were calculated between the language pairs in the language set as a measure of their typological similarities. Fuzzy C-Means Clustering algorithm was applied on these pairwise similarity values in order to divide the languages into clusters. This enabled the low-resource languages in a cluster to take advantage of related higher resource languages in order to create cross-lingual resources as well as inducing the desired cross-lingual embedding space.

2   Identifying Bridge Languages

2.1   Language Set

The languages chosen for the experiment are English, Russian, Belarusian, Bengali and Assamese. The monolingual embedding spaces of these languages were trained on their respective Wikipedias using fastText [9]. These vectors in dimension 300 were trained using the skip-gram method proposed by Bojanowski et al. [10]. Figure 1 shows the vocabulary size of the five embedding spaces. Of these languages, English and Russian are resource-rich languages while Belarusian, Bengali and Assamese are low-resource languages. However, Bengali has an available bilingual lexicon with English unlike Assamese and Belarusian [11].

Apart from the disparity in resources, one other reason that these particular languages were chosen was the typological similarities that Russian-Belarusian and Bengali-Assamese pairs share as languages. Russian and Belarusian are both Slavic languages, Eastern Slavic to be specific. On the other hand, Bengali and Assamese are related Indo-European languages from the Eastern zone. The Russian and Belarusian alphabet are based on the Cyrillic script. Similarly, Bengali and Assamese are both written in the Eastern Nagari script. These two alphabets are almost identical, with only two different letters for the sound /r/ in these two alphabets and an additional letter in the Assamese script for the sound /w/ or /v/ [12].

2.2   Analyzing Pairwise Relational Similarities

Table 1. Pairwise relational similarity values.

	Assamese	Belarusian	Bengali	Russian	English
Assamese	1	0.50946421	0.83918901	0.42186703	0.02149236
Belarusian	0.50946421	1	0.619777	0.56029783	0.03404167
Bengali	0.83918901	0.619777	1	0.43963931	−0.0002599
Russian	0.42186703	0.56029783	0.43963931	1	0.20889628
English	0.02149236	0.03404167	−0.0002599	0.20889628	1

Relational Similarity measure proposed by Vulić et al. [13] was used as the measure for degree of isomorphism between the language pairs in our set of

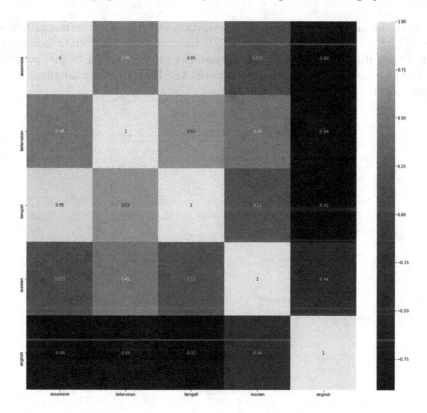

Fig. 2. Heatmap: pairwise relational similarities.

monolingual embedding spaces. This measure is inspired by the intuition that the similarity distributions of translation pairs within their own embedding space should be similar. Cosine similarities are calculated for the words in the translation pairs on both the source and the target side. Pearson correlation coefficient is then calculated between the sorted lists of cosine similarities. Table 1 shows the pairwise Relational Similarity values for five languages in the experiment. As expected, Bengali and Assamese share the highest similarity between themselves. Their individual pairwise similarity with the Russian embedding space is also similar. However, Belarusian and Russian pair do not explicitly reflect their real-world typological similarities. Instead, Belarusian shares the highest relational similarity with Bengali followed by Russian and Assamese. On the other hand, the Russian embedding space reports the highest similarity score with the English embedding space compared to the other three languages in the experiment set. These values may reflect, more than typological similarities, their respective embedding sizes as the Belarusian embedding space is more comparable in size with the Bengali and the Assamese embedding space than it is with Russian.

The correlation heatmap in Fig. 2 produced from pairwise similarity values also does not reveal more than what is evident from the values themselves. The rest of the languages share a negative correlation with English. The magnitude of the positive correlation that Russian and Belarusian share between themselves is higher than Russian-Assamese and Russian-Bengali correlation values, possibly hinting at some typological relation between the Eastern Slavic language pairs being captured by the similarity metric.

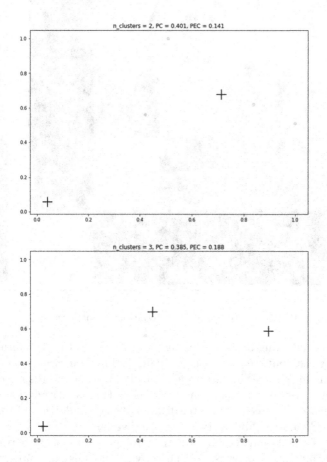

Fig. 3. FCM: optimal cluster number ('+' symbols signify cluster centers).

3   Clustering the Embedding Spaces

Although linguists categorize languages into etymological families, there are other factors that may contribute to similarities among languages. Most notably, the common features between two languages can result from one of the languages borrowing from the other or both of them borrowing from a third language. To an uncritical eye, English may seem to be a Romance language instead of a Germanic one due to its heavy borrowing from French during the middle ages. Fuzzy clustering methods are especially useful for objects that fall on the boundaries of two or more classes. They may provide better insight into the different ways a language can be related to another as it expresses a degree of membership to all the clusters for the data points.

3.1 Fuzzy C-Means Clustering

Fuzzy clustering allows a data element to belong to multiple clusters with a degree of membership. This is in contrast to the hard clustering algorithms where a data point can belong to only one cluster. The fuzzy set theory proposed by Zadeh used a membership function in order to express the uncertainty of belonging [14]. Fuzzy C-Means Clustering is one of the most widely used clustering algorithms. A special case for this algorithm was reported first by Joe Dunn [15] which was later improved by Jim Bezdek [16]. The algorithm requires the user to provide the number of clusters to be created. Partition Coefficient(PC) and Partition Entropy Coefficient(PEC) can be used as validation metrics for determining the optimal number of clusters [17].

After having analyzed the pairwise Relational Similarity Values and the correlation heatmap produced from them, the next step was to apply the Fuzzy C-Means clustering algorithm on the pairwise similarity values. Clustering was done for both n = 2 and n = 3.

Table 2. FCM clusters (n = 2).

Language	Cluster label	0	1
English	0	0.998689	0.001311
Assamese	1	0.049867	0.950133
Belarusian	1	0.062647	0.937353
Bengali	1	0.034730	0.965270
Russian	1	0.187024	0.812976

Figure 3 shows the Partition Coefficient (PC) and Partition Entropy Coefficient (PEC) for n = 2 and n = 3. According to these metrics the optimal number of clusters for our language set is 2, since a higher value of PC and a lower value of PEC signal more uniform clusters. Table 2 shows the cluster formation and membership distribution for n = 2. The languages are clearly clustered into two clusters with English being the sole language in Cluster 0 with almost exclusive membership to its assigned cluster label. Of the languages assigned to Cluster 1, only Russian has a notable degree of membership to Cluster 0 with the degree of membership split at 0.19 and 0.81 between the two clusters. The clusters formed do not align with the real-world typological relationship among the languages. With the Eastern Slavic languages Belarusian and Russian clustered together with Eastern Indo-European Bengali and Assamese.

Table 3. FCM clusters (n = 3).

Language	Cluster label	0	1	2
Assamese	0	0.963607	0.029400	0.006993
Bengali	0	0.975066	0.020458	0.004477
Belarusian	1	0.314450	0.630678	0.054873
Russian	1	0.053230	0.923071	0.023698
English	2	0.000011	0.000016	0.999973

On the other hand, the clusters formed when n = 3, have a lower PC and a higher PEC. Although these two values indicate that the clustering with n = 3 is less uniform than the clusters formed at n = 2, the languages in the three clusters reflect the real-world typological similarities among the languages. Table 3 shows the cluster formation and membership degree of the languages to different clusters. Assamese and Bengali are grouped together in Cluster 0 with more than 0.95 membership to its assigned cluster label. Belarusian and Russian are grouped together in Cluster 1. Belarusian has a 0.63 membership to its assigned cluster label while it belongs to Cluster 0 with a membership degree of 0.31. English, once again, is the only language in its assigned cluster label with exclusive membership to Cluster 2.

4   Cross-Lingual Space Induction and Evaluation

4.1   Bilingual Lexicon Induction Task

Bilingual Lexicon Induction (BLI) is one of the most commonly used intrinsic tasks to evaluate cross-lingual word embeddings. Given N words from the source language, the task aims to determine the most appropriate translation for each source word in the target language. In a shared semantic space, this task is equivalent to finding a target language word that is closest to the source word embedding which we intend to translate. This closest target word is also known as the cross-lingual nearest neighbor. Cosine similarity is usually used to compute the similarity between pairs of word embeddings. This set of learned source-target word pairs are then evaluated against a gold standard. The induced cross-lingual model is evaluated on its ability to predict the translation target word from the gold standard dictionary as the nearest neighbor of the source word in the shared embedding space. The performance of a cross-lingual model at this task is expressed as precision-at-one (P@1). Precision-at-k (P@k) is a more lenient performance metric where the ranked list of k-nearest neighbors from the target language must contain the correct translation where $k \geq 1$.

4.2   Dictionary Construction

Since there is no off-the-shelf dictionary available for Assamese-English and Belarusian-English pairs, it's important to build them. To do that, the triangulation method [18] was followed. This method uses the available resources of a related higher resource language to build a dictionary. Bengali has an off-the-shelf dictionary with English. On the other hand, since Bengali and Assamese share a script, the two embedding spaces share a large number of overlapping words, 22,138 to be precise. These 22,138 words were triangulated separately through the Bengali-English training and testing dictionary, which resulted in Assamese-English training and testing dictionaries with 3,383 training pairs and 560 testing pairs respectively. The reason that the Bengali-English train-test split was preserved is because part of this experiment involves mapping the Assamese and Belarusian embedding spaces through Bengali and Russian respectively. Similarly, training and testing dictionaries were generated for Belarusian by triangulating through the Russian-English dictionary and taking advantage of the 68,554 overlapping words in the Russian and Belarusian embedding spaces. Following the previously described method resulted in a training and a test dictionary for the Belarusian-English pair with 4,582 and 703 word pairs respectively.

4.3   Learning the Mapping

Mapping Method 1: Weak-Supervision. This portion of the experiment is motivated by the work of Smith et al. [19] who use a form of weak supervision to induce the cross-lingual embedding space. The Bilingual Lexicon Induction task performance achieved using this method can serve as the baseline for the rest of the experiments. In their work, Smith et al. [19] used the overlapping words present in pairs of monolingual word embeddings to map 89 embedding spaces onto the English embedding space. Since monolingual spaces are trained on their respective Wikipedias, they often contain many overlapping words, especially with English. There are 16,140 overlapping words between English and Assamese monolingual embedding spaces while there are 35,358 overlapping words between the English and Belarusian embedding spaces. Majority of these are words in the English language. These are words English language words that are also found in Assamese and Belarusian Wikipedia pages.

Table 4. BLI performance: weak supervision.

Language pairs	P@1	P@5	P@10
Assamese - English	0%	0%	0.18%
Belarusian - English	0.14%	0.57%	0.85%

The cross-lingual embedding space induced by mapping Assamese and Belarusian on to the English space using overlapping set of words results in very poor performance on both the test dictionaries constructed using the triangulation method. As shown in Table 4, no correct English translation for the 560 Assamese words could be found among the (1) nearest neighbor and the 5-nearest neighbors in the shared semantic space. The P@10 value is 0.18% since one correct English translation was found when the 10 nearest English translations were retrieved from the shared embedding space. For the Belarusian-English pair, only one correct translation was found as the nearest neighbor among the 703 word pairs in the test dictionary. The number of correct translation when the number of nearest neighbors retrieved was 4 in total. Six correct translations were found among the 10 nearest neighbors retrieved for the 703 Belarusian words in the test dictionary.

The poor performance in the BLI task is commensurate with Relational similarity values calculated between these two pairs of embedding spaces based on these overlapping words. In both cases the degree of isomorphism was small. The correlation heatmap also shows negative correlation values which were produced from the pairwise relational similarity values.

Mapping Method 2: Multiple Pivot Languages. This portion of the experiment is motivated by the Knowledge Distillation idea proposed by Nakashole and Flauger [20]. They map a low-resource language directly to the English embedding space and also learn a triangulated mapping from the low-resource embedding space to English through a language related to the low-resource language but is considerably richer in resources. They distill the knowledge from this triangulated path to learn the direct mapping. They also propose the idea of weighting the triangular path in order to indicate how useful each of these paths is. This experiment also aims to improve Bilingual Lexicon Induction performance for low resource languages by triangulating through a related richer resourced language. But instead of optimizing multiple triangular paths and weight them by their usefulness, the first step tries to identify, without relying completely on pre-existing notion of linguistic similarities, which language would be a good choice to bridge the resource gap between the low-resource languages in the experiment set and English. The largest embedding space in the clusters formed from pairwise Similarity Values are treated as the bridge language while inducing the cross-lingual embedding space. In the experiments that will follow, the final hub language is always the English embedding space as the aim of these experiments is figuring out ways to improve the Bilingual Lexicon Induction task from the low-resource languages to English. In our particular experiment setup, no language was clustered with English. In cases where some languages are clustered with English, they can be directly mapped to the English embedding space, since they are presumably related more to English than any other resource-rich language in the language set.

For this part of the experiment, results will be reported for both cases where cluster number is equal to 2 and 3. As mentioned in Sect. 3, the PC and PEC values for the Fuzzy C-Means clustering indicates a uniform clustering for n = 2 while the language clusters formed reflect the real world typological similarities among languages. Hence, performance for the Bilingual Induction Task will be reported for both the cases.

When n = 2, one of the clusters contain only the English word embedding space while the second one contains Russian, Belarusian, Bengali and Assamese. According to how the methodology is set up, Russian is chosen as the bridge language for Belarusian, Bengali and Assamese, since those languages are lower in resource than the Russian monolingual embedding space. Typologically this clustering does not make sense since Bengali, Assamese and Russian, Belarusian pairs belong to two different language families. Russian is presumably a good bridge language for Belarusian as the two languages are typologically related and their respective alphabets also originate from the same script. The same could not be said for Bengali and Assamese. Since Russian is now the largest embedding space in this cluster of languages containing the four languages. According to the experiment design, Russian will now bridge the resource-gap between the two low-resource languages Belarusian, Assamese and English. Russian first needs to be mapped on to the English space. Our method checks for any available bilingual dictionary between the two languages. Since an off-the-shelf lexicon is available for the Russian-English pair, the mapping is learned using that dictionary. As for Belarusian and Assamese, they will be mapped on to the Russian embedding space. While neither pair shares a bilingual lexicon, Russian and Belarusian embedding space share a large number of overlapping words (68,554) between themselves. Since, the performance will be reported for the Assamese-English and Belarusian-English language pairs, it can be expected that the performance will not improve for the Assamese-English pair as much as it will for the Belarusian-English pair over the performance reported for cross-lingual embedding induced using weak-supervision.

Table 5. BLI performance: bridge language (two clusters).

Source-(Bridge-)Target	P@1	P@5	P@10
Assamese-(Russian-)English	0%	0%	0.18%
Belarusian-(Russian-)English	4.27%	9.25%	10.95%

Table 5 shows the results for the induced cross-lingual space where Russian served the role of the bridge language for both Assamese and Belarusian. As expected, there was no improvement for the Assamese-English BLI task when Russian was used as the bridge language between the source and target embedding space. There was a single correct prediction when 10-nearest neighbors from the English embedding space were fetched for the 560 Assamese words in the test dictionary. The Belarusian-English Bilingual Induction task performance is improved upon using Russian as the bridge language. Thirty correct predictions were made as the nearest neighbors were retrieved from the shared semantic space. As the number of retrieved nearest neighbors increases to 5, the number of correct predictions made goes up to 65. With 10-nearest-neighbors fetched, the number of correctly predicted words is 77.

When n = 3, English, like the previous case, is the only language in its cluster. Bengali, and Assamese are clustered together in one of the other two clusters. Russian and Belarusian are clustered together in the third cluster. as mentioned in Sect. 3, this clustering of languages reflects our real-world knowledge about the typological relationship among these languages. Bengali and Russian are chosen as the languages to bridge the resource gap between Assamese-English and Belarusian-English pairs because of their comparatively richer corpus and readily available off-the-shelf lexicon with English.

Mappings are learned from the Bengali and the Russian embedding space to the English embedding space. Later, the lower-resource languages in the respective clusters learn a mapping to the resource-rich bridge language using a rich set of overlapping words. In this case, The Belarusian space is mapped on to the Russian embedding space learned using the 68,554 overlapping words between the two embedding spaces. Similarly, a mapping is learned from the Assamese to the Bengali embedding space using the 22,138 overlapping words between the Bengali and the Assamese embedding space.

Table 6. BLI performance: bridge language (three clusters).

Source-(Bridge-)Target	P@1	P@5	P@10
Assamese-(Bengali-)English	0.18%	1.48%	2.14%
Belarusian-(Russian-)English	4.27%	9.25%	10.95%

Upon inducing the cross-lingual embedding space using the concept of bridge languages, Bilingual Induction Task tests were carried out for the Assamese-English and Belarusian-English pairs. As seen from Table 5, and Table 6, the results show improvement over simply learning the mappings from Assamese and Belarusian to the English embedding space. There is one correct prediction when the nearest English neighbor was fetched from the shared embedding space. When 5-nearest neighbors are retrieved, the number of correct predictions becomes 8 and a total of 12 correct predictions were made as the 10 nearest English language words were retrieved from the shared embedding space.

Mapping Method 3: Triangulated Training Dictionary. The cross-lingual word embedding space was induced following a third, final method. For this, the training dictionaries obtained using the triangulation method for Assamese-English and Belarusian-English pairs (Sect. 4.2) were utilized. Based on the performance of the cross-lingual embedding space induced using the concept of bridge languages, it was decided to go forward with the 5 languages divided into 3 clusters. All the four languages were independently mapped on to the English embedding space in order to induce the cross-lingual word embedding space.

Table 7. BLI performance: triangulated dictionary.

Language pairs	P@1	P@5	P@10
Assamese-English	0.89%	2.5%	3.9%
Belarusian-English	4.55%	10.24%	10.81%

The results in Table 7 show that there are performance gains for most of the cases. Five nearest-neighbor correct predictions were made among the 560 word pairs in the Assamese-English test dictionary. In total, 14 correct predictions were found among the 5 nearest neighbors retrieved from the shared embedding space, while among the 10 nearest English words retrieved, 22 correct predictions were found. This is a significant improvement over both the previous cases where the Assamese space was simply mapped on to the English embedding space using the overlapping words found in the two spaces and also where Bengali was utilized as a bridge language between the two spaces. The same is true for the BLI performance from Belarusian to English. Except for the P@10 score, this task produced a better performance than when the Russian embedding space was used to bridge the resource gap between Belarusian and English. When the constructed training dictionary was used to map the Belarusian space directly onto the English space, the induced cross-lingual space fetched 32 correct English translations as the nearest neighbor of the 703 Belarusian words in the test dictionary. The number increased to 72 as 5 nearest neighbors were considered instead. A total of 76 correct English translations were retrieved among the 10 nearest neighbors of each of the 703 Belarusian words in the test dictionary.

5 Discussion

This research aims at bridging the resource gap in monolingual embedding spaces to achieve better representation for low-resource languages in a cross-lingual embedding space. The languages chosen for the cross-lingual embedding spaces are English, Russian, Belarusian, Bengali, and Assamese. Of these, English and Russian are resource-rich languages, while the rest are low-resource in terms of their respective monolingual embedding space size. Although disparate in terms of available resources, Russian and Belarusian are typologically related as are Bengali and Assamese. Cross-lingual embedding space was induced with the above mentioned languages using three different form of cross-lingual supervision.

First, the four languages were separately mapped onto the English embedding space using the identically spelled strings that each of the languages shared with the English embedding space. These identically spelled strings resulted from the use of English words in the Wikipedia of other languages. The distribution of these identically spelled strings would be much different in the non-English embedding space and so they served as poor cross-lingual signals. The resulting shared semantic space showed a low score for the Assamese-English and Belarusian-English BLI tasks. Next, the languages were clustered on their pairwise Relational Similarity values. Fuzzy C-Means Clustering algorithm was applied on the relational similarity values calculated for each language pair in our experiment set. This was done with the view to clustering related languages so that the low-resource languages could leverage the available resources of the relatively richer-resource languages in the induced cross-lingual embedding space. The number of cluster was initially set to 2 based on the Partition Coefficient

(PC) and Partition Entropy Coefficient (PEC). This resulted in Russian, Belarusian, Bengali and Assamese in one cluster and English in another. According to the methodology, Russian was chosen as a bridge language to map Belarusian and Assamese on to the English space. The BLI performance improved for the Belarusian-English pair. This can be attributed to Belarusian and Russian being typologically similar and the two languages sharing a large number of identically spelled strings. The same was not true for the Assamese-English pair. It improved when the cluster number was chosen to be 3 and Bengali served as the bridge language between Assamese and English. Further improvements in performance were achieved by first constructing triangulated training dictionaries for the Assamese-English and Belarusian-English pairs by utilizing the available resources of the bridge languages and then directly mapping Assamese and Belarusian on to the English space.

One of the reasons the results are still on the lower side is because the Belarusian and Assamese words in the triangulated training and testing dictionaries are possibly not among the most frequent words in their respective embedding spaces. In future, this issue can be alleviated by constructing refined training dictionaries from the induced cross-lingual space for the Assamese-English and Belarusian-English pairs using the most frequent Assamese and Belarusian words. Also, deciding the optimal number of clusters based on the Partition Coefficient and Partition Entropy Coefficient may not always result in clusters with typologically similar languages. When the optimal number of clusters was chosen to be 2, it resulted in a linguistically diverse cluster with Russian, Belarusian, Bengali and Assamese. While this improved the BLI performance for the Belarusian-English pair, the same did not happen for the Assamese-English pair. Future work may include researching and implementing different indicators for deciding the optimal number of clusters.

This work focused on achieving better results under the existing resource constraints. Transformer based multilingual language models mBERT [21] and XLM-R [22] have shown impressive results for many languages across a wide range of tasks. However, researchers have reported poor performance for low-resource languages in mBERT [23]. Multilingual joint training with related languages have shown to improve low-resource language representation in mBERT. XLM-R uses CommonCrawl as the source of data instead of Wikipedia. The larger amount of data available through CommonCrawl has resulted in performance improvement for low-resource languages. We intend to explore these models in the future.

References

1. Ruder, S., Vulić, I., Søgaard, A.: A survey of cross-lingual word embedding models. J. Artif. Intell. Res. **65**, 569–631 (2019)
2. Mikolov, T., Le, Q.V., Sutskever, I.: Exploiting similarities among languages for machine translation. arXiv preprint arXiv:1309.4168 (2013)
3. Xing, C., Wang, D., Liu, C., Lin, Y.: Normalized word embedding and orthogonal transform for bilingual word translation. In: Proceedings of the 2015 Conference

of the North American Chapter of the Association for Computational Linguistics: Human Language Technologies, pp. 1006–1011 (2015)

4. Artetxe, M., Labaka, G., Agirre, E.: Learning principled bilingual mappings of word embeddings while preserving monolingual invariance. In: Proceedings of the 2016 Conference on Empirical Methods in Natural Language Processing, pp. 2289–2294 (2016)

5. Zhang, M., Liu, Y., Luan, H., Sun, M.: Earth mover's distance minimization for unsupervised bilingual lexicon induction. In: Proceedings of the 2017 Conference on Empirical Methods in Natural Language Processing, pp. 1934–1945 (2017)

6. Søgaard, A., Ruder, S., Vulić, I.: On the limitations of unsupervised bilingual dictionary induction. arXiv preprint arXiv:1805.03620 (2018)

7. Ahia, O., Kreutzer, J., Hooker, S.: The low-resource double bind: an empirical study of pruning for low-resource machine translation. arXiv preprint arXiv:2110.03036 (2021)

8. Oughton, E.: Policy options for digital infrastructure strategies: a simulation model for broadband universal service in Africa. arXiv preprint arXiv:2102.03561 (2021)

9. Joulin, A., Grave, E., Bojanowski, P., Douze, M., Jégou, H., Mikolov, T.: FastText.zip: compressing text classification models. arXiv preprint arXiv:1612.03651 (2016)

10. Bojanowski, P., Grave, E., Joulin, A., Mikolov, T.: Enriching word vectors with subword information. Trans. Assoc. Comput. Linguist. 5, 135–146 (2017)

11. Conneau, A., Lample, G., Ranzato, M.A., Denoyer, L., Jégou, H.: Word translation without parallel data. arXiv preprint arXiv:1710.04087 (2017)

12. Comrie, B. (ed.): The World's Major Languages. Routledge, London (1987)

13. Vulić, I., Ruder, S., Søgaard, A.: Are all good word vector spaces isomorphic?. In: Proceedings of the 2020 Conference on Empirical Methods in Natural Language Processing (EMNLP), pp. 3178–3192 (2020)

14. Zadeh, L. A.: Fuzzy sets. In: Fuzzy Sets, Fuzzy Logic, and Fuzzy Systems: Selected Papers by Lotfi A Zadeh, pp. 394–432 (1996)

15. Dunn, J. C.: A fuzzy relative of the ISODATA process and its use in detecting compact well-separated clusters, pp. 32–57 (1973)

16. Bezdek, J.C., Ehrlich, R., Full, W.: FCM: the fuzzy c-means clustering algorithm. Comput. Geosci. 10(2–3), 191–203 (1984)

17. Bezdek, J.C.: Pattern Recognition with Fuzzy Objective Function Algorithms. Springer, Heidelberg (2013)

18. Anastasopoulos, A., Neubig, G.: Should all cross-lingual embeddings speak English?. In: Proceedings of the 58th Annual Meeting of the Association for Computational Linguistics, pp. 8658–8679 (2020)

19. Smith, S.L., Turban, D.H., Hamblin, S., Hammerla, N.Y.: Offline bilingual word vectors, orthogonal transformations and the inverted softmax. arXiv preprint arXiv:1702.03859 (2017)

20. Nakashole, N., Flauger, R.: Knowledge distillation for bilingual dictionary induction. In: Proceedings of the 2017 Conference on Empirical Methods in Natural Language Processing, pp. 2497–2506 (2017)

21. Devlin, J., Chang, M.W., Lee, K., Toutanova, K.: BERT: pre-training of deep bidirectional transformers for language understanding. arXiv preprint arXiv:1810.04805. Vancouver (2018)

22. Conneau, A., et al.: Unsupervised cross-lingual representation learning at scale. arXiv preprint arXiv:1911.02116 (2019)

23. Wu, S., Dredze, M.: Are all languages created equal in multilingual BERT? arXiv preprint arXiv:2005.09093 (2020)

Classification of Microstructure Images of Metals Using Transfer Learning

Mohammed Abdul Hafeez Khan[1]([✉]) (ID), Hrishikesh Sabnis[2], J. Angel Arul Jothi[1] (ID), J. Kanishkha[1] (ID), and A. M. Deva Prasad[3]

[1] Department of Computer Science Engineering, Birla Institute of Technology and Science, Pilani Dubai Campus, Dubai International Academic City, Dubai, UAE
{f20190091,angeljothi,f20190072}@dubai.bits-pilani.ac.in
[2] Department of Mechanical Engineering, Birla Institute of Technology and Science, Pilani Dubai Campus, Dubai International Academic City, Dubai, UAE
f20170023@dubai.bits-pilani.ac.in
[3] Department of Mechanical Engineering, Rochester Institute of Technology of Dubai, Dubai Silicon Oasis, Dubai, UAE
dxacad@rit.edu

Abstract. This research focuses on the application of computer vision to the field of material science. Deep learning (DL) is revolutionizing the field of computer vision by achieving state-of-the-art results for various vision tasks. The objective of this work is to study the performance of deep transfer learned models for the classification of microstructure images. With light optical microscopes, microstructure images of four different metals were acquired for this task, including copper, mild steel, aluminum, and stainless steel. The proposed work employs transfer learned powerful pre-trained convolutional neural network (CNN) models namely VGG16, VGG19, ResNet50, DenseNet121, DenseNet169 and DenseNet201 to train and classify the images in the acquired dataset into different classes of metals. The results showed that the transfer learned ResNet50 model has obtained the highest accuracy of 99%, outperforming other transfer learning models. This also shows that DL models can be used for automatic metal classification using microstructure images.

Keywords: Microstructure images · Deep learning · Transfer learning · Convolutional neural network · Image classification

1 Introduction

The study of microstructure helps us determine the material's property, performance, and composition. Over the recent years there has been huge focus on microstructure sensitive designs. The idea is to modify or generate microstructures with desired properties. Phase transformation is one of the primary causes of the creation of microstructures [1]. Microstructures can hugely be influenced by the composition and the way in which the materials are processed. Some of the natural ways of processing include heat treatment, forging, extrusion and rolling. Some of the processing techniques being followed

D. Simian and L. F. Stoica (Eds.): MDIS 2022, CCIS 1761, pp. 136–147, 2023.
https://doi.org/10.1007/978-3-031-27034-5_9

in the modern days are selective laser melting (SLM), hot isostatic pressing (HIP) and thermal process [2, 3]. Microstructures are usually observed using the micrometre to nanometre scale length. The varying sizes, structures and composition of the materials have an impact on the properties and performance of the products. Hence, analysing the microstructures is useful for selecting the most suitable and reliable materials according to the requirements [4].

Microstructural examination techniques are widely used in industries, research studies, quality control, performance, and failure investigation. The materials are accepted or rejected based on microstructural analysis. They are compared to select the best one using a number of quantitative measures like grain size, inclusions, impurities, second phases, porosity, segregation, and surface effects [5]. Hence, automatic microstructure image analysis is an emerging research area gaining lot of attention in recent times.

Deep learning (DL) is a subcategory of machine learning (ML) which focuses on algorithms that deals with the neural network architecture. This is one of the upcoming fields in computer science that aims to mimic human brain and its way of thinking. It is composed of several processing layers that help to abstract more and more useful features at each consecutive layer from a dataset [6]. The advantage of DL is that unlike traditional learning methods it can learn features and perform classification together.

DL methods have played a vital role in the domain of material sciences in the past few years. It is used to identify key attributes from the dataset and reduce dimensionality which results in efficient material classification. It is also used to identify the mapping between the structure attributes and property attributes. Understanding the structure is helpful in detecting defects and is useful for aromatic classification and characterization of the materials [7]. Deep convolutional neural network (CNN) can learn patterns from datasets of different scales which is helpful when the materials cannot be analysed at a particular scale. In this work, we experimentally study the performance of several pre-trained deep CNNs for the classification of microstructure images of metals.

The paper is organized as follows: Sect. 2 details the related work on microstructure images. Section 3 explains the dataset. Section 4 outlines the proposed method. Section 5 presents the implementation and metrics while Sect. 6 explicates the results obtained. This is followed by conclusion in Sect. 7.

2 Literature Review

Review of prior work on applying DL and image classification techniques to microstructure images is presented in this section.

Azimi et al. [8] have performed a semantic segmentation-based approach using fully convolutional neural networks (FCNN) which is an extension of CNN accompanied by a max-voting scheme to classify microstructures (MVFCNN). They used scanning electron microscopy (SEM) images over light optical microscopy (LOM) to get better pixelation in images and achieve better accuracy. The method recorded an accuracy of 93.94%.

Balaji et al. [9] introduced a new approach known as DL for structure property interrogation (DLSP) for learning the microstructure-property explorations in photovoltaics. A custom CNN architecture was developed. It is comprised of four blocks of convolution layers with alternating max pooling layers and a batch normalization layer. The model produced an accuracy of 95.80%.

Pazdernik et al. [10] have tested and compared 3 different deep CNN architectures for pixelwise categorization of the microstructure images. (U-Net, SegNet, ResNet50 and Res-Net101). Supervised deep CNN was trained using labelled data obtained from SEM. It was used to classify each pixel as one of the following categories: grain, boundary, voids, precipitates, and impurities. They used a combination of DL, computer vision and spatial statistics to predict the key features like pixel type and location of defects. The best model was SegNet which produced a precision, recall and F1-score of 0.983, 0.991 and 0.987 respectively.

Arun et al. [11] used a DL approach to build a CNN from scratch. It consisted of three convolutional layers followed by a fully connected layer. The model classified the dataset into one of three categories: lamellar, duplex, or acicular. It produced an accuracy of 93.17%.

Shen et al. [12] proposed an electron backscatter diffraction (EBSD) based DL model to establish a generic low-cost model to classify between the microstructures of dual-phase steel (austenite & martensite) and quenching and partitioning (Q&P) steel (ferrite & martensite). The classic U-net architecture was deployed in the process which consisted of four convention layers and four up-convention layers. Batch normalization and dropout regularization with the rate of 50% was performed to avoid overfitting. Adaptive gradient algorithm (AdaGrad) optimizer and mean absolute error (MAE) loss function was used. The model when applied to high-quality images produced a high segmentation accuracy of 82.9%.

Ma et al. [13] used generative adversarial networks to establish a microstructure-processing relationship. Images of depleted U-10Mo (Uranium-Molybdenum) alloy were collected using SEMs (512 × 512). Here progressively growing GAN (pg-GAN) where new layers are added progressively to extract more features was used. Sliced Wasserstein Distance (SWD) has been used between trained images and generated images to evaluate the performance of the model.

Zaloga et al. [14] created a CNN to classify microstructural symmetry from powder x-ray diffraction patterns. The first, second and third convolutional layers had 100 filters of sizes 100, 50, and 25 each. ReLU activation function was applied in all the convolutional layers. Dropout was set to 0.5. The fourth, fifth and sixth layers are fully connected layers with 700, 140, and 7 neurons each. Sigmoid activation function was applied in all the fully connected layers. Adam (Adaptive moment estimation) optimizer was used. The classification accuracy reached 99.57% using the cubic system.

Roberts et al. [15] presented a hybrid CNN algorithm called DefectSegNet for segmentation of defects in stem images of steels. They used transfer learning by using a pre-trained architecture to perform the classification. The three crystallographic defects that were classified using the model are dislocations, precipitates, and voids. The model was trained from high-quality double chain layer, STEM defect images of HT-9(12% chromium & 1% molybdenum steel) martensitic steels (512 × 512). To avoid overfitting L2 regularization and a dropout were applied. Adam optimizer and cross entropy loss function was used. The model DefectSegNet was used to predict the pixel accuracy of the defects. It produced a pixel accuracy of 91.60%, 98.85%, and 59.85% for the line dislocations, voids, and precipitates respectively.

Dong et al. [16] demonstrated a regression CNN for parameter quantification (PQ-Net) for microstructural analysis of powder diffraction data. The model was trained with one-dimensional X-ray diffraction computed tomography (1D XRD-CT) dataset of Ni-Pd or CeO_2-ZrO_2 or Al_2O_3 catalysts. PQ-Net consists of convolution, max pooling, dropout, and fully connected layers. MAE was used as a loss function. Adam optimizer was used as an optimization algorithm and the learning rate was set to 0.0005. The model yields a relative error of less than 5% and Rietveld refinement factor (Rwp) of 3.144%, 1.191%, and 1.835% for scale factor crystallite size and lattice parameter respectively.

Maemura et al. [17] applied local model-agnostic explanations (LIME) technique for the classification of microstructures of low carbon steel. The model was successfully able to classify eight types of low carbon steel namely - upper bainite, lower bainite, martensite and mixed structures of these types. The model produced an accuracy of 97.9% using a majority voting algorithm.

3   Dataset

In this research, a dataset of microstructural images of metals like aluminum, copper, stainless steel, and mild steel was collected manually. The steps followed for the preparation of the metal samples and making them available for the acquisition of images can be found here [28]. The images are RGB and were taken at a magnification of 40×. The dataset consists of a total of 2000 images where for each metal, 500 images in JPG format were captured. Figure 1 demonstrates the microstructure images of the various metals utilized in the dataset.

Fig. 1. Microstructure images of aluminum, copper, stainless steel, and mild steel from the dataset.

4   Methodology

This research seeks to develop an automated system for classification of metals such as copper, aluminum, stainless steel, and mild steel. This is transpired, using feature extraction and classification of micrographs, i.e., microstructure image data, with the aid of DL models, and transfer learning technique. Figure 2 illustrates the steps involved in the proposed method. The input images from the dataset are split in the ratio of 80:20 (train:test). Various pre-trained CNNs are trained using the training data which learn the features from the data automatically. The features learnt by the models can be thought of as an intermediate representation of the input data. Finally, classification results are obtained using the intermediate representation for the test data.

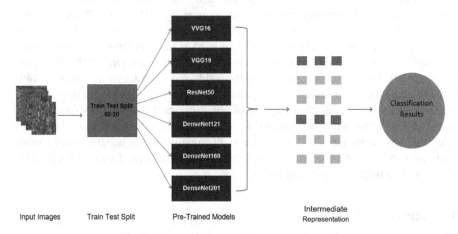

Fig. 2. The architecture of the proposed model.

Usually, a large amount of data is required for training the DL models. Additionally, deep neural networks take days or even weeks to be trained from scratch. These problems could be overcome by employing transfer learning. In the case of transfer learning, a new DL model for a current problem can be developed from an already existing DL model (pre-trained CNN). The already existing DL model is trained on large amount of data for another similar problem previously. These pre-trained models tend to provide better performance and guarantees the new DL model uses a comparatively less amount of training data and training time.

In this work, six popular pre-trained models, that are currently being used for research and commercial purposes are applied to the acquired dataset. The different pre-trained CNNs used in this work are: VGG16, VGG19, ResNet50, DenseNet121, DenseNet169 and DenseNet201 [18–24]. Table 1 provides the comparison of all the pre-trained models used in this work with respect to their architecture. The following paragraphs provide a brief description about each model.

4.1   VGG16 and VGG19

VGG16 is a CNN architecture having 16 trainable layers consisting of replicative structure of convolution, ReLU and pooling layers [18]. The first convolutional layer receives
an image with a size of 224×224 as the input. Consequently, the input image is propa-
gated through a set of convolutional layers and pooling layers. The convolutional layers
have three-dimensional receptive field. VGG16 have convolutional filter size of 3×3
and stride of 1 pixel. Five max-pooling layers with a stride of 2 and size of 2×2 pixel
window are used. Three fully connected layers with neuron sizes 4096, 4096, and 1000
follows the set of convolutional and pooling layers. Activations of the neurons in the
previous layer are reflected in each neuron in the fully connected layer. The softmax
layer is the final layer which is used for classification. As opposed to VGG16, VGG19
has 19 trainable layers [17] and provides better accuracy of classification. VGG19 has
few more convolution layers when compared with the VGG16.

4.2   ResNet50

CNN's must have at least one layer of convolution, wherein a convolution operation is
performed on the input matrix for learning distinct low-level and high-level features of
an image [25]. However, increasing the depth of the network by increasing the number
of layers results in problems of vanishing gradients and degradation [26]. Hence, the
Residual Neural Networks (ResNet) was developed for easing the training of deep CNN
models. It resulted in easier optimization of the network, and attained a higher accuracy.
In this network architecture, there is a residual block that adds the output of a layer with
the following layers through a skip connection by skipping some layers [20]. ResNet50 is
a ResNet model that is 50 layers deep. The number of trainable parameters in ResNet50
are about 23 million.

4.3   DenseNet

Dense convolutional network (DenseNet) connects each layer in a CNN to every other
layer in a feed-forward fashion. It alleviates the vanishing-gradient problem, encourages
the reuse of features, while strengthening feature propagation, and substantially reducing
the total number of parameters. Furthermore, DenseNet identifies that integrating all
available feature maps can reduce the fusion effect due to fusion redundancy [21]. Three
different architectures of the DenseNet were applied to the dataset, i.e., DenseNet121
which uses only 12 filters with a small set of new feature maps, Dense-Net169 which has
a capability of accessing feature maps from all its preceding layers, and DenseNet201
which exploits the condensed network providing easy to train and highly parametrically
efficient models [22–24].

Table 1. Comparison of the pre-trained models in terms of architecture.

Features	VGG16	VGG19	Densenet201	DenseNet169	DenseNet121	ResNet50
Convolutional layers	13	16	200	168	120	48
Max pool layers	5	5	1	1	1	1
Trainable parameters	134.26 M	143.67 M	20.1 M	12.8 M	7.2 M	23.9 M

5 Implementation and Evaluation

The transfer learning models were implemented using Jupyter Notebook with python version of 3.9.2. All the models were trained for 50 epochs with a learning rate of 0.01 and categorical cross entropy loss function. It was observed that the validation loss increased after 50 epochs, which resulted in a reduction of the validation accuracy. Due to this, the epochs were set to 50 for preventing the models from overfitting. The models were tested using 2 different optimizers: RMSprop and Adam's optimizer. Adam's optimizer produced better results in terms of accuracy and loss than RMSprop, therefore Adam's optimizer was used to optimize the transfer learning models.

Four metrics are used for measuring the performances of the transfer learning methods as shown in Table 2. They are: Precision (PPV), Sensitivity (TPR), F1 score (F1), and Accuracy (ACC). In Table 2, TP, FP, TN, and FN represent the number of True Positive, False Positive, True Negative and False Negative, respectively.

Table 2. Metrics used for evaluating the performance of the models.

Metric	Derivations
Precision	$PPV = TP/(TP + FP)$
Sensitivity	$TPR = TP/(TP + FN)$
F1 Score	$F1 = 2TP/(2TP + FP + FN)$
Accuracy	$ACC = (TP + TN)/(TP + TN + FP + FN)$

6   Results and Discussions

This section gives the classification results of the pre-trained deep CNN models on the acquired microstructure dataset. Table 3 shows performance of the pre-trained models measured in terms of precision, sensitivity, F1 score, and accuracy. It is evident from Table 3 that the pre-trained ResNet50 model has acquired PPV, TPR, F1, and ACC of 99% for the metal detection and classification. It is superior to the VGG19, DensNet201, VGG16, DensNet169, and DensNet121 models where they have acquired an ACC of 94%, 94%, 95%, 96%, and 97% respectively. ResNet50 performs efficiently due to its architecture that helps in maintaining a low error rate much deeper in the network. This is because even though ResNet is a much deeper network, its layers are able to approximate identity mappings through shortcuts. This enables the model to have lower error rates and is able to address the degradation problem. It also handles the vanishing gradient problem in a much better way compared to other models [29].

Table 3. Performance metrics for the DL models.

Classifier	Precision (PPV)	Sensitivity (TPR)	F1 Score (F1)	Accuracy (ACC)
VGG19	0.95	0.94	0.94	0.94
DensNet201	0.95	0.94	0.94	0.94
VGG16	0.96	0.95	0.95	0.95
DensNet169	0.97	0.96	0.96	0.96
DensNet121	0.98	0.97	0.97	0.97
ResNet50	**0.99**	**0.99**	**0.99**	**0.99**

It can be seen from the Fig. 3 that the precision of aluminum and sensitivity of mild steel is low for all the classification models. This anomaly occurs due to the misclassification of mild steel images as aluminum images. When further investigated, it was observed that some of the images in the dataset belonging to these classes looked very similar. Figure 4 depicts sample images from the dataset for which the deep neural networks failed to classify the images correctly. As a result, the number of false positives increase in the case of aluminum, while the number of false negatives increase in the case of mild steel. Hence, the F1-score of both the metals are affected and it results in a lower F1-score for both aluminum and mild steel.

The proposed work is compared with previous work [28] on the dataset. Table 4 shows the comparison. In [28] the authors extracted texture features from the images and used ML classifiers like Support Vector Machines, Naïve Bayes, K-Nearest Neighbor, Decision Tree, Random Forest, and XGBoost to classify the images into four classes. The results showed that the XGBoost classifier achieved the highest classification accuracy of 98% among all the other experimented ML models. Though DL models are better at classification tasks when compared with traditional ML models, it could be understood from the results in Table 4 that for the dataset under study there is no significant difference between the performance of the ML model (XGBoost) and the highest performing DL

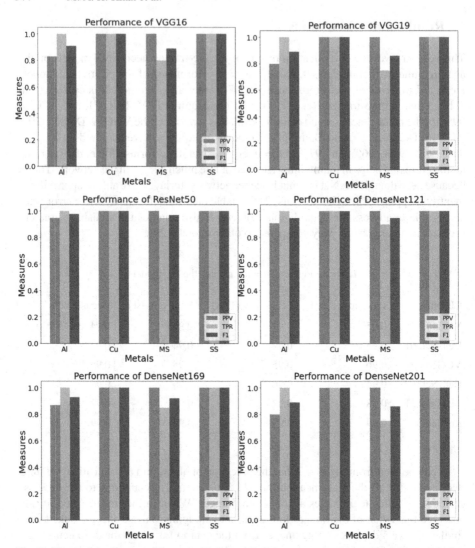

Fig. 3. Illustration of the precision, sensitivity and the F1 score attained by each of the models for the respective metals that were detected and classified, i.e., Aluminum (Al), Copper (Cu), Mild Steel (MS), and Stainless Steel (SS).

model (ResNet50). This might be because, the hand-crafted features extracted by the study in [28] was able to capture the necessary features to discriminate the metal samples perfectly. Another possible reason for the comparable performance of the ML models might be because the dataset is of small size.

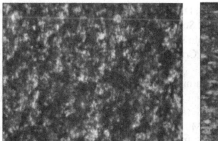 

Fig. 4. Similar looking aluminum (left) and mild steel (right) images from the dataset.

Table 4. Comparison of the proposed work with previous work.

Classifier	Precision (PPV)	Sensitivity (TPR)	F1 score	Accuracy
XGBoost [26]	0.98	0.98	0.98	0.98
ResNet50 (proposed work)	0.99	0.99	0.99	0.99

7 Conclusion

This paper proposed the use of DL and transfer learning models for the microstructure image classification of four types of metals. The metal detection and classification experiments were conducted using six pre-trained deep CNN models, and their results were compared. Based on the experimental results, the transfer learning model using ResNet50 showed the highest performance in terms of precision, sensitivity, F1 score and accuracy. A variety of model parameters were also analysed and applied in order to better understand which features were critical for classifying the different types of metals. The presented model can be expanded and used in future research and development applications to assist researchers and developers in designing reliable systems for automated metal classification. This can greatly reduce the time and costs involved in commercial and expensive production processes. In future, researchers and data scientists can use the results of this research to further improve the DL solutions. Specifically, future research can focus on the disadvantages of using DL for microstructural analysis which is the difficulty of accessing the features that the deep neural network model is based on [27]. This can have a greater impact in the field of material sciences.

References

1. Bhadeshia, H., Honeycombe, R.: Steels: Microstructure and Properties. Butterworth-Heinemann, Oxford (2017)
2. Lin, C.Y., Wirtz, T., LaMarca, F., Hollister, S.J.: Structural and mechanical evaluations of a topology optimized titanium interbody fusion cage fabricated by selective laser melting process. J. Biomed. Mater. Res. **83**(2), 272–279 (2007)
3. Kruth, J.P., Mercelis, P., Van Vaerenbergh, J., Froyen, L., Rombouts, M.: Binding mechanisms in selective laser sintering and selective laser melting. Rapid Prototyp. J. **11**(1), 26–36 (2005)

4. Chowdhury, A., Kautz, E., Yener, B., Lewis, D.: Image driven machine learning methods for microstructure recognition. Comput. Mater. Sci. **123**, 176–187 (2016)
5. German, R.M.: Coarsening in sintering: grain shape distribution, grain size distribution, and grain growth kinetics in solid-pore systems. Crit. Rev. Solid State Mater. Sci. **35**(4), 263–305 (2010)
6. LeCun, Y., Bengio, Y., Hinton, G.: Deep learning. Nature **521**(7553), 436–444 (2015)
7. https://www.nasa.gov/centers/wstf/supporting_capabilities/materials_testing/microstructu ral_analysis.html. Accessed 05 Dec 2022
8. Azimi, S.M., Britz, D., Engstler, M., Fritz, M., Mücklich, F.: Advanced steel microstructural classification by deep learning methods. Sci. Rep. **8**(1), 1–14 (2018)
9. Pokuri, B.S.S., Ghosal, S., Kokate, A., Sarkar, S., Ganapathysubramanian, B.: Interpretable deep learning for guided microstructure-property explorations in photovoltaics. NPJ Comput. Mater. **5**(1), 1–11 (2019)
10. Pazdernik, K., LaHaye, N.L., Artman, C.M., Zhu, Y.: Microstructural classification of unirradiated LiAlO2 pellets by deep learning methods. Comput. Mater. Sci. **181**, 109728 (2020)
11. Baskaran, A., Kane, G., Biggs, K., Hull, R., Lewis, D.: Adaptive characterization of microstructure dataset using a two stage machine learning approach. Comput. Mater. Sci. **177**, 109593 (2020)
12. Shen, C., Wang, C., Huang, M., Xu, N., van der Zwaag, S., Xu, W.: A generic high-throughput microstructure classification and quantification method for regular SEM images of complex steel microstructures combining EBSD labeling and deep learning. J. Mater. Sci. Technol. **93**, 191–204 (2021)
13. Ma, W., et al.: Image-driven discriminative and generative machine learning algorithms for establishing microstructure–processing relationships. J. Appl. Phys. **128**(13), 134901 (2020)
14. Zaloga, A.N., Stanovov, V.V., Bezrukova, O.E., Dubinin, P.S., Yakimov, I.S.: Crystal symmetry classification from powder X-ray diffraction patterns using a convolutional neural network. Mater. Today Commun. **25**, 101662 (2020)
15. Roberts, G., Haile, S.Y., Sainju, R., Edwards, D.J., Hutchinson, B., Zhu, Y.: Deep learning for semantic segmentation of defects in advanced STEM images of steels. Sci. Rep. **9**(1), 1–12 (2019)
16. Dong, H., et al.: A deep convolutional neural network for real-time full profile analysis of big powder diffraction data. NPJ Comput. Mater. **7**(1), 1–9 (2021)
17. Maemura, T., et al.: Interpretability of deep learning classification for low-carbon steel microstructures. Mater. Trans. **61**(8), 1584–1592 (2020)
18. Tammina, S.: Transfer learning using VGG-16 with deep convolutional neural network for classifying images. Int. J. Sci. Res. Publ. **9**(10), 143–150 (2019)
19. Mateen, M., Wen, J., Song, S., Huang, Z.: Fundus image classification using VGG-19 architecture with PCA and SVD. Symmetry **11**(1), 1 (2018)
20. Wen, L., Li, X., Gao, L.: A transfer convolutional neural network for fault diagnosis based on ResNet-50. Neural Comput. Appl. **32**(10), 6111–6124 (2019). https://doi.org/10.1007/s00 521-019-04097-w
21. Huang, G., Liu, Z., Van Der Maaten, L., Weinberger, K.Q.: Densely connected convolutional networks. In: Proceedings of the IEEE Conference on Computer Vision and Pattern Recognition, pp. 4700–4708, IEEE, Honolulu, HI, USA (2017)
22. Huang, G., et al.: Densely connected convolutional networks. In: Proceedings of the IEEE Conference on Computer Vision and Pattern Recognition (2017)
23. Varshni, D., Thakral, K., Agarwal, L., Nijhawan, R., Mittal, A.: Pneumonia detection using CNN based feature extraction. In: 2019 IEEE International Conference on Electrical, Computer and Communication Technologies (ICECCT), pp. 1–7. IEEE (2019)

24. Jaiswal, A., Gianchandani, N., Singh, D., Kumar, V., Kaur, M.: Classification of the COVID-19 infected patients using DenseNet201 based deep transfer learning. J. Biomol. Struct. Dyn. **39**(15), 5682–5689 (2021)

25. Medium Homepage: https://towardsdatascience.com/a-comprehensive-guide-to-convolutional-neural-networks-the-eli5-way-3bd2b1164a53. Accessed 05 Dec 2022

26. Medium Homepage: https://medium.com/@shaoliang.jia/vanishing-gradient-vs-degradation-b719594b6877. Accessed 05 Dec 2022

27. Gola, J., et al.: Objective microstructure classification by support vector machine (SVM) using a combination of morphological parameters and textural features for low carbon steels. Comput. Mater. Sci. **160**, 186–196 (2019)

28. Sabnis, H., Angel Arul Jothi, J., Deva Prasad, A.M.: Microstructure image classification of metals using texture features and machine learning. In: Patel, K.K., Doctor, G., Patel, A., Lingras, P. (eds.) Soft Computing and its Engineering Applications. icSoftComp 2021. CCIS, vol. 1572, pp. 235–248. Springer, Cham (2022). https://doi.org/10.1007/978-3-031-05767-0_19

29. Medium Homepage: https://towardsdatascience.com/resnets-why-do-they-perform-better-than-classic-convnets-conceptual-analysis-6a9c82e06e53. Accessed 05 Dec 2022

Generating Jigsaw Puzzles and an AI Powered Solver

Stefan-Bogdan Marcu[1]([envelope]) [ID], Yanlin Mi[1], Venkata V. B. Yallapragada[4] [ID],
Mark Tangney[3] [ID], and Sabin Tabirca[1,2] [ID]

[1] School of Computer Science and Informational Technology,
University College Cork, Cork, Ireland
{120227147,120220407,Tabirca}@umail.ucc.ie
[2] Faculty of Mathematics and Informatics, Transilvania University of Braşov,
Braşov, Romania
Tabirca@unitbv.ro
[3] Cancer Research@UCC, APC Microbiome Ireland, SynBioCentre, iEd Hub,
University College Cork, Cork, Ireland
M.Tangney@ucc.ie
[4] Centre for Advanced Photonics and Process Analytics,
Munster Technological University, Cork, Ireland

Abstract. This paper tackles the problem of assembling a jigsaw puzzle, starting only from a picture of the scrambled jigsaw puzzle pieces on a random, textured background. This manuscript discusses previous approaches in dealing with the jigsaw puzzle problem and brings two contributions: an open source tool for creating realistic scrambled jigsaw puzzles meant to serve as a foundation for further research in the field; and an end to end AI based solution taking advantage of the convolutional neural network architecture, capable of solving a scrambled jigsaw puzzle of unknown pictorial and with an unknown, uniformly textured, background. The lessons and techniques learned in engaging with the jigsaw puzzle problem can be further used in approaching the more general and complex problem of Protein-Protein interaction prediction.

Keywords: Jigsaw puzzle · Deep neural network · AI

1 Introduction

1.1 Description of the Jigsaw Puzzle Challenge

The jigsaw puzzle problem is familiar to most people since childhood. For more than 3 centuries [26] people have been entertained by tackling the challenge of reconstructing a picture from it's interlocking components. The ability of solving pictorial jigsaw puzzles is an index of meta-cognitive development [5].

A jigsaw puzzle is a tiling puzzle with interlocking, irregular shaped pieces, often with a pattern printed on them such that, upon the correct assembly, an image is revealed (Fig. 2D). In the classical, most common version of the puzzle, the pieces have a shape similar to one of the 18 generic types of puzzles pieces (Fig. 1) joined together in a grid layout with the help of pegs and sockets.

D. Simian and L. F. Stoica (Eds.): MDIS 2022, CCIS 1761, pp. 148–160, 2023.
https://doi.org/10.1007/978-3-031-27034-5_10

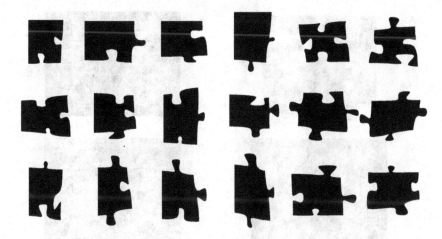

Fig. 1. Illustration of a possible versions for each of the 18 types of jigsaw puzzle pieces.

The objective of the puzzle is to find the correct assembly configuration and reveal the message or image composed by the pieces starting from a set of scrambled pieces of unknown orientation and position (Fig. 2C).

The first record of a jigsaw puzzle dates back to 1760. It's manufacturer was a map maker from London named John Spilsbury [26]. Jigsaw puzzles were first created by cutting a map printed on a wooden board in order to teach geography.

The problem, as it is attempted to be solved in this paper, represents the most common encountered rendition of the Jigsaw puzzle.

The input is an image of the puzzle with all the pieces visible, facing with the pictorial side up, non-overlapping, scrambled both translationaly and rotationaly on a uniform textured background.

The puzzle is a pictorial, connected, rectangular shaped puzzle, of unknown length and width, with no missing or extraneous pieces, with no holes, for which the correct orientation is unknown at the start of the assembly and the correct assembled configuration is unique.

Each piece is unique with small derivations from the 18 standard types Fig. 1. All the internal pieces have four connecting neighbours, the edge pieces have three connecting neighbours and the corner pieces two connecting neighbours. In the solved configuration all of the pieces are arranged in a grid-like layout such that all of the internal junctions form a + shaped quadradial intersection and the edge junctions form a T shaped triradial intersection.

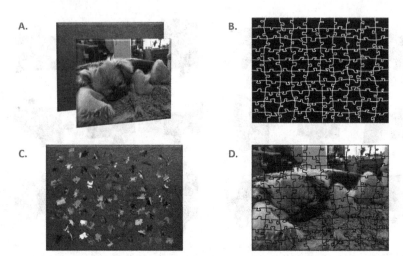

Fig. 2. A. - The pictorial of the Puzzle and the Background Texture B. - The cutout template created during the generation of the puzzle C. - The output of the puzzle generator, the scrambled jigsaw puzzle on its background D. - A correctly assembled Jigsaw Puzzle.

All of the reviewed techniques fail at least one of the assumptions listed above, even when the equivalence between the classical jigsaw puzzle and the box puzzle variant [3] is taken in consideration.

1.2   Motivation on Tackling the Problem

The jigsaw puzzle is a fun, natural and challenging problem familiar to most people. The development of a system capable of solving this type of puzzles has fascinated scientists for over half a century. It is a hard problem, proven to be NP-complete [3], for which children as young as 4 can find a correct solution [5]

At it's core the Jigsaw puzzle is a problem of segmentation and efficient application of pattern recognition using geometric and chromatic clues provided by the piece. The knowledge and solutions gained from the development and design of a puzzle solver can be applied to multiple different scientific fields like archaeology, art, image reconstruction and manipulation, structural biology [22], etc.

The problem, as formulated in this paper, is very similar to the problem of finding pairs of proteins that can interact with each other given just their sequences. The extraction of the puzzle piece from their regularly patterned background is comparable to extracting meaningful information from the proteins sequence and the matching step is similar to finding which proteins would interact in a basic lock-and-key model. The 3D shape of the protein surface together with its surface properties, such as electric charge can be accurately modelled as a jigsaw puzzle with 3D pieces and connections where the chromatic hints represent the protein's surface properties. This is illustrated in Fig. 3; Illustration A depicts a Gaussian surface representation [19, 25] of the 2JEL [21] protein complex. B depicts the same protein complex in an x-ray like view, allowing for a better visualisation of the topology of the inner connecting surfaces.

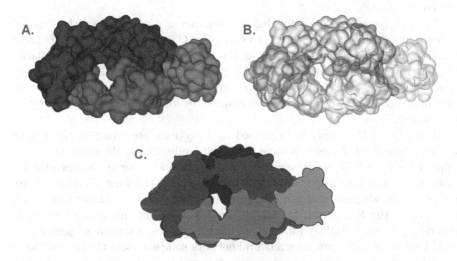

Fig. 3. Images of 2JEL [21] protein, visualised with the help of MOL [25] on the RCSB website [19], that illustrate the similarity between the jigsaw puzzle problem and Protein-Protein docking. A. - the Gaussian surface representation of the protein. B. - the x-ray visualisation of the Gaussian surface. C. - Internal view of the Gaussian surface using a clipping plane.

The same as in the case of a jigsaw puzzle, the number of useful interactions between each component vastly outnumbers the number of possible interactions. Furthermore segmenting the piece from it's background and retaining only the relevant information of a puzzle piece, such as shape and boundary colour, is similar to detecting which portion of a protein surface is going to be active, which specific characteristics, such as polarity of the surface, would be relevant.

1.3   Our Contribution

In this paper, the jigsaw puzzle piece detection is included as an integral part of the puzzle solving problem.

A Jigsaw Puzzle generator is introduced that is capable of creating realistic images of scrambled Jigsaw puzzles and an AI powered solver is proposed that, with no prior knowledge of the puzzle, is capable of correctly and accurately indicating the correct piece connections starting only from a realistic image of a scrambled puzzle.

Making the generator publicly available aims to create the foundation for further development of neural networks-based solutions for real life-like puzzles.

A deep neural network architecture is proposed, this architecture shows promise in extracting useful information from the provided input picture and has the capability to model the interaction between the pieces in order to find the correct match.

The 2D jigsaw problem is tackled in this paper as a precursor for the Protein-Protein interaction problem. Figure 3 sub-point C exemplifies how the 2D problem is a more constrained version of the 3D jigsaw puzzle and by similarity of the Protein-Protein interaction. The 2D version is tackled in order to learn what approaches and what types of architectures are capable of learning a solution and show promise to be used as candidates for predicting Protein-Protein interaction.

The rest of the paper is organised in 4 sections according to the following structure: Sect. 2 presents some historical landmarks in the evolution of the approaches for solving the jigsaw puzzle before presenting contemporaneous solutions and discussing their assumptions in tackling the challenge. Section 3 introduces the Puzzle generator, an open source jigsaw puzzle generator tool meant to serve as the foundation for further developments in this area of research, describing its methodology for creating life-like images of scrambled jigsaw puzzles. This section also proposes a neural network solution using the convolutional neural network architecture that starts with nothing more than an image of a scrambled jigsaw on a textured surface, as produced by the generator. Section 4 constitutes the conclusion of the paper and elaborates on possible future directions the research can be further developed.

2 Evolution of the Different Approaches in Tackling the Jigsaw Puzzle

The first attempt [24] at a computerised solver was in 1964 as part of a research study in pattern analysis supported by the Air Force Office of Scientific Research. H. Freeman et al. introduce a technique of assembling puzzles solely based on shape [6]. This type of approach with small improvements was able to tackle puzzles with as many as 200 pieces [8].

In 1994 D.A. Kosiba et al. introduces the Automatic Puzzle Solver, a system that uses a combination of shape features along with the colour information as a matching criteria [11].

M Sağiroğu et al. flips the previous approaches and disregards the boundary shape, taking in consideration only the colour and textures. An in-painting synthesis is used to forecast the texture of a band outside the piece's boundary. The forecast is used to calculate feature values and in the end align and match with other pieces [23].

With the increase in computing power, more and more emphasis is placed on the contents of each piece and the importance placed on the edge information is diminished to the boundary shape being reduced to a square [2,7,9,14,17,20,

26, 27]. Even though this seems an important distinction Demaine, Erik D et al. [3] have shown that the classical jigsaw puzzle and box puzzle variant are equivalent.

Most state-of-the-art solutions to the jigsaw puzzle revolve around variants of dissimilarity/compatibility function over different aspects of the pieces, such as: sum-of-squared differences for patches along the piece boundary [2], changes in the intensity of the gradient for patches along the pieces edge [7] or checking the line of pixels at the edge [9, 17, 20, 26].

A common approach is to solve the puzzle in a greedy fashion, choosing a starting piece and recursively choosing neighbours based on the dissimilarity/compatibility function [2, 7, 17, 20, 27].

One technique that uses the dissimilarity/compatibility score in a non-greedy way is presented by Sholomon, Dror et al. [26]. In their genetic algorithm approach to solving jigsaw puzzles the dissimilarity/compatibility function is summed over neighbouring pieces for all the pieces in order to compute the fitness score of each individual. This technique was capable of solving puzzles of impressive dimensions with the caveat that the piece orientation has to be known. This assumption together with knowing the size of the puzzle, reduces the puzzle solving problem to finding the correct permutation of the pieces.

In their paper, Marie-Morgane Paumard et al. introduce Deepzzle, a system that tackles a jigsaw-like problem using deep neural networks. They illustrate the ability of deep learning techniques to tackle the puzzle [18]. In their paper they decide to focus on an archaeological perspective, choosing to force their network to ignore edge information such as shape and colour by cutting pieces of square shape that have a gap between them. This is useful in archaeology because edge information might be lost due to manipulation and erosion.

A majority of the proposed solutions [2, 7, 9, 10, 14, 17, 18, 26, 27] start with the assumption that the pieces are already correctly segmented from their background, a non-trivial task.

3 Proposed Method

3.1 Puzzle Generation

In order to have a result that is comparable with a real picture of a scrambled Jigsaw puzzle, the generating process closely follows the real fabrication process of Jigsaw puzzles.

The process of generating a scrambled puzzle starts with selecting an image that is going to be the pictorial of the puzzle (Fig. 2A). After the pictorial of the puzzle is selected and therefore the dimensions of the puzzle are known, a new cutout image of the same dimensions is created.

The next step consists in the design of the pieces shape. The image is sectioned in squares, for which the size is dictated by the desired number of rows and columns. Each of these squares is going to have a horizontal edge with a peg/socket joint running through the middle of it. This approach can be taken because the most common version of the jigsaw puzzle is arranged in a grid layout.

The boundary of the piece is defined with the help of Bézier Curves. The Bézier curves used in this project are a parametric curves defined by the formula for Bézier curves with n control points (1). The curve is parameterised by parameter t, taking values in the interval $[0, 1]$, that defines the interpolation distance between the start and the end of the curve. The curve is defined by n control points $P_0, P_1, ..., P_n$. This type of parametric curve has been chosen because of its ability to model a variety of complex shapes and for three of its properties: the first and last control points are the start and end points of the curve; a curve can be split into arbitrarily many sub-cuvres, each of which is also a Bézier curve; the resulting curve will not extend outside the convex envelope of its control points.

$$B(t) = \sum_{i=0}^{n} \binom{n}{i} (1 - t)^{n-i} t^i P_i \tag{1}$$

The first property assures that the edges can have arbitrarily complex shapes while keeping precise control over the start and end of the curve, this means that the edge is going to be continuous between consecutive pieces as long as the last point of the previous edge is taken to be the first point in the creation of the new edge.

The second and third property together assure that the peg/socket is not going to go outside the perimeter defined by the control points, the socket joint is not able to go outside the square being worked on.

The same process is repeated for the vertical edges with a different offset for the partition into rectangles and with a check for overlapping of the peg/socket joint with the horizontal boundary, producing the digital equivalent of the cutout template of the puzzle (Fig. 2B).

In the next step a background image is selected that has enough space to fit all of the scrambled pieces in an non overlapping manner. Experimentally results showed that a dimension twice the size of the puzzle face is sufficient. If the background image is bigger than the required size, it is randomly cropped to the desired dimensions and randomly flipped and mirrored in order to increase the diversity of the backgrounds. On this surface the picture sections are pasted according to the cutout template at a random, non-overlapping, position and applying a random rotation.

A final step of blurring is applied over the final scrambled puzzle picture in order to dampen the sharp transition at the edge of each piece and to better mimic the softness of the transition of real life pictures (Fig. 2C).

The generator is available for research purposes on the following github page: https://github.com/Bogst/JigsawPuzzleGenerator.

3.2   Proposed Puzzle Solver Solution

In order to solve the Jigsaw puzzle problem a solution that incorporates the use of a deep convolutional neural network architecture is proposed. This approach has been selected because of the necessity to accurately distinguish the puzzle pieces from their background, task for which deep neural networks are producing state of the art results [16].

Even though at a first glance the problem of detecting puzzle pieces on a background image seems similar to the problem of segmenting different objects

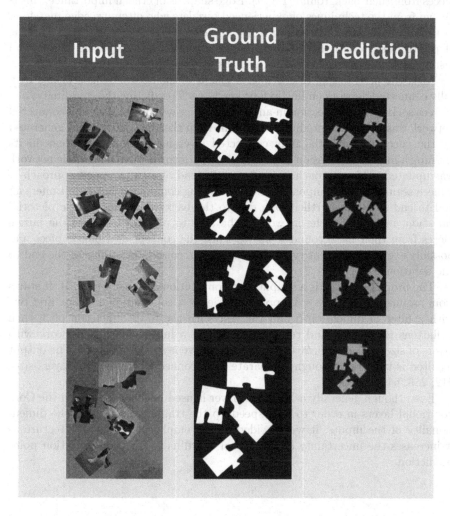

Fig. 4. Input, Ground Truth and Prediction of the encoder-decoder training, green represents places where the network missed predicting, red places where it labelled wrongfully as piece and yellow where it correctly predicted a piece. (Color figure online)

in a picture, it is subtly different because of the pictorial of the puzzle. In the case of object detection, the network can learn abstract representation of physical objects. While in the case of puzzle pieces, the detection should not get misled by the graphical details withing the puzzle piece itself. This constitutes enough reason in order to train the network from scratch and not attempt to fine tune an existing state of the art model.

In order to help the model learn that edge shape as an important factor in the determination of matching pieces as well as to ignore the textured background, the network was pre-trained on an encoder-decoder task of segmenting the puzzle pieces from their background (Fig. 4). Edge shape is of crucial importance, this is known from the techniques of experienced puzzlers who often use edge matching as the ultimate test in determining if two pieces are true neighbours. This style of pre-training forces the network to pay attention to the puzzle shapes and disregard useless information such as background colour and texture.

After the pretraining, the decoder segment of the network was replaced by a fully connected network ending in four heads, each predicting the coordinates of a correct socket/peg connection in an four piece jigsaw puzzle. The fully connected network was trained to predict connections from the embedding space generated by the encoder. After the fully connected layer has reached a plateau in its training a fine-tuning session of training was performed where the whole network was updated. This approach was selected instead of a two stepped approach of a piece segmentation component and a matching component because it offers an end to end solution. Furthermore, due to the background and puzzle pictorial variation, in order for the network to be able to properly segment the puzzle pieces from their background it has to learn the concept of a puzzle piece and possibly even the different types. Taking this in consideration makes the end to end solution preferred.

The proposed model is a simple deep convolutional network (Fig. 5). It starts from an image of $98 \times 98 \times 3$, applies a series of 4 convolutional layers, first two with a filter of dimension 5 and the rest with the filter dimension of 3, then it flattens the output and runs them through a fully connected network with layers of sizes 8192, 4096, 2048 and 3 consecutive layers of size 1024. The output resulted is then ran through 4 separate fully connected heads with layers sizes 512, 256, 64, 16 and 4.

Even though generally a Max-Pool layer is used in connection with the Convolutional layers in order to help speed up the training and reduce the dimensionality of the inputs, it was decided not to employ it in the architecture as it increases the uncertainty of the actual coordinates for the connection point prediction.

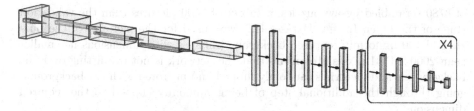

Fig. 5. Architecture of the network, 4 convolutional layers, first 2 with filter size of 5 and the last 2 with size 3, 6 fully connected layers of sizes 8192, 4096, 2048 and 3 layers of size 1024 followed by 4 heads with fully connected layers of sizes 512, 256, 64, 16 and 4. Diagram generated with the help of NN-SVG [13].

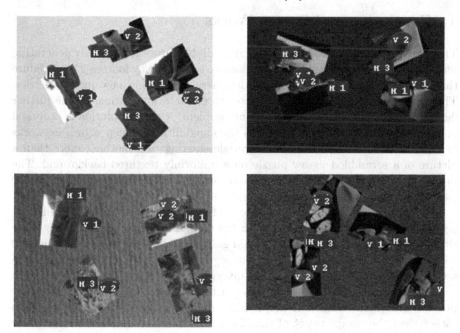

Fig. 6. Examples of the networks prediction for connection points (blue) over the ground truth (green). (Color figure online)

The network was trained using euclidean pixelwise distance from the true centre of peg/socket connection as a loss function, using the Adam optimizer. The best result achieved a loss of 423.859px. Figure 6 illustrates a subset of these results.

Initially the generator was directly linked to the training of the network providing it with on the spot new samples. Due to the fact that a non-trivial portion of the training time was spent generating new puzzles the decision to save puzzles to disk was made, this significantly improved training times. In order to generate a training set of 100,000 scrambled jigsaw puzzles and a validation set of 8780 scrambled jigsaw puzzles, a subset of 1000 pictures from the validation stage of the Open Images Dataset [12] were used for the pictorial of the puzzle and 100 uniformly textured background images with dimensions no smaller than 2000×1331 pixels. To ensure that the network is not overfitting over the backgrounds and pictorials randomly flipped and mirrored with the background going through the additional step of being randomly clipped to the required dimension.

The model creation and training were done with the help of the Tensorflow [4] library. All the experiments were ran on a machine with an Intel i7-10750h CPU, 16 GB of RAM and an NVIDIA RTX 2060 graphics card.

4   Conclusion and Future Works

This paper presented and made publicly available a tool for digitally generating scrambled jigsaw puzzles that are comparable to real life by following a procedure that mimics the workflow of puzzle designers in creating new jigsaw puzzles. The generator is open source and meant to create a foundation for the further development of the jigsaw puzzle solving problem. A simple solution is proposed in the form of a solver that leverages the capabilities of deep neural networks in order to solve a four piece jigsaw puzzle starting from nothing more than a picture of a scrambled jigsaw puzzle on a uniformly textured background. The proposed solution includes the segmentation problem and does not impose the use of a known, or plain, background.

The generator has the potential to be further extended with the ability to generate non traditional puzzles. With the addition of different sectioning algorithms for the pieces the generator could gain the ability to create jigsaw puzzles with intersections that have a different degree of radiality, with different intersecting angles and with a layout that diverges from the grid-like pattern.

One direction in which the solver can be improved to generalise on jigsaw puzzles with more than four pieces is to predict neighbours of a given piece rather than the complete set of coordinates for the solution. This would allow the model to tackle puzzles of unknown piece number. Different architectures and training techniques can be potentially employed to improve the performance of the network. One such technique that shows potential is the attention mechanism. A further path for improvement could be constituted by the usage of meta-heuristics and swarm intelligence approaches [1] in order to further refine the model and it's hyper-parameters. Feature selection techniques [15] could be successfully employed in tackling the Protein-Protein interactions.

Lessons learned in tackling this problem can be extrapolated onto the problem of Protein-Protein interaction due to the similarity of the two problems. The lock and key interaction model of proteins can be seen as a generalised 3D version of the jigsaw puzzle problem. The results of the proposed model are a strong indicator that a bigger model could tackle the Protein-Protein interaction problem. Such a model would prove valuable in fields that deal with Protein-Protein interaction such as structural-biology.

References

1. Bacanin, N., Stoean, R., Zivkovic, M., Petrovic, A., Rashid, T.A., Bezdan, T.: Performance of a novel chaotic firefly algorithm with enhanced exploration for tackling global optimization problems: application for dropout regularization. Mathematics **9**(21) (2021). https://doi.org/10.3390/math9212705, https://www.mdpi.com/2227-7390/9/21/2705

2. Cho, T.S., Avidan, S., Freeman, W.T.: A probabilistic image jigsaw puzzle solver. In: 2010 IEEE Computer Society Conference on Computer Vision and Pattern Recognition, pp. 183–190. IEEE (2010). https://doi.org/10.1109/CVPR.2010.5540212

3. Demaine, E.D., Demaine, M.L.: Jigsaw puzzles, edge matching, and polyomino packing: connections and complexity. Graphs Comb. **23**(S1), 195–208 (2007). https://doi.org/10.1007/s00373-007-0713-4

4. Developers, T.: Tensorflow (2022). https://www.tensorflow.org

5. Doherty, M.J., Wimmer, M.C., Gollek, C., Stone, C., Robinson, E.J.: Piecing together the puzzle of pictorial representation: how jigsaw puzzles index metacognitive development. Child Dev. **92**(1), 205–221 (2021)

6. Freeman, H., Garder, L.: Apictorial jigsaw puzzles: the computer solution of a problem in pattern recognition. IEEE Trans. Electron. Comput. **EC-13**(2), pp. 118–127 (1964). https://doi.org/10.1109/PGEC.1964.263781

7. Gallagher, A.C.: Jigsaw puzzles with pieces of unknown orientation. In: 2012 IEEE Conference on Computer Vision and Pattern Recognition, pp. 382–389. IEEE (2012). https://doi.org/10.1109/CVPR.2012.6247699

8. Goldberg, D., Malon, C., Bern, M.: A global approach to automatic solution of jigsaw puzzles. In: Proceedings of the Eighteenth Annual Symposium on Computational Geometry, SCG 2002, pp. 82–87. Association for Computing Machinery, New York (2002). https://doi.org/10.1145/513400.513410

9. Gur, S., Ben-Shahar, O.: From square pieces to brick walls: the next challenge in solving jigsaw puzzles. In: 2017 IEEE International Conference on Computer Vision (ICCV), pp. 4049–4057. IEEE (2017). https://doi.org/10.1109/ICCV.2017.434

10. Khoroshiltseva, M., Vardi, B., Torcinovich, A., Traviglia, A., Ben-Shahar, O., Pelillo, M.: Jigsaw puzzle solving as a consistent labeling problem. In: Tsapatsoulis, N., Panayides, A., Theocharides, T., Lanitis, A., Pattichis, C., Vento, M. (eds.) CAIP 2021. LNCS, vol. 13053, pp. 392–402. Springer, Cham (2021). https://doi.org/10.1007/978-3-030-89131-2_36

11. Kosiba, D., Devaux, P., Balasubramanian, S., Gandhi, T., Kasturi, K.: An automatic jigsaw puzzle solver. In: Proceedings of 12th International Conference on Pattern Recognition, vol. 1, pp. 616–618. IEEE Computer Society Press (1994). https://doi.org/10.1109/ICPR.1994.576377

12. Krasin, I., et al.: OpenImages: a public dataset for large-scale multi-label and multi-class image classification (2017). https://github.com/openimages

13. LeNail, A.: NN-SVG: publication-ready neural network architecture schematics. J. Open Sour. Softw. **4**(33), 747 (2019). https://doi.org/10.21105/joss.00747

14. Li, R., Liu, S., Wang, G., Liu, G., Zeng, B.: JigsawGAN: auxiliary learning for solving jigsaw puzzles with generative adversarial networks. IEEE Trans. Image Process. **31**, 513–524 (2022). https://doi.org/10.1109/TIP.2021.3120052

15. Malakar, S., Ghosh, M., Bhowmik, S., Sarkar, R., Nasipuri, M.: A GA based hierarchical feature selection approach for handwritten word recognition. Neural Comput. Appl. **32**(7), 2533–2552 (2019). https://doi.org/10.1007/s00521-018-3937-8

16. Minaee, S., Boykov, Y., Porikli, F., Plaza, A., Kehtarnavaz, N., Terzopoulos, D.: Image segmentation using deep learning: a survey. IEEE Trans. Pattern Anal. Mach. Intell. **44**(7), 3523–3542 (2022). https://doi.org/10.1109/TPAMI.2021.3059968

17. Paikin, G., Tal, A.: Solving multiple square jigsaw puzzles with missing pieces. In: 2015 IEEE Conference on Computer Vision and Pattern Recognition (CVPR), pp. 4832–4839. IEEE (2015). https://doi.org/10.1109/CVPR.2015.7299116

18. Paumard, M.M., Picard, D., Tabia, H.: Deepzzle: solving visual jigsaw puzzles with deep learning and shortest path optimization. IEEE Trans. Image Process. **29**, 3569–3581 (2020). https://doi.org/10.1109/TIP.2019.2963378

19. PDB, R.: RCSB PDB. https://www.rcsb.org/

20. Pomeranz, D., Shemesh, M., Ben-Shahar, O.: A fully automated greedy square jigsaw puzzle solver. In: CVPR 2011, pp. 9–16. IEEE (2011). https://doi.org/10.1109/CVPR.2011.5995331

21. Prasad, L., Waygood, E.B., Lee, J.S., Delbaere, L.T.J.: The 2.5 å resolution structure of the jel42 fab fragment/HPr complex. J. Mol. Biol. **280**(5), pp. 829–845 (1998). https://doi.org/10.1006/jmbi.1998.1888

22. Priatama, A.R., Setiawan, Y., Mansur, I., Masyhuri, M.: Regression models for estimating aboveground biomass and stand volume using landsat-based indices in post-mining area. J. Manajemen Hutan Tropika **28**(1), 1–14 (2022). https://doi.org/10.7226/jtfm.28.1.1

23. Sagiroglu, M., Ercil, A.: A texture based matching approach for automated assembly of puzzles. In: 18th International Conference on Pattern Recognition (ICPR 2006), vol. 3, pp. 1036–1041. IEEE (2006). https://doi.org/10.1109/ICPR.2006.184

24. Sahu, E., Mishra, G., Singh, H., Kumar, V.: A review on the evolution of jigsaw puzzle algorithms and the way forward. In: 2022 9th International Conference on Computing for Sustainable Global Development (INDIACom), pp. 281–287. Institute of Electrical and Electronics Engineers (IEEE) (2022). https://doi.org/10.23919/INDIACom54597.2022.9763234

25. Sehnal, D., et al.: Mol* viewer: modern web app for 3D visualization and analysis of large biomolecular structures. Nucleic Acids Res. **49**(W1), W431–W437 (2021). https://doi.org/10.1093/nar/gkab314

26. Sholomon, D., David, O., Netanyahu, N.S.: A genetic algorithm-based solver for very large jigsaw puzzles. In: 2013 IEEE Conference on Computer Vision and Pattern Recognition, pp. 1767–1774. Institute of Electrical and Electronics Engineers (IEEE) (2013). https://doi.org/10.1109/CVPR.2013.231

27. Son, K., Moreno, D., Hays, J., Cooper, D.B.: Solving small-piece jigsaw puzzles by growing consensus. In: 2016 IEEE Conference on Computer Vision and Pattern Recognition (CVPR), pp. 1193–1201. IEEE (2016). https://doi.org/10.1109/CVPR.2016.134

Morphology of Convolutional Neural Network with Diagonalized Pooling

Roman Peleshchak⬤, Vasyl Lytvyn⬤, Oleksandr Mediakov⬤,
and Ivan Peleshchak(✉)⬤

Lviv Polytechnic National University, 12 S. Bandera Str., Lviv, Ukraine
`rpele@ukr.net`, `oleksandr.mediakov.sa.2019@lpnu.ua`,
`peleshchakivan@gmail.com`

Abstract. In the paper, we introduce the structure of a convolutional neural network with diagonalized pooling (DiagPooling). Existing deterministic poolings, like max or Average, cause loss of some critical information, which may be helpful for a further classification task. To minimize that loss, we propose an algorithm that considers information from all values over the pooling window via eigendecomposition. Eigendecomposition changes the basis to one in which the matrix of the pooling window has a diagonal form. Then DiagPooling yields the greatest diagonal element of the matrix, which corresponds to the greatest eigenvalue of the pooling window. The main difference between our idea and classical or PCA pooling methods is a change of basis without returning to it. Also, we suggest combinations of diagonalized pooling with normalization methods that increase models' performance compared to CNNs that use classical pooling layers.

Keywords: Diagonalized pooling · Convolutional neural network · Matrix diagonalization · Eigendecomposition · Normalization methods

1 Introduction

The problem of high-accuracy spatial and spectral image recognition with optimal computing resources using convolutional neural networks requires the development of new pooling methods. The process of pooling can be seen as the downsampling of the activation values of convolved images. Pooling layers are usually placed between sequential convolutions. Those layers are necessary because they perform several essential tasks: (1) pooling the most valuable and discarding less valuable or irrelevant information, (2) lower spatial dimensions of convolved tensors, (3) gradually reduce the number of weights, and (4) stabilize the learning process by preventing overfitting, along with other regularization methods, like a dropout [2].

However, classical poolings cause a loss of information, which might be helpful for eventual classification. We introduce a new diagonalized pooling method that overcomes that issue.

© The Author(s), under exclusive license to Springer Nature Switzerland AG 2023
D. Simian and L. F. Stoica (Eds.): MDIS 2022, CCIS 1761, pp. 161–172, 2023.
https://doi.org/10.1007/978-3-031-27034-5_11

The paper is organized as follows. Firstly we present the result of an analysis of some existing pooling methods in Sect. 2. Then, we thoroughly describe an algorithm of our method, its place in CNN, and its combination with normalization techniques in Sect. 3. Last but not least, in Sect. 4, the results of validating the theoretical superiority of diagonalized pooling over classical methods on experimental data are demonstrated.

2   Related Works

Different methods and operations characterize existing types of poolings. For instance, there are some classical pooling layers, like max, min, average [7], or stochastic [15].

Still, also researchers have created other interesting methods based on classical or independent ones. Regardless of the type, the pooling layer is expected to extract the essential information while down-sampling the feature maps [4].

For example, max pooling losses information while discarding all values except the maximum one. The average pooling method considers all elements of the convolved and activated image, so minimal pooling areas reduce maximum activations, causing the loss. Those are the drawbacks of classical pooling.

Ideas from [3,8,12,13,15] are able to partially minimize that loss. In particular, a PCA pooling from PCANet [3], in [8] proposed "responsible" poolings: gated or tree functions, [13] uses L_p pooling, in [15] idea of stochastic pooling was developed, and [12] introduced spectral pooling.

For instance, pooling strategies from [8] propose mixed, gated poolings yield a masked and weighted combination of average and max, and so take into consideration the biggest values and all values with some proportion. Simple mixed pooling can be represented with Eq. 1.

$$f_{mixed}(x) = \omega f_{max}(x) + (1 - \omega) f_{avg}(x) \tag{1}$$

where ω defines the use of maximum and average pooling. The value of ω is chosen randomly between 0 and 1 for each layer, usually lambda is a trainable parameter. In the case of $\omega = 0$, we get the average pooling method, and with $\omega = 1$ - the maximum.

But, for so-called "responsible" pooling methods authors proposed to learn a "gating mask" with parameter ω of the same size as pooling window. The gated pooling can be expressed with Eq. 2.

$$f_{gated}(x) = \sigma(\omega^\top x) + (1 - \sigma(\omega^\top x)) f_{avg}(x) \tag{2}$$

where ω is trainable tensor, $\sigma(\cdot)$ is a sigmoid function.

The L_p pooling method proposed in [13] has better generalization properties than max polling. This method can be described with Eq. 3, so it can be seen as a weighted average of inputs from the pooling window matrix.

$$f_{L_p}(x) = \left(\sum_{a \in x} a \right)^{\frac{1}{p}} \tag{3}$$

Here x is current pooling window, and p is taken from the interval $[1, \infty]$. Apparently, when $p = \infty$, L_p pooling is a maximum pooling.

Both gated and L_p operations are a kind of trade-off between average and max pooling.

Another approach proposed in [15] is stochastic pooling, which uses a multinomial distribution to select a pooled value, ensuring the non-maximal elements of the pooling feature map will also be considered (see Fig. 1). Firstly, for each pooling window, the probabilities p_i are computed according to the normalization Eq. 4. Those probabilities are considered as event probabilities of the multinomial distribution.

$$p_i = \frac{a_i}{\sum_j a_j}, \quad a_{i,j} \in x \tag{4}$$

where x is a pooling window.

After, the stochastic method yields value at position l of the pooling window where index l is taken from multinomial distribution (see Eq. 5). It gives high chances to stronger activations and suppresses the weaker activations.

$$f_{stoch}(x) = a_l \in x, \quad l \sim P(p_1, ..., p_i, ...) \tag{5}$$

However, that pooling method has different behaviors in training and interference modes. Because stochasticity may generate noise in the testing or prediction mode, the authors described the idea of a probabilistic form of mean value, namely $f_{stoch}(x) = \sum_i a_i p_i$, which can be seen as a model averaging [15] rather than elements one.

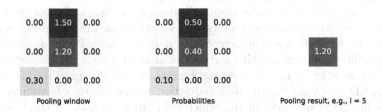

Fig. 1. Example of stochastic pooling on concrete pooling window.

Still, it can be seen that the pooling method has information loss in both modes.

Yet another method of pooling, proposed in [3] is PCA poolingwhere the principal component analysis is used at a pooling stage. Created pooling layers with PCA are used to learn multistage filter banks.

Lastly, in [12] authors have proposed a method of spectral pooling based on dimensionality reduction via cropping a spectral image of input data. Let $x \in R^{m \times m}$ be an input feature map, and $h \times w$ will be the new lower dimension of the map. The first step of spectral pooling is the discrete Fourier transform (DFT) of the input map, and the second step — is the created spectral image is cropped from the center into the submatrix of hv order. The last step uses inverse DFT on the cropped map, getting it back to a spatial basis. Formally those steps can be expressed with Eq. 6.

$$f_{spectral}(x) = \text{iFFT}\Big(\text{Crop}\left[\text{FFT}(x),\ h \times w\right]\Big) \tag{6}$$

where $\text{FFT}(\cdot)$ is a fast Fourier transform algorithm, and $\text{Crop}(\cdot)$ is a central cropping.

Spectral pooling keeps significantly more information for the same number of parameters. This is achieved by exploiting the non-uniformity of typical input signals in their signal-to-noise ratio as a function of frequency. Spectral pooling is based on natural images having a spectrum with power primarily concentrated in lower frequencies, while higher frequencies tend to encode noise [14]. But, even though, the truncation of spectral matrix will cause information loss.

Spectral pooling enables to specify dimensionalities of the output feature map, which results in the gradual and controlled reduction. In that case, it can be treated as a function of the neural network depth. It is important to mark that use of spectral pooling is not an extremely expensive computation in deep neural networks because of the algorithm of fast transform.

Thus, an analysis of different pooling methods allows us to assert that the use of these methods leads to some kind of information loss. In this paper, we introduce a different approach to solving the problem based on one of the most crucial ideas from linear algebra - eigendecomposition (ED).

Performing ED leads to the diagonalization of a matrix, finding its eigenvalues and corresponding eigenvectors. ED is widely used in machine learning. From feature extraction and dimensionally reduction with PCA to more complex tasks, like quantum machine learning models, and even has connections to proper initialization of neural networks weights, as was discussed in [11].

Also, ED was used for the diagonalization of the matrix of synaptic connections in neural networks [9,10], which leads to a decrease in their number since only diagonal elements remain. Those diagonal elements contain complete information about the vector of input signals and are directly used for fitting a neural network with reduced computing resources.

In our work, we have developed a concept of decoding the most significant information from the pooling window to minimize the possible loss of information. And for that purpose, the ED is a nice tool. The main difference between our idea and classical pooling methods is a change of basis without returning to it (in contrast to PCA). As a result of developing this idea, we created an algorithm for diagonalized pooling - DiagPooling.

3   DiagPooling

Pooling of features via DiagPooling is based on changing the basis for the pooling window via eigendecomposition. Performing ED results in a diagonalized matrix of the pooling window and, importantly, in a set of its eigenvalues. After ED, our algorithm yields the biggest diagonal element, corresponding to the matrix's greatest eigenvalue. Pooled eigenvalues of matrices form new feature maps of lower dimensionality (see Fig. 2).

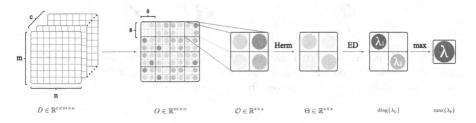

Fig. 2. DiagPooling scheme with kernel size $s = 2$.

Lets $D \in \mathbb{R}^{c \times m \times n}$ corresponds to the result of the convolution layer, with c beings a number of output channels or a number of filters, and $m \times n$ is the shape of the image or a feature map.

With the given size of pooling kernel s, the method considers each feature map $O_k \in \mathbb{R}^{m \times n}$ from every channel of D.

Remark 1. There is a requirement that m and n are divisible by s.

Each matrix O_k is divided into square blocks (or submatrices) $\mathcal{O}_{ij}$, of the size $s \times s$. Those blocks aren't overlapping, i.e., strides of pooling are equal to the kernel size s, so calculations for each block are independent.

In order to prevent appearance of complex eigenvalues and make ED operation a little faster, one could transform each block $\mathcal{O}_{ij}$ into a Hermitian matrix Θ_{ij}. In out method this step is done by operation $\Theta_{ij} = \mathcal{O}_{ij}\mathcal{O}_{ij}^{\top}$.

Then, by performing ED of a matrix, one finds such operator T and set of eigenvalues $\{\omega_l^{ij} \,|\, l = \overline{1,s}\}$ that $\mathrm{diag}(\{\omega_l^{ij} \,|\, l = \overline{1,s}\}) = T^{-1}\Theta T$. After, the greatest eigenvalue $\omega_m^{ij} = \max\{\omega_l^{ij} \,|\, l = \overline{1,s}\}$ should be selected.

The selected max eigenvalue ω_m^{ij} is yielded from DiagPooling over the current block $\mathcal{O}_{ij}$.

Repeating such calculations for each block $\mathcal{O}_{ij} \subset O_k$ and each feature map $O_k \subset D$, one forms resulting tensor $\mathcal{D} \in \mathbb{R}^{c \times p \times q}$ where $p = \frac{m}{s}$, $q = \frac{n}{s}$ are the new spatial dimensions of feature maps.

Like other deterministic pooling layers, our can be used for image down-sampling or as image transformation with dimensionality reduction [4]. For instance, single-channeled MNIST images of size 28×28 can be lowered to 14×14 or 7×7 with DiagPooling, as shown in Fig. 3.

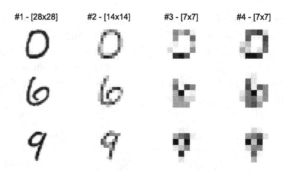

Fig. 3. Transformation of MNIST images with DiagPooling. # 1 images are the original digits, # 2 - transformed with DiagPooling with $s = 2$, # 3 - down-sampling from # 2 with same pooling, and # 4 - pooled from # 1 with $s = 4$.

It's common to use intermediate normalization of data in CNN. The most usual case is using batch normalization [6], but in a combination with DiagPooling, this method won't give the expected increase in the model's performance. However, we discovered that using layer normalization [1] or transposed instance normalization has a great impact on the performance of CNN with DiagPooling.

Using layer normalization means that one should consider all eigenvalues from each feature map from the diagonalized pooling. Formally, an operation is given in Eq. 7. In that case, mean and variance are scalars used to normalize all values of the resulting tensor $\mathcal{D}$.

$$\widehat{\mathcal{D}} \leftarrow \frac{\mathcal{D} - \mathbb{E}[\mathcal{D}]}{\sqrt{\mathrm{Var}[\mathcal{D}] + \varepsilon}} \tag{7}$$

The idea of transposed instance normalization is to normalize eigenvalues from every channel of a single pixel. For pooled tensor $\mathcal{D} \in \mathrm{R}^{c \times p \times q}$, this method individually normalizes $m \cdot n$ vector $\mathcal{C}_{i,j} \in \mathrm{R}^{c}$ by formula 8.

$$\widehat{\mathcal{D}}_{[:,i,j]} \leftarrow \frac{\mathcal{D}_{[:,i,j]} - \mathbb{E}[\mathcal{D}_{[:,i,j]}]}{\sqrt{\mathrm{Var}[\mathcal{D}_{[:,i,j]}] + \varepsilon}} \; \Bigg| \; i = \overline{1, p}, \; j = \overline{1, q} \tag{8}$$

Regardless of the normalization method, we suppose to use scaling and centering from [6] with trainable per-channel parameters alpha and beta. The final result of DiagPooling with normalization is a tensor $\alpha \widehat{\mathcal{D}} + \beta$ (operations of element-wise multiplication and addition are performed with the compatibility of tensor's shapes).

We recommend using intermediate normalization right after the pooling of eigenvalues. That is why we include the normalization method into the full algorithm of the proposed pooling method formed in Table 1.

Table 1. Algorithm of DiagPooling with normalization.

Algorithm 1: DiagPooling

Input	: Result of convolution as tensor, with c channels and $n \times m$ spatial dimensionalities $D \in \mathbb{R}^{c \times m \times n}$
Input	: Pooling kernel size s
Input	: Normalization method Norm, e.g. (7), (8)
Require:	$m \mod s = 0$, $n \mod s = 0$
Output	: Resulting tensor $\widehat{\mathcal{D}}$

$p \leftarrow \dfrac{m}{s}, \; q \leftarrow \dfrac{n}{s}$

init $\mathcal{D} \in \mathbb{R}^{c \times p \times q}$

for $k = 1$ **to** c :

$\quad O^{[k]} \leftarrow \mathcal{D}_{[k,:,:]}$

$\quad$ **for** $i = 1$ **to** p :

$\quad\quad$ **for** $j = 1$ **to** q :

$\quad\quad\quad \mathcal{O}_{i,j} \leftarrow O^{[k]}_{[(i-1)s..is,(j-1)s..js]}$ $\qquad\qquad$ ▷ $\mathcal{O}_{ij} \in \mathbb{R}^{s \times s}$

$\quad\quad\quad \Theta_{i,j} \leftarrow \mathcal{O}_{i,j}\mathcal{O}^{\top}_{i,j} + \varepsilon I_s$

$\quad\quad\quad \mathrm{diag}(\{\omega_k^{ij}\}) = T^{-1}\Theta_{i,j}T$ $\qquad$ ▷ $\{\omega_k^{ij}\}$ - **eigenvalues of** Θ

$\quad\quad\quad \mathcal{D}_{[k,i,j]} \leftarrow \max\{\omega_k^{ij}\}$

$\widehat{\mathcal{D}} \leftarrow \mathrm{Norm}(\mathcal{D})$ $\qquad\qquad$ ▷ **normalization with scaling and centering**

Even though DiagPooling was designed as an alternative or a replacement for standard pooling layers in a convolutional block of CCN, it also can be used as a transformation or preprocessing layer. For that, our solution could be used in combination with already defined poolings. We consider only a few numbers of those combinations, shown in Fig. 4.

For instance, as mentioned, DiagPooling can transform input images with a reduction of input dimensionalities and eventual normalization. Another idea - use DiagPooling instead of global pooling methods for processing convolved data before feeding it to the classification head. Of course, one can use both of described technics simultaneously.

4 Computer Experiments

For the experiment, we taught several different models on an MNIST dataset. The dataset contains 60000 training single-channeled images and 10000 validation images. All performances were measured on the validation data.

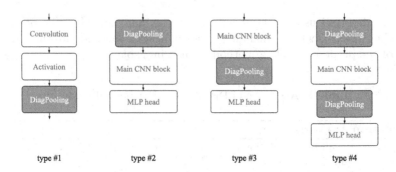

Fig. 4. Usage types of DiagPooling within a CNN. Type # 1 - replace {max, avg, min}Poolings with DiagPooling in the convolutional block, type # 2 - input transformation, # 3 transformation of convolution output, and type # 4 - combination of # 2 and # 3.

We used TensorFlow for the definition and training of the models. All models have many equal hyperparameters, namely: (1) activation function after convolution layer - LeakyReLU, (2) kernel size of convolution layers is 3, number of filters - 16, (3) head's hidden layer has 64 units with elu activation function, (4) optimizer - Adam, (5) batch size is 64, (6) number of epochs is 15.

We created two types of models. The first one has more parameters and a classic CNN structure. In contrast, the second one uses poolings as a dimensionality reduction layer or global pooling, collapsing the spatial axis, and therefore has less trainable parameters. There are five models of the first type and three of the second.

Each model has two convolutional blocks, and each block has a convolutional layer, activation layer, pooling, and normalization. After those blocks, goes operation of flattening and classification head.

The $M1$ model uses our DiagPooling and layer normalization. The $M2$ model uses DiagPooling and transposed instance normalization method. Model $M3$ and $M4$ have max pooling, but the $M3$ uses batch normalization, while $M4$ - layer normalization. Lastly, the $M5$ model uses Average pooling with batch normalization.

Models of the second type also have two convolutional blocks. Models $m1$ and $m3$ use pooling layers with kernel size two as a dimensionality-reducing layer and has a convolutional block with max pooling and batch normalization. The difference is that $m1$ uses DiagPooling, and $m3$ uses max pool. The last model - $m2$ doesn't transform the input images, but it uses our pooling method as a global pooling of the features after two convolutional blocks.

The number of trainable parameters, the resulting accuracy of the models, and also time of training per epoch on an MNIST dataset can be seen in Table 2. Also, the learning curves of the models on validation data are shown in Fig. 5.

As it's seen, models $M1$ and $M2$ have better performance than others ($M3 - 5$). Also models with our solution, except for the $m2$, keep "stable" learn-

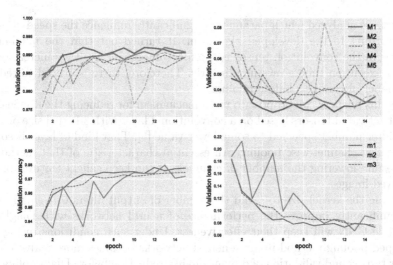

Fig. 5. Learning curves of models on validation MNIST data.

ing behavior longer the classic models. Because of that "stability" and constant superiority for almost all epochs, they require a lesser number of epochs to get a desirable result.

There is an important note on the models of the second type. The main purpose of testing those models is to get the benefits of DiagPooling without making computations much heavier.

It's obvious, that diagonalized pooling has a worse time complexity than max pooling. But, as it's seen from the results, we are able to insert DiagPooling only in the specific places within a CNN to get increase the performance (e.g., an advantage of m1 over the m3).

However, to use DiagPooling in CNN entirely instead of max pooling, there is a need to build a mechanism for reducing the necessary computing resource.

Table 2. Performance and number of trainable parameters of the defined models Mi and mi.

Model	Epoch 1	Epoch 5	Epoch 15	# of parameters	Time per epoch
m3 (max reducing pool)	94.3	96.6	97.5	$4.282 \cdot 10^3$	11.0 s
m2 (our global pool)	94.4	93.6	97.2	$4.284 \cdot 10^3$	29.9 s
m1 (our reducing pool)	94.4	97.3	97.8	$4.284 \cdot 10^3$	33.3 s
M5 (avg+BN)	97.6	98.8	98.9	$1.9642 \cdot 10^4$	24.5 s
M4 (max+LN)	98.0	98.7	98.9	$1.9642 \cdot 10^4$	28.8 s
M3 (max+BN)	98.3	98.7	98.9	$1.9642 \cdot 10^4$	24.6 s
M2 (our+TrIN)	98.4	98.9	**99.1**	$1.9642 \cdot 10^4$	593.3 s
M1 (our+LN)	**98.5**	**99.2**	99.0	$1.9642 \cdot 10^4$	591.2 s

First, we analyzed which parameters significantly influence the speed of pooling according to the developed algorithm. It turned out that the number of channels in the convolution tensor is the most important. As they increase, the pooling execution time grows non-linearly. Therefore, there is a need to reduce the number of channels before pooling.

For this purpose, we propose to use a mechanism for reducing the dimensionality of activated tensors based on a convolutional layer with a kernel size of 1×1 and a filter number specified to some lower value [5]. That kind of layer is sometimes called channel-wise pooling. An essential characteristic of that method is that despite lowering the number of channels, it can keep salient features of the convolved images.

Setting the kernel size to 1 and the number of output filters to a small value will significantly improve the performance of neural networks with DiagPooling, while CNN will keep their effectiveness. Under such conditions, using the developed pooling method in large neural networks becomes more realistic.

For testing and validation of dimensionality reduction before DiagPooling, we built and trained a CNN for image classification based on the CIFAR10 dataset.

Similar to the previously discussed Mi models, we developed networks with DiagPooling, maximum and average poolings with different normalization methods. A total of 5 models were created - $C1 - 5$.

Similar to the previous models, the new ones also have the same set of specific hyperparameters, particular: (1) activation function after convolution layer - LeakyReLU, (3) head's hidden layer has 256 units with *softplus* activation function, (4) optimizer - Adam, (5) batch size is 128, (6) number of epochs is 25.

The designed networks have simple morphologies. We included in those models several convolutional blocks, each of them, in general, consists of: (1) two convolution layers connected in series, with the exact kernel sizes but different output filter sizes, (2) a convolutional layer with a single kernel designed to reduce dimensionality (optional layer and can be omitted), (3) pooling layer, (4) normalization layer (optional layer and can be omitted).

In the C1 model, we use our pooling algorithm with layer normalization. It consists of three convolution blocks: the first two use dimensionality reduction and normalization, and the last block does not reduce dimensionality and does not use normalization. We designed the last block so that DiagPooling plays the role of global pooling (see type #3 from Fig. 4).

The C2 model is a copy of the C1 but with a different normalization method - TrIN.

The C3 and C4 models are analogs of the C1 model. However, the pooling method is max, and the normalization methods are layer and batch, respectively. Apparently, global max pooling is used after the last block in those models.

The C5 model is an analog of the C4 model, but the chosen pooling method is the average one.

The resulting accuracy of the models, and also time of training per epoch on an MNIST dataset can be seen in Table 3. Also, the learning curves of the models on validation data are shown in Fig. 6. The number of trainable parameters were equal.

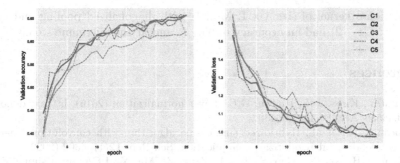

Fig. 6. Learning curves of models on validation CIFAR10 data.

Table 3. Performance and training time estimation of the defined models Ci.

Model	Epoch 10	Epoch 15	Epoch 20	Epoch 25	Time per epoch
C5 (avg+BN)	61.2	62.2	64.3	61.3	66.6 s
C4 (max+LN)	58.0	59.1	61.3	62.1	69.7 s
C3 (max+BN)	60.4	63.5	64.4	65.5	65.9 s
C2 (our+TrIN)	61.3	**63.7**	64.5	**65.7**	309.7 s
C1 (our+LN)	**61.5**	63.5	**64.7**	**65.7**	273.2 s

As it can be seen, again, models with DiagPooling $(C1 - 2)$ have better performance than others $(C3 - 5)$. Learning curves also show similar behavior to MNIST ones, as models with our pooling method seem to have a bit more stable process of learning.

The most significant result from experiments on the CIFAR10 dataset is the success of using dimensionality reduction methods. More complicated and more extensive models $(C1 - 2)$ have shorter training time than models $(M1 - 2)$.

5   Conclusions

A new diagonalized pooling algorithm was developed. We described its interaction with normalization methods and possible positions in CNNs.

We showed that the diagonalized pooling method reduces convolutional feature maps' dimensionality with minimal information loss. In contrast, the max and average poolings reduce the size of images with an inevitable loss.

It was found that if the diagonalized pooling layer is placed after the activation function, the model's performance $(M1 - 2)$ is 99.2%, while the max pooling is 98.7% for the MNIST dataset.

We established that to reduce the training time of the CNN with diagonalized pooling, it is necessary to reduce the number of convolutional channels using the convolutional kernel of size 1×1. Models with diagonalized pooling and such methods $(C1 - 2)$ had an accuracy of 65.7%, and at the maximum - 65.5%.

References

1. Ba, J.L., Kiros, J.R., Hinton, G.E.: Layer normalization (2016). https://doi.org/10.48550/ARXIV.1607.06450
2. Bacanin, N., et al.: Hybridized sine cosine algorithm with convolutional neural networks dropout regularization application. Sci. Rep. **12**(1), 6302 (2022)
3. Chan, T.H., Jia, K., Gao, S., Lu, J., Zeng, Z., Ma, Y.: PCANet: a simple deep learning baseline for image classification? IEEE Trans. Image Process. **24**(12), 5017–5032 (2015). https://doi.org/10.1109/TIP.2015.2475625
4. Gholamalinezhad, H., Khosravi, H.: Pooling methods in deep neural networks, a review (2020). https://doi.org/10.48550/ARXIV.2009.07485
5. He, K., Zhang, X., Ren, S., Sun, J.: Deep residual learning for image recognition. CoRR abs/1512.03385 (2015). http://arxiv.org/abs/1512.03385
6. Ioffe, S., Szegedy, C.: Batch normalization: accelerating deep network training by reducing internal covariate shift (2015). https://doi.org/10.48550/ARXIV.1502.03167
7. LeCun, Y., et al.: Handwritten digit recognition with a back-propagation network. In: Touretzky, D. (ed.) Advances in Neural Information Processing Systems, vol. 2. Morgan-Kaufmann (1989). https://proceedings.neurips.cc/paper/1989/file/53c3bce66e43be4f209556518c2fcb54-Paper.pdf
8. Lee, C.Y., Gallagher, P.W., Tu, Z.: Generalizing pooling functions in convolutional neural networks: mixed, gated, and tree (2015). https://doi.org/10.48550/ARXIV.1509.08985
9. Peleshchak, R., Lytvyn, V., Peleshchak, I., Doroshenko, M., Olyvko, R.: Hechth-Nielsen theorem for a modified neural network with diagonal synaptic connections. Math. Model. Comput. **6**(1), 101–108 (2019). https://doi.org/10.23939/mmc2019.01.101
10. Peleshchak, R., Lytvyn, V., Peleshchak, I., Olyvko, R., Korniak, J.: Decision making model based on neural network with diagonalized synaptic connections. In: Świątek, J., Borzemski, L., Wilimowska, Z. (eds.) ISAT 2018. AISC, vol. 853, pp. 321–329. Springer, Cham (2019). https://doi.org/10.1007/978-3-319-99996-8_29
11. Pennington, J., Schoenholz, S., Ganguli, S.: Resurrecting the sigmoid in deep learning through dynamical isometry: theory and practice. In: Guyon, I., et al. (eds.) Advances in Neural Information Processing Systems, vol. 30. Curran Associates, Inc. (2017). https://proceedings.neurips.cc/paper/2017/file/d9fc0cdb67638d50f411432d0d41d0ba-Paper.pdf
12. Rippel, O., Snoek, J., Adams, R.P.: Spectral representations for convolutional neural networks (2015). https://doi.org/10.48550/ARXIV.1506.03767
13. Sermanet, P., Chintala, S., LeCun, Y.: Convolutional neural networks applied to house numbers digit classification. In: Proceedings of the 21st International Conference on Pattern Recognition (ICPR 2012), pp. 3288–3291 (2012)
14. Torralba, A., Oliva, A.: Statistics of natural image categories. Netw. Comput. Neural Syst. **14**(3), 391–412 (2003). https://doi.org/10.1088/0954-898X_14_3_302. pMID: 12938764
15. Zeiler, M.D., Fergus, R.: Stochastic pooling for regularization of deep convolutional neural networks (2013). https://doi.org/10.48550/ARXIV.1301.3557

Challenges and Opportunities in Deep Learning Driven Fashion Design and Textiles Patterns Development

Dana Simian[✉][iD] and Felix Husac

Faculty of Sciences, Research Center in Informatics and Information Technology,
Lucian Blaga University of Sibiu, Sibiu, Romania
dana.simian@ulbsibiu.ro

Abstract. Creative industries were thought to be the most difficult avenue for Computer Science to enter and to perform well at. Fashion is an integral part of day to day life, one necessary both for displaying style, feelings and conveying artistic emotions, and for simply serving the purely functional purpose of keeping our bodies warm and protected from external factors. The Covid-19 pandemic has accelerated several trends that had been forming in the clothing and textile industry. With the large-scale adoption of Artificial Intelligence (AI) and Deep Learning technologies, the fashion industry is at a turning point. AI is now in charge of supervising the supply chain, manufacturing, delivery, marketing and targeted advertising for clothes and wearable and could soon replace designers too. Clothing design for purely digital environments such as the Metaverse, different games and other on-line specific activities is a niche with a huge potential for market growth. This article wishes to explain the way in which Big Data and Machine Learning are used to solve important issues in the fashion industry in the post-Covid context and to explore the future of clothing and apparel design via artificial generative design. We aim to explore the new opportunities offered to the development of the fashion industry and textile patterns by using of the generative models. The article focuses especially on Generative Adversarial Networks (GAN) but also briefly analyzes other generative models, their advantages and shortcomings. To this regard, we undertook several experiments that highlighted some disadvantages of GANs. Finally, we suggest future research niches and possible hindrances that an end user might face when trying to generate their own fashion models using generative deep learning technologies.

Keywords: GAN · Fashion industry · Deep learning

1 Introduction

In the last decade, Machine Learning (ML) techniques were successfully used to solve complex classification and regression problems in different fields. Like

D. Simian and L. F. Stoica (Eds.): MDIS 2022, CCIS 1761, pp. 173–187, 2023.
https://doi.org/10.1007/978-3-031-27034-5_12

the optimization field, ML is governed by a No Free Lunch theorem, meaning that the average of computational performances of ML algorithms on all classes of problems is the same [30]. Therefore many research articles are devoted to propose new ML algorithms for specific type of data or to compare the performances of different existing ML algorithms on different classes of problems. The Covid 19 pandemic leads to many studies related to Covid 19 diagnosis and prediction [19]. Many ML applications can be found in computer vision [29], health, business, social media, cyber security [26], etc. The Covid-19 pandemic also accelerated the adoption of the Artificial Intelligent (AI) technologies in the fashion industry: in the beginning only for analytics and afterwards for virtualization of fashion collections and events [27] and for trend fashion forecasting [36].

Artificial Neural Networks (ANNs) simulate the human brain behaviour. The backforward ANNs [22] evolved as a result of the increase in the complexity of the problems that wanted to be solved, the availability of an increased volume of data, often unstructured and of high complexity, and of the increase in the computing power of computers. Thus, more efficient ANN architectures were developed, leading to the apparition of a new field in ML known as Deep learning. Initially Deep Neural Networks (DNN) were ANNs with a large number of hidden layers, allowing the complexity of the trained models to increase. Currently there is a wide range of DNN architectures [15,33] including Convolutional Neural Networks (CNN), Recurrent Neural Networks (RNN), Graph Neural Networks (GNN) Generative Adversarial Networks, etc. The superiority of DNNs over ANNs is given by several capabilities than can be highlighted in different DNN architectures:

- They are scalable (their performances does not reach a flat plateau when the volume of date is increasing)
- They can handle raw data performing feature learning (automatic features extraction)
- They are able to learn hierarchical representations

Another direction of evolution for ANN architectures was oriented through generative models. In 2013 Kingma and Welling proposed a generative model named Variational Auto Encoders (VAE) [14]. An autoencoder consists of two neural networks, an encoder and a decoder. The encoder reduces the dimensionality of the data into a latent space representation. The decoder rebuilds the data, given the latent representation. VAEs are used with success for image processing [8] and natural language processing [2]. Another generative neural network is represented by the Generative Adversarial Networks (GAN), introduced by Ian Goodfellow et al. [7]. A GAN train simultaneously a generative network and a discriminative network using a min-max two-player game model.

One of the important application of the generative models consists in artificial data synthesis. Image, text or audio generation can be obtained based on a training set of examples [2,4,5,9,32].

Any domain that can benefit from artificial data synthesis is a good candidate for the use of deep learning generative models.

The fashion and textile industries are two connected fields that aim to design clothing along with the specific pattern and colors of the materials used. The design relies on the fashion trend, on the already existing collections of clothing and materials and on the customers preferences. Therefore, the use of generative models in fashion design increased in interest nowadays.

The aim of this article is to investigate different generative models that were used or could be used for transposing clothing design in a virtual environment. We also provide an overview on the use of deep learning generative models for clothing design, focusing especially on the main types of GANs used for this purpose. Section 2 of the article briefly present different GAN architectures and their advantages over variational autoencoders. Some state of the art deep learning generative models already implemented in fashion design are presented in Sect. 3. Thus we can have an overview of the particular requirements imposed by the clothing design process and at the same time we can follow the trend of the involvement of generative models in this process.

Another goal of the article is to identify possible issues that an individual fashion creator can encounter in applying generative models on his own. To this end, in Sect. 4, we present a series of our own experiments, highlighting the obtained results as well as some drawbacks when trying to create clothing item designs using freely available platforms and technologies.

We consider our study of interest from many reasons. First of all, the involvement of generative models in apparel design paves the way for AI approaches into human creativity tasks, being a sensitive subject for artists in general.[1] Secondly, the fashion industry has the necessary funds that can accelerate research in the mentioned direction, opening a niche with potential for future development. In recent years, the fashion industry had a good financial trend. In 2019, the fashion industry was valued at 1.5 trillion EUR. At the same time fashion industry development negatively affects the environment. The information provided by the Institute for Sustainable Development and the Environment form Poland (https://izrs.eu/) shows that it is the second most polluting industry worldwide after the oil industry, being responsible of about 10% of the total global greenhouse gas emissions. The use of AI and deep generative models to eliminate some operations from the clothing design flow, responsible for increasing pollution, are also of great interest.

2   Generative Adversarial Networks

ANNs can be used both for discriminative and generative tasks. The classical ANNs are used for solving classification and regression tasks and falls in the class of discriminative models. A discriminative model return a label or a prediction for an unseen example based on the training data. A generative model returns

---

[1] Nikoleta Kerinska, professor, PhD in art sciences considers that the computer can simulate creativity using AI, but does not have from the beginning an artistic goal, as humans have. https://www.heuritech.com/articles/fashion-solutions/artificial-intelligence-fashion-creativity/.

a probability for a given example based on a distribution of a dataset. Generative models are statistical models that are able to generate new instances. One important step toward generative models based on ANN is represented by the Variational Auto Encoders (VAE). VAE are obtained by using an ANN as probabilistic encoder in the Auto-Encoding Variational Bayes (AEVB) algorithm introduced in [14]. AEVB performs an efficient inference using the Stochastic Gradient Variational Bayes (SGVB) estimator to optimize a recognition model. At the same time the algorithm allows learning the model parameters more efficient than other previous inference scheme. A VAE consists of two neural networks, an encoder and a decoder. The encoder maps the samples in the dataset in a lower dimensionality latent space. The decoder, decompress a latent space sample, called a noise vector, into the original feature space. The VAE aims for a low reconstruction error of the data.

GANs are deep learning generative models [7]. A GAN architecture is composed by two DNNs, a Generator (G) and a Discriminator (D), trained together in a two players zero-sum game (Fig. 1). The generator generates fakes images. The Discriminator receive both real and fake images and need to discriminate between the two classes. The Generator goal is to full the Discriminator such that more fake sample be classified as real ones. Both the Discriminator and the Generator used the same loss function. The loss function can be the absolute error or the binary cross entropy loss function of the discriminator. When training the discriminator we aim to minimize this error and when training the generator we aim to maximize it. As a result, training a GAN architecture is comparable to a zero-sum game in which players try to minimize their own losses or penalties while maximizing the enemy's loss.

In an ideal case the model should reach convergence, or Nash equilibrium. This means that the generated sample cannot be distinguished from a real world sample and the the the synthetic samples produced by the Generator, when run through the discriminator, receive a 50% probability score.

Convergence can occasionally be faked by the Generator. This phenomenon is called mode collapse. Mode collapse is one of the most common problems with GAN design in general. The generator repeatedly provides the same images for which it received a high accuracy score as feedback, the game stops and no network can make progress.

GANs have enjoyed tremendous popularity from the research community, with more than 360 papers published on GANs every month in 2018 and the publication trend in the field of GANs continues to rise also in present. This sustained interest in adversarial networks brought along many improvements to the GAN architecture. In less than a decade since their invention, GANs have evolved from creating 28×28 pixel grayscale images to pictures of over 1MP [18]. Several GAN architectures that helped push the research forward in the generative AI field are briefly mentioned.

Conditional GANs were introduced in 2014 [20]. They can generate examples from an imposed domain. To this regard, additional information has to be provided to both Generator and Discriminator. By example, the input data of

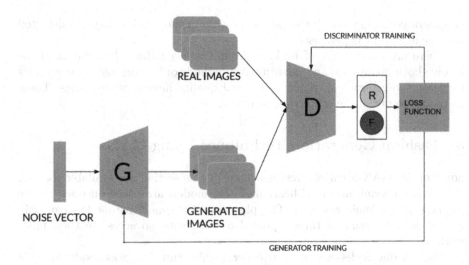

Fig. 1. Classic GAN architecture.

the Discriminator are labeled and the Generator provides synthetic or fake data together with their labels [37]. Conditional GANs can be used in image-to-image translation, like changing the season, or changing an image from day to night. The Deep Convolutionary GAN (DC-GAN) improves the classical GAN architecture by using convolutional neural networks as discriminator and generator, and batch normalization in between each convolution layer except the first one [21]. Wasserstein GAN (WGAN) [1] uses the earth mover's distance as its loss function which describes the gap between two probability distributions. The Wasserstein GAN model's loss function is highly correlated to the quality of the synthetic images. The Wasserstein GAN approach is still considered state-of-the-art even today, because its loss function behaves similar to a human critic. Cycle-GAN [37] was a notable improvement in the world of image-to-image transfer, making possible the training of the image-to-image transfer model without paired images. The idea of progressively growing both the generator and the discriminator led to the development of Progressive GAN [12]. In a Progressive GAN, the training process starts from a low resolution, adding new layers during the training, to obtain finer and finer details. In 2018, a notable contribution was made by Nvidia's StyleGAN architecture, which is based on a Progressive GAN model, capable to train on smaller sized datasets [13]. StyleGAN was released as an open-source project one year later.

More recently, researchers focused on models such as Evolutionary and Energy-Based GAN architectures in order to escape mode collapse. Evolutionary GAN relies upon creating populations of generators and choosing the fittest one. The Energy-Based approach, namely EBGAN-PT (Energy-Based GAN with Pulling-away Term) stops the discriminator from training faster than the generator and avoids mode collapse by computing a vector cosine similarity between all generated images [35]. If the values get close to one, mode collapse is detected and the generator is penalised.

There are many ways of ranking generative algorithms. Ranking methods involve both qualitative and quantitative methods of judging synthetic samples for feature diversity, coherence and visual quality among other things. These methods are out of the scope of this article.

3   Fashion Generation Techniques Using GANs

Some of the GAN architectures presented in this section are available for use, reverse engineering and modification. These models are based on open source projects or academic research. The plug-and-play generative models can only be reviewed in terms of their reported output, with no access to their inner workings.

The Nvidia StyleGAN style transfer architecture [13] was modified by a research team from Berlin-based Zalando, a startup focused on fashion and AI. The aim of their research was to create photorealistic virtual models, with accurate body poses, that showcase the clothing in the best way. The researchers created their own dataset, containing over 380000 image sets. Each image set is composed from a picture of a posing model and six other pictures of individual clothing items worn by the model. They tested both a conditional and an unconditional approach, with the latter producing better results [39]. Zalando intends to employ this technique to provide virtual try-on environments for customers.

A research team at Amazon used a conditional Stack GAN based architecture called Reccurrent Stackable GAN (ReSt-GAN) to create images of pants, shorts and jeans. A Stack GAN is made up of two completely functional GAN models. One GAN produces a low resolution seed image, and the other is tasked to scale the image to a higher resolution and to add the requested features. Human agents can request different features to be present in the synthetic images by feeding a text prompt to the algorithm. The training process was unsupervised, the dataset used containing no further labelling of the items other than the product title. To improve the visual performance of the model, researchers introduced a secondary classifier that checked if the generated images matched the textual description in type, gender and colour. In order to improve the representation of color shades such that they match their textual representation, a new colorspace was developed [31].

Attribute GAN (AttGAN) is a great candidate for facial editing applications. This architecture allows finer granularity control for the human agent to change specific features and influence the desired outcome of the synthetic image. C. Yuan and M. Moghaddan applied the AttGAN approach to fashion image editing. The results they witnessed were less than satisfactory. The authors hypothesised that the reason for which the architecture fails to edit fashion images and excels at editing faces is the relatively large area of a garment in comparison to the size of a human face. To improve on their shortcoming, they created Design-AttGAN. Design-AttGAN outperformed their previous experiment in term of results. Design-AttGAN inherits one severe limitation from its predecessor: the performance of the model is highly dependent on the frequency with which a certain feature repeats itself in the dataset [40].

The conditional AttGAN architecture was also employed by L. Liu et al., with the purpose of integrating different types of clothing items and creating an appealing, harmonious outfit. The model contains two discriminators: a colocation discriminator and an attribute discriminator. The output of their algorithm is a collection of pictures showing what the GAN has learned to be harmonious, coherent and tasteful pairs of fashion items [17].

Attention GAN (AttnGAN) is an improvement of AttGAN and focuses on finding visual elements that correspond to keywords given as a prompt [38]. Staked Generative Adversarial Networks (StackGAN) is another GAN architecture proposed to generate 256×256 artificial photos conditioned on their textual descriptions [34]. The model is based on a new Conditioning Augmentation technique.

Fayyaz et al. [6] proved that the performance of generative models can be enhanced by cleaning an existing dataset. They first test their hypothesis on a classification task, where the cleaned and pseudo-annotated dataset provided a 2% increase in prediction accuracy. They proposed another variant of the dataset as a training base for state-of-the-art generative algorithms such as DC-GAN, Wasserstein GAN with gradient penalty (WGAN-GP) and a convolutional variational autoencoder. The best synthetic results in their study belong to the WGAN-GP architecture.

Shawon et al., in [28] used a state-of-the-art conditional image to image translation architecture GAN called Pix-2-Pix [11] to create traditional Bangladesh Jamdani textures from a rough, pencil and paper sketch. Scarfs and dresses are decorated with the vibrant woven silk fabric motifs known as jamdani. To train the Pix-2-Pix algorithm, the researchers generated a sizable collection of Jamdani patterns. With this method, new designs can be tested quickly and as proof-of-concepts using only a quick hand drawn sketch as the starting point.

Cui et. al. [3], proposed a GAN architecture named FashionGAN to provide a virtual garment display method based on the end users preferences. Fashion-Gan uses a Cycle GAN. The end users need only to chose and introduce as input a desired sketch of their desired outfit and a texture sample. FashionGAN was tested against five other cutting-edge GAN models, receiving an equal or occasionally lower FID (Frechet Inception Distance) score. Human judges were requested to assign a score of 0 (least satisfactory) to 5 (most satisfactory) for each of the synthetic image batches produced by each of the tested GAN architectures. The human participants consistently appreciated the results obtained using FashionGAN with a better score.

J. Liu et al. built a Multi-modal Generative Compatibility Modelling system (MGCM) based on the GAN architecture, in order to assemble complementary pairings between the top portion of an outfit and the bottom part (for example matching shorts with a t-shirt). The proposed architecture solves an image to image translation task, and is based on a conditional GAN architecture. Conditions are described both visually and textually in the descriptions of each article of apparel. The researchers used compatibility restrictions between items and between items and templates, highlighting high-level similarities for the fashion items in the subdomain divided dataset (the top part of the outfit and the bottom part). Their method provides accurate visual representations of the compatible object and also impresses with how stylistically similar the generated samples are to the original image pair. Comparisons were carried out for 8 state-of-the-art generative approaches, across two datasets [16].

Ready-made generative technologies are powerful solutions that are designed to be easy to use even by everyday, non technical people. They rely on conditional algorithms and the output can be controlled via a text prompt.

OpenAI released DALL-E in 2021 [23], a conditional generative model that users could interact with and request images by typing short descriptions. Several months later, in the spring of 2022, DALL-E2 was launched, bringing significant improvements over the previous architecture [24]. One month later, Google launched Imagen [25], a generative solution based on diffusion models and VAEs. Imagen was proven to be better at understanding text prompts and at generating correct specular highlights and translucent materials than previous attempts. As a complementary research, two months later, Google released Parti (Pathways Autoregressive Text-to-Image) [41]. Parti was ranked even higher than Imagen, and it was based on a Vector Quantised GAN (VQGAN). While Google will not be releasing its models to the public, OpenAI opened their models to researchers and faculty, before planning to roll them out as a subscription service to broader consumers. A very impressive plug and play solution is a diffusion model called Stable Diffusion. It is impressive not necessarily by the stunning images it produces, but by the fact that the researchers have made it available free and open source (https://stablediffusionweb.com/). Also, Stable Diffusion can be run on a consumer machine, which is a novelty in the world of synthetic data generation.

4   Own Experiments

The big fashion houses have the financial resources and the potential to access the latest solutions offered by IT companies in the field of generative models for design or virtual visualization. We want to present some options that an ordinary user (consumer or individual fashion creator) has to create his own fashion style or even just a few clothing items using GANs and other generative models. For this, we considered some experiments carried out by us previously [10] and analyzed them from this perspective. We have also highlighted the most important difficulties that those who try to use the options proposed by us could face.

We focused on existing open-source web tools and free online work environments that can allow users to create their own fashion designs.

To create personalised designs, the end user needs to compile a dataset with the desired clothing items, or to pick an existent one. The most straightforward way of gathering a large number of images and creating a new dataset is through a technique called webscraping. Webscraping can be done manually, or via robots - scripts that automatically filter pages of a website for a specific kind of information. The target website for our data acquisition test was online retail giant's Amazon men's clothing page. When designing a webscraping robot, one has to take into account the robots.txt file of a domain in order to safely and legally gather data from that website. The approach used to create a webscraping script in this instance uses the Scrapy Python library. This method is the quickest and simplest to implement and deploy. The advantages of using a designated scraping library is that Scrapy's user agent lets the webserver know that its not a regular user and the server can decide how many requests per second the script can make to access the needed data. Scrapy also tightly enforces robots.txt compliant behaviour out-of-the-box. A Google Chrome plugin called CSS Selector was used to highlight the relevant information the webscraping script needs to gather from the webpage, namely the image URL and item name. The data collecting script used by the web scraper looks for the CSS selectors previously discovered by the browser plugin and stores the information contained within the HTML elements into a python list that can be displayed to the console or stored in a .CSV file. The file containing the scraped image URLs is then loaded in another Python script that will download every picture locally. The user can then sort through and clean the dataset, or add new desired images to it. In order to create a dataset compatible with the Google Colab online working environment, Kaggle's automatic dataset creation feature was used to make a private dataset.

In order to gather accurate results and properly interpret them, we chose a common benchmark dataset for preliminary training of the GAN algorithms. The chosen dataset is the Fashion_MNIST dataset, created by Zalando. It contains over 70.000 grayscale images of different fashion items and clothes, 28×28 pixels each. We chose three unconditional GAN models: the classic GAN (with binary cross entropy loss function), Deep Convolutional GAN (DC-GAN) and the Wasserstein GAN (WGAN). As training environment we used Google Colab Free, and we trained the models for 20.000 epocs for each model. We tuned by hand only the number of epochs and the batch size, leaving the other hyperparameters at their default values.

Training the algorithms on a smaller scale revealed some shortcomings of this approach. Problems arise both from the generative models themselves and from the nature of the training environment. The three GAN models were trained on the same hardware configuration provided by Google Collab - Nvidia Tesla K80, 12 GB RAM. The average training time for each model was about 5 h. Classic GAN was the fastest in training, followed by DC-GAN and finally WGAN. The classic GAN architecture was also tested using a CPU-only training run, however the training time was significantly longer. Figure 2 shows some of the results returned, after training, from the three models. The visual results are quite acceptable given the training environment constraints that needed to be followed, and the fact that we did not work much on the hyperparameters optimisation.

Analysing all the images generated during our experiment, we observed that the classic GAN architecture was more stable than its counterparts. Training

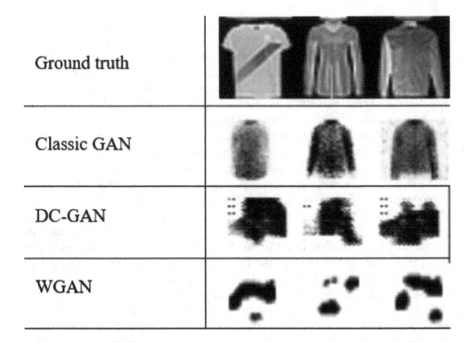

Fig. 2. Selection of visual results produced by the three tested GAN architectures.

the classic GAN took a little less time than DC-GAN and considerably less time than the WGAN approach. The classic GAN model shows higher visual quality output than the other two models, but it is obvious that it suffers from mode collapse. The Fashion_MNIST dataset contains images of shoes, boots, t-shirts, jeans, handbags etc., features not present in the output of the synthetic results of the classic GAN. DC-GAN generated images that do not belong to the real distribution. We can find by example we find sweaters with one sleeve. The model overgeneralizes. Sensitivity studies on the influence of the training time and on the influence of different hyperparameters on DC-Gan performances could lead to finding solutions to improve the results. The WGAN approach suffered from non convergence, giving incoherent results across its whole training. It becomes obvious that further work is still required to transform it into a more stable variant.

Next we will summarize the main drawbacks of using the generative models from our experiment in the simple chosen environment.

– Mode collapse - meaning that the distributions of the output features do not match the distribution of real data features. This means that the Generator network has learned a shortcut to achieving a good probability result for its synthetic images by repeating the same image over and over again.
– Slow convergence - long training times combined with GAN's increased sensitivity to hyperparameters can often lead to a slow convergence and even

non convergence over the span of a training run. Finding the ideal set of hyperparameters for each model requires more work.
- Overgeneralization - takes place when the generator creates feature pairs that do not coexist in the training dataset, such as creating a t-shirt with five arms or a half a hoodie.

Lack of standardised testing can be perceived as a standalone problem for ranking not only GAN models, but any generative deep learning techniques. With multiple evaluation metrics to choose from, it is difficult for researchers to find and use the correct ranking methods for their algorithms.

Finding a new cloud computing platform to execute the training sequences on is a much needed enhancement that will considerably speed up and, ideally, improve the performance of the GAN models. The main disadvantages regarding the training environment come from the lack of alternatives for a simple end user with a modest machine. Online server renting services are expensive, and they require additional knowledge to set up and run. Google Collab is a good alternative to expensive services, but it also has its downsides. The free version of this service has limited time sessions and can even disconnect a running session prematurely if it detects no user interaction with the page in which it is running. Paid tiers of Google Collab, like Google Collab Pro or Pro+ are affordable to buy, but are available only in select countries and do not offer a compelling set of advantages over the free version.

5   Other Directions of Study

Generative Adversarial Neural Networks took less than ten years since their introduction in 2014 to significantly impact the artificial intelligence research community. GANs are used to create synthetic training data for other machine learning models where real world data is insufficient, nonexistent or of poor quality [10]. The years 2021 and 2022 have seen a tremendous leap forward in the visual quality, sample diversity and a deeper understanding of natural language by out-of-the-box generative solutions. There are plenty of practical research avenues in data augmentation and restoration that can benefit from the implementation of GANs in their workflow. This is a turning point in the art world, with many artists seeing themselves under threat from such technologies. GANs inherent shortcomings still make its behaviour unpredictable, but theoretical research into famous problems such as slow convergence and mode collapse will prove to be an important contribution to generative algorithms.

In artistic research, synthetic images can represent a starting point for new designs that can be further expanded upon by a human expert. Deep learning techniques such as those discussed in this paper could also be directly integrated in a design production workflow, such as a print-on-demand business model. GANs could also be used in most interior design applications, from wallpaper designs to car interior designs and upholstery. Generative approaches can help create unique custom digital assets for the gaming industry such as more clothing options for non playable characters (NPCs) and for the crypto market via non fungible tokens (NFTs) [10].

Another industry that can greatly benefit from adopting generative technologies is the image editing software industry. Computer editing programs can benefit from implementing generative algorithms directly inside the editing program. Texture generators and texture brushes created by a deep learning algorithm can mimic a given sample, and can help replace current content-aware fill methods. Generative image inpainting techniques avoid common content-aware fill problems such as smeared edges or repeated objects because the deep learning algorithm takes into account the whole image and not only the pixels in the immediate vicinity of zone to be retouched. Techniques such as super resolution GANs (SRGAN) can help upscale an image both in size and resolution, and image-to-image translation can help morph objects between pictures. Style transfer techniques could then be applied to the overall image in order to give it a look based on a single reference image.

It would be of interest to bring together the achievements in the field of generative models and the ideas and beliefs of the researchers in art sciences. The latter claim that the generative models can only simulate human creativity, but they are far from the way a human artist creates his work, because they cannot envision their final goal of their creative endeavour and blindly combine feature samples. Creating a generative algorithm capable of being more than a simulator of creativity could prove to be a rewarding task and a good starting point for ethical and philosophical debate. Form the legislative point of view, the European Union will impose digital watermarks on all artificially generated content by 2025, and projections predict that by then, 10% of data available on the Internet will have been synthetically generated. Synthetic art, as it stands today, is a worthy opponent to human creatives simply because of the speed with which a deep learning generative algorithm comes up with a novel design or image. Plug-and-play solutions such as the algorithms developed by Google and OpenAI and briefly enumerated in the paper enable every user to come up with their own artistic creations, regardless if they are familiar with machine learning concepts or not.

In the end, these deep learning technologies seek to provide the customer the option of a higher level of design personalisation, whether for branding reasons, such as logo designs, or for home or fashion use [10].

6   Conclusions

In this article, we made a short overview of deep learning generative models, and especially GANs and the current solutions for using these models in fashion design and textile patterns generation. This article presents the evolution and comparison of the main generative models based on GANs, as well as models specially adapted and evolved for the requirements of the fashion industry. We identified opportunities, pitfalls and challenges in the use of generative models based on GANs both for large fashion companies and for individual designers and simple consumers.

In this article we briefly summarised the emergence of generative technologies and the fashion industry's shift towards workflow digitisation, we presented and compared generative adversarial neural networks to variational autoencoders and outlined several important GAN architectures as they evolved along the years. We also discussed the ways in which GANs are being used to aid designers in the fashion industry and we looked into the opportunities that a member of the broader public has when it comes to generating their own fashion designs, while also taking into account the common pitfalls of such an approach. Such breakthroughs and advancements in creating generative technologies that are easy to use by the broader public will be the step to democratising art and design. These technologies could empower end users to create their own fashion designs, among other things, and thus help make the fashion industry more accessible to the broader public.

Such breakthroughs and advancements in creating generative technologies that are easy to use by the broader public will be the step to democratising art and design.

Acknowledgement. The first author, Dana Simian, was supported from the project financed by Lucian Blaga University of Sibiu through the research grant LBUS-IRG-2022-08.

References

1. Arjovsky, M., Chintala, S., Bottou, L.: Wasserstein GAN. arXiv:1701.07875v3 [stat.ML] (2017)
2. Bowman, S.R., et al.: Generating sentences from a continuous space. arXiv:1511.06349 (2016)
3. Cui, Y.R., Liu, Q., Gao, C.Y., Su, Z.: FashionGAN: display your fashion design using conditional generative adversarial nets. Comput. Graph. Forum **37**(7), 109–119 (2018)
4. Donahue, C., McAuley, J., Puckette, M.: Adversarial audio synthesis. In: Proceedings of the International Conference of Learning Representation, ICLR 2019 (2019). arXiv:1802.04208v3 [cs.SD]
5. Engel, J., et al.: GANSynth: adversarial neural audio synthesis. In: Proceedings of the International Conference of Learning Representation, ICLR 2019 (2019). arXiv:1902.08710 [cs.SD]
6. Fayyaz, R.A., Raja, A., Maqbool, M.M., Hanif, M.: Textile design generation using GANs. In: Proceedings of the 2020 IEEE Canadian Conference on Electrical and Computer Engineering (CCECE), pp. 1–5 (2020)
7. Goodfellow I., et. al.: Generative adversarial nets. In: NeurPS Proceedings. Part of Advances in Neural Information Processing Systems, vol. 27 (2014)
8. Gregor, K., et al.: A recurrent neural network for image generation. In: Proceedings of the 32nd International Conference on Machine Learning, in Proceedings of Machine Learning Research, vol. 37, pp. 1462–1471 (2015)
9. Gupta, C., Kamath, P., Wyse, L.: Representations for synthesizing audio textures with generative adversarial networks. arXiv:2103.07390 [eess.AS] (2021)

10. Husac, F.: Using GANs to innovate creative industries: fashion and textile design. In: Proceedings of the International Conference on Applied Informatics, ICDD 2022, Sibiu (2022)
11. Isola P., Zhu J.-Y., Zhou, T., Efros, A.: Image-to-image translation with conditional adversarial networks. In: Proceedings of the 2017 IEEE Conference on Computer Vision and Pattern Recognition, Honolulu, pp. 5967–5976 (2017)
12. Karras, T., Aila, T., Laine, S., Lehtinen, J. : Progressive growing of GANs for improved quality. Stability, and Variation. arXiv:1710.10196 [cs.NE] (2017)
13. Karras, T., Laine, S., Aila, T.: A style-based generator architecture for generative adversarial networks. arXiv:1812.04948 [cs.NE] (2018)
14. Kingma, D.P., Welling, M.: Auto-encoding variational bayes. arXiv:1312.6114 (2014)
15. LeCun, Y., Bengio, Y., Hinton, G.: Deep learning. Nature **521**, 436–444 (2015)
16. Liu, J., Song, X., Chen, Z., Ma, J.: MGCM: multi-modal generative compatibility modeling for clothing matching. Neurocomputing **414**, 215–224 (2020)
17. Liu, L., Zhang, H., Ji, Y., Wu, Q.M.J.: Toward AI fashion design: an attribute-GAN model for clothing match. Neurocomputing **341**, 156–167 (2019)
18. Marchesi, M.: Megapixel size image creation using generative adversarial networks. arXiv:1706.00082 [cs.CV] (2017)
19. Meraihi, Y., Gabis, A.B., Mirjalili, S. et al.: Machine learning-based research for COVID-19 detection. Diagnosis, and prediction: a survey. SN Comput. Sci. **3**, 286 (2022)
20. Mirza, M., Osindero, S.: Conditional generative adversarial nets. arXiv:1411.1784 [cs.LG] (2014)
21. Radford, A., Metz, L., Chintala, S.: Unsupervised representation learning with deep convolutional generative adversarial networks. arXiv:1511.06434 [cs.LG] (2015)
22. Rumelhart, D., Hinton, G., Williams, R.: Learning representations by back-propagating errors. Nature **323**(6088), 533–536 (1986)
23. Ramesh A., et. al.: Zero-shot text-to-image generation. arXiv:2102.12092 (2021)
24. Ramesh, A., et. al: Hierarchical text-conditional image generation with CLIP latents. arXiv:2204.06125 [cs.CV] (2022)
25. Saharia, C., et. al.: Photorealistic text-to-image diffusion models with deep language understanding. arXiv:2205.11487 [cs.CV] (2022)
26. Sarker, I.H.: Machine learning: algorithms, real-world applications and research directions. SN Comput. Sci. **2**, 160 (2021)
27. Särmäkari, N.: Digital 3D Fashion Designers: Cases of Atacac and The Fabricant, Fashion Theory. Francis&Taylor (2021)
28. Shawon, M.T.R., Tanvir, R., Shifa, H.F., Kar, S., Jubair, M.I.: Jamdani motif generation using conditional GAN. In: Proceedings of the 23rd International Conference on Computer and Information Technology (ICCIT), pp. 1–6 (2020)
29. Shekhar, H., Seal, S., Kedia, S., Guha, A.: Survey on applications of machine learning in the field of computer vision. In: Mandal, J.K., Bhattacharya, D. (eds.) Emerging Technology in Modelling and Graphics. AISC, vol. 937, pp. 667–678. Springer, Singapore (2020). https://doi.org/10.1007/978-981-13-7403-6_58
30. Sterkenburg, T.F., Grünwald, P.D.: The no-free-lunch theorems of supervised learning. Synthese **199**, 9979–10015 (2021)
31. Surya, S., Setlur, A., Biswas, A., Negi, S.: ReStGAN: a step towards visually guided shopper experience via text-to-image synthesis. In: Proceedings of the 2020 IEEE Winter Conference on Applications of Computer Vision (WACV), pp. 1189–1197 (2020)

32. Wang, L., et al.: A state-of-the-art review on image synthesis with generative adversarial networks. IEEE Access **8**, 63514–63537 (2020)

33. Wang, X., Zhao, Y., Pourpanah, F.: Recent advances in deep learning. Int. J. Mach. Learn. Cyber. **11**, 747–750 (2020)

34. Zhang, H., et al.: StackGAN: text to photo-realistic image synthesis with stacked generative adversarial networks. arXiv:1612.03242 [cs.CV] (2016)

35. Zhao, J., Mathieu, M., LeCun, Y.: Energy-based generative adversarial network. arXiv:1609.03126 (2016)

36. Zhao, L., Li, M., Sun, P.: Neo-Fashion: a data-driven fashion trend forecasting system using catwalk analysis. Clothing Tex. Res. J. OnlineFirst, 1–16 (2021). https://doi.org/10.1177/0887302x211004299

37. Zhu, J.-Y., Park, T., Isola, P., Efros, A.: Unpaired image-to-image translation using cycle-consistent adversarial networks. In: IEEE International Conference on Computer Vision (ICCV) (2017). arXiv:1703.10593 [cs.CV]

38. Xu, T., et. al.: AttnGAN. Fine-grained text to image generation with attentional generative adversarial networks. In: Proceedings of the IEEE/CVF Conference on Computer Vision and Pattern Recognition (CVPR), pp. 1316–1324 (2018)

39. Yildirim, G., Jetchev, N., Vollgraf, R., Bergmann, U.: Generating high-resolution fashion model images wearing custom outfits. In: International Conference on Computer Vision, ICCV 2019, Workshop on Computer Vision for Fashion, Art and Design (2019). arXiv:1908.08847

40. Yuan, C., Moghaddam, M.: Attribute-aware generative design with generative adversarial networks. IEEE Access **8**, 190710–190721 (2020)

41. Yu, J., et. al.: Scaling autoregressive models for content-rich text-to-image generation. arXiv:2206.10789 [cs.CV] (2022)

Feature Selection and Extreme Learning Machine Tuning by Hybrid Sand Cat Optimization Algorithm for Diabetes Classification

Marko Stankovic[1], Nebojsa Bacanin[1(✉)], Miodrag Zivkovic[1],
Dijana Jovanovic[2], Milos Antonijevic[1], Milos Bukmira[1,2],
and Ivana Strumberger[1]

[1] Singidunum University, Danijelova 32, 11010 Belgrade, Serbia
{marko.stankovic.201,milos.bukmira.20}@singimail.rs,
{nbacanin,mzivkovic,mantonijevic,istrumberger}@singidunum.ac.rs
[2] College of Academic Studies "Dositej", 11000 Belgrade, Serbia
dijana.jovanovic@akademijadositej.edu.rs

Abstract. It is vitally important to establish a system that is able to provide an early detection of diabetes as a stealth disease of modern era. In order to achieve this goal, this manuscript proposes a novel framework for feature selection and extreme learning machine (ELM) hyper-parameter optimization applied to diabetes diagnostics. Feature selection and hyper-parameter optimization are two of the most important challenges in the domain of machine learning and they both belong to the group of NP-hard challenges. An upgraded version of the newly suggested sand cat swarm optimization (SCSO) is developed to address these issues, and adapted for ELM hyper-parameter tuning and feature selection. A preliminary set of biases and weights are established using the proposed approach, as well as the optimal (sub-optimal) no. of neurons for the ELM hidden layer, as well as to establish initial set of biases and weights. Furthermore, each swarm individual also tries to select the most relevant features for classification tasks against a widely utilized diabetes dataset. The performance of proposed methods was compared to other well-known state-of-the-art swarm intelligence algorithms in terms of accuracy, precision, recall and f1 score. Experimental findings demonstrate that the improved SCSO is more efficient than other algorithms in addressing both challenges.

Keywords: Machine learning · Extreme learning machine · Sand cat optimization algorithm · Metaheuristic algorithms · Diabetes classification

1 Introduction

Diabetes mellitus is a class of metabolic disorders that compromise an individual's ability to process glucose in the bloodstream. This disease is characterized

D. Simian and L. F. Stoica (Eds.): MDIS 2022, CCIS 1761, pp. 188–203, 2023.
https://doi.org/10.1007/978-3-031-27034-5_13

by chronic hyperglycemia (high levels of blood glucose) resulting from defects in insulin secretion, insulin action, or both [19]. This is an indispensable anabolic hormone created in the pancreas and responsible for the metabolism of protein, fat and sugar. It allows glucose from the bloodstream to enter the cell where it is converted into usable energy or stored. This finely balanced feedback between cells and insulin is disrupted when there isn't a sufficient amount of insulin produced or the cells don't respond to it, indicating the onset of diabetes.

Unfortunately, despite of many medical advancements, diabetes persists and is one of the most common endocrine diseases of modern times. Major risk factors include diseases, pregnancy, genetics, obesity, drugs, and chemical agents. Diabetes affects many systems in the body, and if left untreated can lead to blindness, loss of limbs, and ultimately death. Early detection can greatly improve the quality of life of patients, as with appropriate treatment many severe complications may be avoided.

In the modern literature, many researches were published that deal with early detection of diabetes from both classification and regression perspectives [6,19] by using machine learning models and techniques (ML). The ML is a category of artificial intelligence (AI) capable of accurately predicting outcomes without explicit programming. Classification in ML is a process of categorizing a set of input data into classes. Many ML methods excel at classification through a supervised learning approach. Algorithms are capable of learning from available data to make more accurate decisions. Recent technological advances in the field of ML have proven to be invaluable to modern medicine, addressing complex tasks. However, despite many advantages, ML methods are not without their shortcomings.

Some of the most notable challenges and flaws of ML models are feature selection and hyper-parameter optimization. Most of the datasets consist of large number of attributes, where some of them are noisy and do not improve classification/regression performance. In order to enhance the classification accuracy for learning algorithms, the goal of feature selection is to identify the relevant subset from high-dimensional data sources by excluding the irrelevant features. According to G. Chandrashekar et al., there are three feature selection strategies: filter, wrapper, and embedded approaches [7].

In general, every ML model utilizes two types of parameters, trainable and non-trainable. As the name suggests, trainable parameters are being adjusted during the training phase, while the non-trainable parameters, which are also known as the hyper-parameters, need to be set manually by the researcher for every particular task. Additionally, the best set of hyper-parameter does not exist and these values must be determined for every task separately. In the literature, this problem is known as the hyper-parameter optimization.

Both above mentioned challenges belong to the group of non-deterministic polynomial hard (NP-hard) problems and it was shown that they can be successfully tackled by using swarm intelligence [3,27,28]. Therefore, in the research presented in this manuscript both feature selection challenge and hyper-parameter optimization by using the efficient extreme learning machine (ELM) model are tackled.

The ELM is generally burdened with two types of challenges: determining the hidden layer's optimal (or nearly optimal) number of neurons for every practical task, and determining the initial values for weights and biases since they have a large influence on the model's performance [15]. To tackle these challenges, this research proposed an enhanced version of recently developed sand cat swarm optimization (SCSO) meta-heuristics, that falls into the family of swarm intelligence. Proposed method was adopted for feature selection and ELM model tuning and validated against well-known diabetes dataset [8] with the goal of improving early prediction of this dangerous state.

The contributions of this paper may be summarized as the following:

- A proposal of a novel SCSO, that overcomes weaknesses of the basic algorithm by using hybridization technique;
- First application of SCSO for optimizing ELM hyper-parameter and feature selection and
- Improving early diagnosis of diabetes by innovative improved HSCSO-ELM framework.

The remainder of this paper is organized as follows: Sect. 2 presents basic background data along with literature review. Basic SCSO is discussed initially in Sect. 3, then an improved strategy is suggested, along with modifications for the ELM tuning task. Diabetes dataset overview, comparative analysis and discussion is demonstrated in Sect. 4, while Sect. 5 is the epilogue of this paper and contains final remarks.

2   Background and Related Works

This section provides some basic background relevant to the proposed research. First, mathematical formulation of ELM model is given and afterwards, basics of swarm intelligence algorithms are given along with respective literature review.

2.1   Extreme Learning Machine

Huang et al. [13] first unveiled the Extreme Learning Machine (ELM) with the intention of training single-hidden layer feed-forward neural networks (SLFNs). Training the network connotes determining the weights in its layers which enables the network to best approximate the desired solution. A major setback in the application of feed-forward networks for past decades has been their sluggish learning speed. According to Huang et al. [14], there are two potential explanations for this: 1) slow gradient-based learning algorithms, which are widely employed to train neural networks, and 2) the iterative tuning of all network parameters by employing such learning algorithms.

Truly, traditional algorithms such as back propagation (BP) tend to fall into so-called local traps and miss out on the global optimum altogether. BP is susceptible to the complexity of the feature space and the initialization of the parameters, making a local optimum more likely to be reached. ELM, on the other hand, selects hidden layer biases and input-to-hidden layer weights at random, and these parameters don't fluctuate throughout training. The only parameters SLFN has to learn in this case are output weights, which are determined analytically by utilizing Moore-Penrose (MP) generalized inverse [22]. ELM has shown the ability to produce better generalization performance and ensure thousand times higher learning speeds compared to traditional feed-forward neural network learning algorithms [14]. Moreover, this model has proven to be superior to support vector machines in both regression and classification problems [11]. Because it learns without iteration, ELM converges significantly more rapidly than conventional algorithms.

Given a training set $\{(x_j, t_j)\}_{j=1}^{N}$ with N samples and m classes, the SLFN with L hidden nodes and activation function g(x) is expressed in equation [13]:

$$\sum_{i=1}^{L} \beta_i g(we_j \cdot x_j + bi_i) = t_j, j = 1, 2, \ldots, N \tag{1}$$

where $we_i = [we_{i1}, \ldots, we_{in}]^T$ is the input weight, bi_i is the bias of the i-th hidden node, $\beta_i = [\beta_{i1}, \ldots, \beta_{im}]^T$ signifies the weight vector that links the ith hidden node and the output nodes, $we_i \cdot x_j$ describes the inner product of we_i and x_j, and t_j stands for the network output regarding input x_j. The Eq. 1 has the following form:

$$H\beta = T \tag{2}$$

where

$$H = \begin{bmatrix} g(we_1 \cdot x_1 + b_1) & \ldots & g(we_L \cdot x_1 + b_L) \\ \vdots & \ldots & \vdots \\ g(we_1 \cdot x_N + b_1) & \ldots & g(we_L \cdot x_N + b_L) \end{bmatrix}_{NxL}, \beta = \begin{bmatrix} \beta_1^T \\ \vdots \\ \beta_L^T \end{bmatrix}_{Lxm}, T = \begin{bmatrix} t1^T \\ \vdots \\ tN^T \end{bmatrix}_{Nxm} \tag{3}$$

In Eq. (3), H signifies the NN's hidden layer output matrix [12], with β being the output weight matrix.

2.2   Swarm Intelligence

Beni and Wang first proposed the concept of swarm intelligence in 1989 in relation to intelligent behavior in cellular robotic systems [4]. This discipline studies both artificial and natural systems where many individuals utilize decentralized control and self-organization to achieve benefiting results for the entire population.

A typical swarm intelligence system is comprised out of a set of individual units or "boids" that act locally and with their environment. Boid is a coin from "bird-like object" and it was introduced by Craig Reynolds in his paper in 1987 as a part of the artificial life program [21]. The inspiration came from observing the behavior of bird flocks seeking food. Boids are an example of emergent behavior: the emergence of intelligent global behavior, unbeknownst to these individuals, is the outcome of seemingly random behaviors of individual agents abiding to a set of simple principles without a centralized framework to manage their coordination. In the aforementioned paper, Reynolds presented some of the simplest rules that must be followed in a boid universe: separation (manoeuvre to prevent collisions with nearby flock members), alignment (manoeuvre towards the average heading of the local flock mates), and cohesion (navigate in the direction of the flock's "center of mass").

Nowadays, there are many swarm intelligence approaches and some of the most notable examples include particle swarm optimization (PSO) [9], firefly algorithm (FA) [25], artificial bee colony (ABC) [17], elephant herding optimization (EHO) [24], etc. There are many swarm intelligence metaheuristic applications for both challenges addressed in this paper, feature selection [3,28] and hyper-parameter optimization [1,2], however literature lacks approaches that tackle both feature selection and hyper-parameter tuning challenges.

3   Proposed Method

The fundamental sand cat swarm optimization (SCSO) overview is presented in this part, succeeded by the enhanced approach and its modifications for ELM tuning that are suggested in the study.

3.1   Elementary Sand Cat Swarm Optimization Algorithm

The SCSO, an innovative metaheuristic algorithm introduced by Seyyedabbasi and Kiani [23], is based on the feeding and prey-seeking behaviors of the sand cat (Felis margarita). The researchers depicted sand cats as herds despite the fact that these feral cats dwell in solitude in the wilderness to emphasize the concept of swarm intelligence. These creatures can also sense low-frequency emissions up to 2 kHz, that might allow them to swiftly and easily catch distant prey. Sand cats are also remarkably adept at digging up their catch. They are certain to have a special affinity for locating and preying in the outdoors owing to these two exceptional traits.

For a number of reasons, the SCSO algorithm has shown to be superior than the most recent metaheuristic algorithms. Particularly noteworthy is that it's characterized by a more advantageous time complexity of $O(n^2)$. The SCSO also exhibits an adequate and harmonious response between the exploitation and exploration stages, managing to avoid straying into the local optimum trap. Thirdly, compared to other metaheuristic algorithms, the SCSO has fewer operators and parameters. The implementation of SCSO is also simpler.

Initializing the population and establishing the optimization task is the first stage in the algorithm. The quantities of the variables are known as "sand cats", in line with the way the actual problem is solved (e.g. in PSO, this variable is called particle position [18]). In an optimization process with d dimensions, each of these felines is specified as a vector, or a $1 \times d$ array. The formula for individual sand cat's solution is given as $C_i = (c_{i1}, c_{i2}, c_{i3}, \ldots, c_{id})$. Each c in this case must be located inside the specified limits ($\forall c_i \in [$ lower, upper $]$). The fitness function defines the pertinent problem parameters, and the SCSO will determine the optimal parameter values. A value will be generated for each sand cat's corresponding function. The feline with the best cost at the end of an iteration is chosen, and the remaining cats try to transition in the general area of this best-chosen cat in the following iteration. The cat with greatest proximity to the target is represented by the best solution in each iteration. The solution is not saved if a more suitable value is not discovered in the succeeding iterations, ensuring efficient memory use.

The exploration step is based on low-frequency vibration that sand cats detect during hunting, which correlates to their hunting for prey. As a result, the sensitivity range for each cat is provided. These organisms can sense weak signals up to 2 kHz, as was already reported. This overall sensitivity range is quantitatively given as $(\vec{s_R})$, where the value will gradually fall from 2 to 0 over the span of iterations.

$$\vec{s_R} = H_A - \left(\frac{2 \times H_A \times \text{curr}_I}{\text{max}_I + \text{curr}_I} \right), \tag{4}$$

where H_A signifies the quantity influenced by the sand cat's perceptive abilities, and is constant with the value of 2, curr_I is the current iteration, and max_I is maximum number of iterations in the run.

Additionally, the SCSO makes use of the parameter M, which stands for the important variable that dictates how the exploration and exploitation stages are merged. The vector in question is generated via Eq. (5). The two phases' transitions will be more evenly balanced as a result of this adaptive method.

$$\vec{M} = 2 \times \vec{s_R} \times \text{random}(0, 1) - \vec{s_R}, \tag{5}$$

The search space is randomized within the given bounds. Throughout the search step, a random position is used to update each agent's placement. The search agents can therefore discover different domains inside the search space. Every cat has a unique sensitivity spectrum that is specified by Eq. 6 in order to escape the local optimum pitfall.

$$\vec{s} = \vec{s_R} \times \text{random}(0, 1). \tag{6}$$

According to the best-candidate location $\left(\overrightarrow{\text{Loc}_{bc}} \right)$, current location $\left(\overrightarrow{\text{Loc}_c} \right)$, and sensitivity spectrum $(\vec{s})$, each search agent modifies its own location. Consequently, sand cats are able to identify prospective locations for additional prey, as shown in Eq. (7).

$$\overrightarrow{\text{Loc}}(t + 1) = \vec{s} \cdot \left(\overrightarrow{\text{Loc}_{bc}}(t) - \text{random}(0, 1) \cdot \overrightarrow{\text{Loc}_c}(t) \right). \tag{7}$$

Equation 8 is used to determine the gap between the sand cat's best location $\left(\overrightarrow{Loc_b}\right)$ and current location $\left(\overrightarrow{Loc_c}\right)$ in order to describe the striking (exploitation) phase of SCSO quantitatively. The orientation of the motion is determined by an arbitrary angle (α) on the circle, which is also how the sensitivity spectrum of the sand cat is represented. Given that the angle chosen falls within range $[0, 360]°$, the value of the cosine parameter will fluctuate between -1 and 1. This allows for the circular movement of the search agents. The SCSO chooses a random angle for each sand cat using the Roulette Wheel Selection (RWS) technique. In this way, the cat will be capable of approaching the hunting stance. This arbitrary angle guarantees avoidance of the local optimum trap, and is responsible for the guidance of the search agents, as shown in Eq. 8. A random parameter designated as $\left(\overrightarrow{Loc_{rand}}\right)$ steers cats nearer the prey.

$$\overrightarrow{Loc}_{rand} = \left|random(0, 1) \cdot \overrightarrow{Loc_b}(t) - \overrightarrow{Loc_c}(t)\right|$$
$$\overrightarrow{Loc}(t + 1) = \overrightarrow{Loc_b}(t) - \vec{s} \cdot \overrightarrow{Loc}_{rand} \cdot \cos(\alpha). \tag{8}$$

The adaptive values of the $(\overrightarrow{s_R})$ and M parameters ensure both exploration and exploitation, and enable SCSO to easily change between two stages. The M's fluctuation range will be reduced because it depends on $(\overrightarrow{s_R})$. As previously mentioned, the M value will likewise be evenly distributed when the values of the $(\overrightarrow{s_R})$ are. When $M \leq 1$, the SCSO algorithm forces the felines to exploit; alternatively, the search agents are directed towards exploration.

Throughout this hunt's prey-seeking process (exploration), each cat's individual radius aids in escaping the local optimum pitfall. Equation 9 demonstrates the revised positions of each feline during the two stages.

$$\vec{C}(t+1) = \begin{cases} \overrightarrow{Loc_b}(t) - \overrightarrow{Loc}_{rand} \cdot \cos(\alpha) \cdot \vec{s} & |M| \leq 1; \text{ exploitation} \\ \vec{s} \cdot \left(\overrightarrow{Loc_{bc}}(t) - random(0, 1) \cdot \overrightarrow{Loc_c}(t)\right) & |M| > 1; \text{ exploration} \end{cases} \tag{9}$$

3.2   Enhanced SCSO Strategy and ELM Tuning Challenge Modifications

The SCSO is still being fully evaluated for NP-hard optimization problems because it is a relatively new method. Due to this, fundamental SCSO was rigorously assessed using the Congress on Evolutionary Computation (CEC) test suite's traditional bound-constrained (unconstrained) benchmarks as a component of the study proposal. Although the SCSO generates outcomes that are largely reliable and robust, it was observed that there is need for enhancement.

In particular, it was seen in certain runs that the algorithm became stuck in the less-than-ideal regions of the search space and was unable to locate the optimum region. Given the stochastic character of SCSO, when a suitable domain is found, the search quickly converges to an optimum, and as a result, the method performs below average. The approach suggested in this publication integrates the idea of "exhausted solutions" from the famous ABC methodology [17] to address this problem. According to this concept, a solution is replaced with an arbitrary one generated within the search space's bounds if it cannot be enhanced in the predetermined number of runs.

When a solution cannot be improved upon, the suggested technique specifies each solution with an additional parameter $trial$, which is increased with each iteration. Ultimately, a pseudo-random solution is used in place of the solution when the $trial$ parameter reaches a predefined $limit$ value, and it is generated as:

$$\vec{C} = \vec{lb} + \beta \cdot (\vec{ub} - \vec{lb}), \tag{10}$$

with β being a pseudo-random integer selected from a uniform distribution, and vectors $\vec{lb}$ and $\vec{ub}$ represent the lower- and upper-boundaries of the solution's parameters, respectively.

However, based on empirical data, it becomes considerably more difficult to refine current solutions in consecutive iterations once the seeking procedure has converged towards an optimum area. For this reason, proposed metaheuristics incorporate an flexible $limit$ variable that is being increased in subsequent runs beginning from its original value $limit_0$.

The suggested method was given the moniker hybrid SCSO (HSCSO) as it integrates exploration techniques from the ABC metaheuristics. Although the recommended adjustments are quite straightforward, they are extremely effective as demonstrated in Sect. 4, and more significantly, HSCSO does not complicate the standard SCSO in any way.

The suggested HSCSO is modified to address the concerns of feature selection and ELM tuning. Each HSCSO unit is represented by a one-dimensional L-sized array, with $L = nf + 1 + nsf \cdot nn + nn$. First nf parameters represent the net no. of features in the dataset. Parameter on the position $nf + 1$ is integer by type and it encodes the no. of neurons (nn) in the hidden ELM's layer, and the nsf represents the no. of selected features. The product $nsf \cdot nn$ is the size of parameters that hold ELM's weights, while nn stores biases for the hidden ELM layer.

The objective is expressed as the following, taking into consideration both the number of chosen features (nsf) and the classification error rate (cer):

$$O = \theta \cdot cer + \phi \cdot \frac{nsf}{nf}, \tag{11}$$

where θ and ϕ denote weight coefficients dictating the relative importance of cer to the nsf. It should be emphasized that the literature frequently uses this objective function formulation for feature selection problems [5].

Furthermore, every HSCSO individual is composed of binary, discrete and continuous parameters. For transforming continuous into binary search space, sigmoid function is employed, while the discrete argument is derived by simple rounding operation to the nearest integer. Proposed method is in native adopted for continuous search space, therefore last $L - (1 + nf)$ parameters do not need transformation.

The suggested HSCSO algorithm's pseudocode is presented below, taking into consideration everything stated previously.

Algorithm 1. The HSCSO pseudo-code

Define the population and *trial* for each search agent to 0
Define s, s_R, M and *limit*
Utilizing the objective function, calculate the fitness function.
while $t \leq$ T **do**
    for each sand cat **do**
        Using the RWS, obtain a random angle ($0° \leq \alpha \leq 360°$)
        if ($abs(M) <= 1$) **then**
            Adjust the sand cat's location in accordance with Eq. (8)
            Between the original and adjusted agent locations, execute greedy selection
        else
            Adjust the search agent's location in accordance with Eq. (7)
            Between the original and adjusted agent locations, execute greedy selection
        end if
        if (result remains unchanged) **then**
            Increase cat's *trial*
        end if
    end for
    Each agent where the *trial* $\geq$ *limit* is substituted with a random value generated by Eq. (10).
    $t = t + +$
end while

---

4   Experiments and Discussion

This chapter introduces details of the employed dataset along with performance metrics description, followed by experimental setup and comparative analysis between proposed HSCSO and other SOTA metaheuristics adapted for ELM tuning, and validated against the diabetes dataset.

In all experimental findings tables, best result for each metric is marked with bold style.

4.1   Dataset Description and Performance Metrics

Research proposed in this manuscript uses the PIMA Indian Diabetes (PID) dataset, generated by the National Institute of Diabetes and Kidney Diseases center, available in the UCI machine learning repository [8]. The dataset is composed of 768 instances with 8 features for healthy and female individuals older than twenty-one age that suffer from diabetes. Dataset includes the following features: number of pregnancies, blood glucose levels, measurements of blood pressure, bicep skin fold thickness, insulin levels, body mass index (BMI), age and diabetes pedigree function describing their hereditary tendency towards developing the condition.

The diabetes dataset falls into the group of binary classification challenges and whether the patient is diagnosed as a diabetic is shown as a binary value given as the outcome target feature. Additionally, besides diagnosing diabetic patients application, this data can also be used to predict the chances of a patient developing the condition within the next four years. However, this research employs this dataset only for a classification task. Visualization of the exploratory data analysis (EDA) for the employed diabetes dataset is provided in Fig. 1.

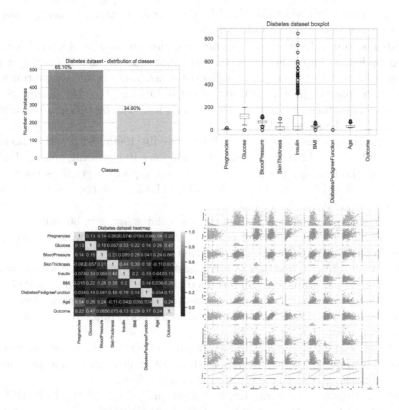

Fig. 1. Diabetes dataset visualization.

From the provided figure it can be seen that the dataset is moderately imbalanced with more negative classes. Also, from the box plot diagram, distribution of values per each feature can be seen, while the correlation between features can be observed from the provided heatmap and pairplot diagrams.

For the purpose of comparative analysis, the following metrics were taken into account: accuracy (classification error), precision, recall, f1 score, as it is common practice in the modern literature. Formulation of these metrics can be found in [16].

4.2   Experimental Setup and Comparative Analysis

The ELM models generated by proposed HSCSO (ELM-HSCSO) were compared by other ELM models evolved by the following SOTA metaheuristics: basic SCSO [23], bat algorithm (BA) [26], artificial bee colony (ABC) [17], harris hawks optimization (HHO) [10] and salp swarm algorithm (SSA) [20]. The recommended control parameter settings from the algorithms' original papers were used to test each algorithm. Initial value for ELM-HSCSO *limit* parameter ($limit_0$) was set to 2 and it was determined empirically.

Python's core and data science libraries: numpy, pandas, and scikit-learn are used to implement all methods. Since ELM is not available in scikit-learn, it was developed from the scratch. All metaheuristics are applied in a single run with the solution space of ($SN = 20$) and no. of iterations ($T = 15$). Furthermore, because of their stochastic nature, all methods were run 50 times in total, with average metrics being recorded.

All simulations were executed on Intel Core i9-11900K with turbo mode at 5.3 GHz and 64 GB of RAM platform and Windows 10 operating system. Unfortunately, Python numpy library, which does not support a graphics processing unit (GPU) execution, is used to implement ELM, therefore the framework could not be executed on GPU.

Search space boundaries are defined as follows: $nn_{lb} = 30$, $nn_{ub} = 150$, $w_{lb} = -1$, $w_{ub} = 1$, $b_{lb} = -1$ and $b_{ub} = 1$, where b and w represent biases and weights in the hidden ELM layer respectively.

Moreover, objective function as formulated in Eq. (11) is used with $alpha = 0.95$. Again, it was empirically determined that if $\alpha = 0.95$ and $\beta = 0.05$ satisfying balance between the number of selected features and classification error rate can be obtained. Results are captured in for both, objective functions and classification error.

In Tables 1 and 2, basic metrics, averaged over 50 runs, in terms of accuracy and objective function value, are reported. Additionally, the nn and percentage of selected features of the best solution is also provided.

From the reported basic metrics it can be concluded that the ELM-HSCSO managed to obtain in average best results. Also, all developed models, except ELM-SCSO acquired the best objective and the best accuracy (error) in the same run. At the other hand, for the ELM-SCSO the best accuracy (error) is captured with 87.50% of selected features and 35 neurons, while the best objective is accomplished with 30 hidden layer neurons ELM model and 75.00% of selected features. It is also noteworthy that the ELM-ABC managed to acquire satisfying accuracy with only 50.00% of selected features.

Figure 2 reports convergence speed graphs and box plot diagrams for error and objective, while Fig. 3 shows confusion matrix, precision-recall (PR), and

Table 1. Comparative analysis of accuracy - general metrics

Metaheuristic						
	ELM-HSCSO	ELM-SCSO	ELM-BA	ELM-ABC	ELM-HHO	ELM-SSA
Best (%)	**85.71**	83.77	85.06	85.06	85.06	85.06
Worst (%)	83.12	82.47	82.47	83.12	81.82	83.12
Mean (%)	**84.03**	83.12	83.64	83.77	83.38	83.90
Median (%)	83.77	83.12	83.77	83.77	83.12	83.77
std	0.0108	**0.0064**	0.0116	0.0079	0.0117	0.0071
nn	150	35	150	54	91	67
Features (%)	75.00	87.50	87.50	**50.00**	100.00	62.5

Table 2. Comparative analysis of objective - general metrics

Metaheuristic						
	ELM-HSCSO	ELM-SCSO	ELM-BA	ELM-ABC	ELM-HHO	ELM-SSA
Best	**0.1550**	0.1741	0.1639	0.1564	0.1664	0.1589
Worst	0.1780	0.1868	0.1793	0.1830	0.1857	0.1780
Mean	**0.1705**	0.1800	0.1734	0.1731	0.1774	0.1718
Median	0.1741	0.1780	0.1766	0.1780	0.1805	0.1741
std	0.0091	**0.0054**	0.0071	0.0106	0.0074	0.0076
nn	150	30	150	54	91	67
Features (%)	75.00	75.00	87.50	**50.00**	100.00	62.50

receiver operating characteristics (ROC) curves with the aim of visualizing the presented results.

From the presented convergence speed graph on Fig. 2 can be undoubtedly concluded that the proposed HSCSO manages to converge must faster to optimum regions of search space than the basic SCSO and other cutting-edge algorithms. Also, provided box plot diagrams prove the stability of proposed approaches over conducted runs.

For unbalanced datasets like a diabetes one, measuring accuracy alone does not provide enough information to understand how well the method performs. Therefore, Table 3 also includes comprehensive performance characteristics for the run that had the lowest error. In the presented table, metrics per each class are given along with the weighted (micro) averaged metric values.

Similarly, like in the case of basic metrics, detailed indicators also prove that the ELM generated by proposed HSCSO achieves better performance than all other methods included in comparison. Additionally, notable improvements over basic SCSO are accomplished by the proposed HSCSO algorithm.

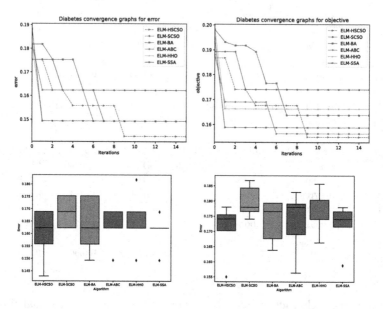

Fig. 2. Convergence speed graphs and box plot diagrams for error and objective.

Table 3. Comparative analysis - detailed metrics.

Performance metrics										
	Accuracy (%)	P0	P1	W.Avg.P	R0	R1	W.Avg.R	F1S0	F1S1	W.AvgF1S
ELM-HSCSO	**85.71**	**0.8750**	0.8200	**0.8557**	0.9100	0.7592	**0.8571**	**0.8921**	**0.7884**	**0.8557**
ELM-SCSO	83.77	0.8571	0.7959	0.8356	0.9000	0.7222	0.8376	0.8780	0.7572	0.8357
ELM-BA	85.06	0.8737	0.8039	0.8492	0.9000	0.7592	0.8506	0.8867	0.7809	0.8496
ELM-ABC	85.06	0.8598	0.8297	0.8492	0.9200	0.7222	0.8506	0.8888	0.7722	0.8479
ELM-HHO	85.06	0.8598	0.8297	0.8492	0.9200	0.7222	0.850649	0.8888	0.7722	0.8479
ELM-SSA	85.06	0.8666	0.8163	0.8490	0.9100	0.7407	0.8506	0.8878	0.7766	0.8488

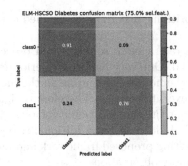
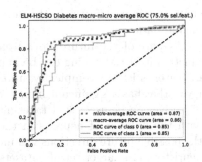

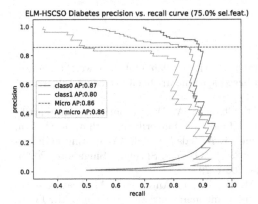

Fig. 3. The ELM-HSCSO confusion matrix, PR and ROC AUC.

5   Conclusion

In this research, a novel framework utilizing an enhanced version of SCSO algorithm, adapted for ELM hyper-parameter tuning and feature selection for diabetes classification was proposed. Additionally, introduced improved SCSO metaheuristics overcomes drawbacks of its original implementation by incorporating exploration procedure from the well known ABC algorithm.

Research shown in this manuscript tackles two of the most significant obstacles in the field of machine learning - feature selection and hyper-parameters tuning, which are both NP-hard in nature. Proposed SCSO is used to simultaneously perform feature selection and determining the number of neurons in a single hidden layer of an ELM, as well as computing the initial weights and biases values of each neuron in the hidden layer. Metaheuristics is validated against well-known diabetes classification datasets, which is frequently used for benchmark purposes.

In terms of accuracy, precision, recall, and f1 score, the performance of the proposed method was compared to that of other well-known, cutting-edge swarm intelligence algorithms. The suggested improved SCSO is able to address both challenges more effectively than other algorithms, according to reached experimental data.

As part of the future work from this area, a more sophisticated objective function for ELM tuning can be implemented. One very important performance indicator is ELM's complexity, in terms of the number of neurons in the hidden layer, that has a significant impact on the execution speed. Therefore, weighted objective can be used, that takes into account the classification error rate along with the number of neurons in the ELM's hidden layer. Additionally, one of the possible future directions will include testing proposed framework and metaheuristics on a larger set of datasets with more observations.

References

1. Bacanin, N., et al.: Artificial neural networks hidden unit and weight connection optimization by quasi-refection-based learning artificial bee colony algorithm. IEEE Access **9**, 169135–169155 (2021)
2. Bacanin, N., Bezdan, T., Zivkovic, M., Chhabra, A.: Weight optimization in artificial neural network training by improved monarch butterfly algorithm. In: Shakya, S., Bestak, R., Palanisamy, R., Kamel, K.A. (eds.) Mobile Computing and Sustainable Informatics. LNDECT, vol. 68, pp. 397–409. Springer, Singapore (2022). https://doi.org/10.1007/978-981-16-1866-6_29
3. Bacanin, N., Petrovic, A., Zivkovic, M., Bezdan, T., Antonijevic, M.: Feature selection in machine learning by hybrid sine cosine metaheuristics. In: Singh, M., Tyagi, V., Gupta, P.K., Flusser, J., Ören, T., Sonawane, V.R. (eds.) ICACDS 2021. CCIS, vol. 1440, pp. 604–616. Springer, Cham (2021). https://doi.org/10.1007/978-3-030-81462-5_53
4. Beni, G., Wang, J.: Swarm intelligence in cellular robotic systems. In: Dario, P., Sandini, G., Aebischer, P. (eds.) Robots and Biological Systems: Towards a New Bionics? NATO ASI Series, vol. 102, pp. 703–712. Springer, Heidelberg (1993). https://doi.org/10.1007/978-3-642-58069-7_38
5. Bezdan, T., Zivkovic, M., Bacanin, N., Chhabra, A., Suresh, M.: Feature selection by hybrid brain storm optimization algorithm for COVID-19 classification. J. Comput. Biol. **29**, 515–529 (2022)
6. Butt, U.M., Letchmunan, S., Ali, M., Hassan, F.H., Baqir, A., Sherazi, H.H.R.: Machine learning based diabetes classification and prediction for healthcare applications. J. Healthc. Eng. (2021)
7. Chandrashekar, G., Sahin, F.: A survey on feature selection methods. Comput. Electr. Eng. **40**, 16–28 (2014)
8. Dua, D., Graff, C.: UCI machine learning repository (2017). http://archive.ics.uci.edu/ml
9. Eberhart, R., Kennedy, J.: Particle swarm optimization. In: Proceedings of the IEEE International Conference on Neural Networks, Australia, vol. 1948 (1942)
10. Heidari, A.A., Mirjalili, S., Faris, H., Aljarah, I., Mafarja, M., Chen, H.: Harris hawks optimization: algorithm and applications. Future Gener. Comput. Syst. **97**, 849–872 (2019)
11. Huang, G.B., Zhou, H., Ding, X., Zhang, R.: Extreme learning machine for regression and multiclass classification. IEEE Trans. Syst. Man Cybern. Part B (Cybern.) **42**, 513–529 (2012)

12. Huang, G.B.: Learning capability and storage capacity of two-hidden-layer feed-forward networks. IEEE Trans. Neural Netw. **14**, 274–281 (2003). https://doi.org/10.1109/TNN.2003.809401
13. Huang, G.B., Zhu, Q.Y., Siew, C.K.: Extreme learning machine: a new learning scheme of feedforward neural networks. In: 2004 IEEE International Joint Conference on Neural Networks (IEEE Cat. No.04CH37541), vol. 2, pp. 985–990 (2004). https://doi.org/10.1109/IJCNN.2004.1380068
14. Huang, G.B., Zhu, Q.Y., Siew, C.K.: Extreme learning machine: theory and applications. Neurocomputing **70**, 489–501 (2006)
15. Jovanovic, D., Antonijevic, M., Stankovic, M., Zivkovic, M., Tanaskovic, M., Bacanin, N.: Tuning machine learning models using a group search firefly algorithm for credit card fraud detection. Mathematics **10**, 2272 (2022). https://doi.org/10.3390/math10132272
16. Jovanovic, D., Antonijevic, M., Stankovic, M., Zivkovic, M., Tanaskovic, M., Bacanin, N.: Tuning machine learning models using a group search firefly algorithm for credit card fraud detection. Mathematics **10**(13) (2022). https://doi.org/10.3390/math10132272, https://www.mdpi.com/2227-7390/10/13/2272
17. Karaboga, D., Akay, B.: A comparative study of artificial bee colony algorithm. Appl. Math. Comput. **214**(1), 108–132 (2009)
18. Kennedy, J., Eberhart, R.: Particle swarm optimization. In: Proceedings of ICNN 1995 - International Conference on Neural Networks, vol. 4, pp. 1942–1948 (1995). https://doi.org/10.1109/ICNN.1995.488968
19. American Diabetes Association: Diagnosis and classification of diabetes mellitus. Diabetes Care **37**(Suppl 1), 81–90 (2014)
20. Mirjalili, S., Gandomi, A.H., Mirjalili, S.Z., Saremi, S., Faris, H., Mirjalili, S.M.: Salp swarm algorithm: a bio-inspired optimizer for engineering design problems. Adv. Eng. Softw. **114**, 163–191 (2017)
21. Reynolds, C.W.: Flocks, herds and schools: a distributed behavioral model, vol. 21, pp. 25–34. Association for Computing Machinery, New York (1987). https://doi.org/10.1145/37402.37406
22. Serre, D.: Matrices. Graduate Texts in Mathematics, 2nd edn. Springer, New York (2010)
23. Seyyedabbasi, A., Kiani, F.: Sand cat swarm optimization: a nature-inspired algorithm to solve global optimization problems. Eng. Comput. 1–25 (2022). https://doi.org/10.1007/s00366-022-01604-x
24. Wang, G.G., Deb, S., Coelho, L.D.S.: Elephant herding optimization. In: 2015 3rd International Symposium on Computational and Business Intelligence (ISCBI), pp. 1–5. IEEE (2015)
25. Yang, X.-S.: Firefly algorithms for multimodal optimization. In: Watanabe, O., Zeugmann, T. (eds.) SAGA 2009. LNCS, vol. 5792, pp. 169–178. Springer, Heidelberg (2009). https://doi.org/10.1007/978-3-642-04944-6_14
26. Yang, X.S., Gandomi, A.H.: Bat algorithm: a novel approach for global engineering optimization. Eng. Comput. **29**, 464–483 (2012)
27. Zivkovic, M., et al.: COVID-19 cases prediction by using hybrid machine learning and beetle antennae search approach. Sustain. Urban Areas **66**, 102669 (2021)
28. Zivkovic, M., Stoean, C., Chhabra, A., Budimirovic, N., Petrovic, A., Bacanin, N.: Novel improved salp swarm algorithm: an application for feature selection. Sensors **22**(5), 1711 (2022)

Enriching SQL-Driven Data Exploration with Different Machine Learning Models

Sabina Surdu[✉]

Faculty of Mathematics and Computer Science, Babeș-Bolyai University,
Cluj-Napoca, Romania
sabina.surdu@ubbcluj.ro

Abstract. Nowadays we rely heavily on technology to understand the world around us. Databases and machine learning are key components in this endeavour. In the data exploration realm, users can have difficulties formulating the right queries for various reasons, e.g., the dataset has a large number of attributes, the user doesn't have a clear idea of what they're looking for in their data. In our previous work, we aimed to bridge the gap between SQL and machine learning by providing the user who poses an SQL query with an answer set and a reformulation of their query, generated using the C4.5 decision tree algorithm. We now investigate the use of three different machine learning models in a new experimental study and interaction paradigm: LightGbm, FastTree, and GAM. Upon posing an SQL query, the user is presented with: the answer set, three trained models along with a rich set of new metrics that assess the models' quality, and the most important features for each model, computed with Permutation Feature Importance. By analyzing the metrics' results, the user can decide for themselves which model(s) they'll use further in their exploratory quest. Once a model is chosen, the user can formulate new SQL queries using some of the most important features for the model.

Keywords: SQL queries · Machine learning · Data exploration

1 Introduction

We have always been on a quest to understand the world around us. Nowadays we rely heavily on technology to figure out physical phenomena, health and education related issues, trends in the financial sector, business opportunities, and more.

Databases play a key role in this context, as they can be loaded with enormous quantities of data and queried in a timely fashion. The user can thus gain valuable insights into a particular organization, field of interest, etc.

Relational Database Management Systems (DBMSs) come first in DB-Engines' ranking of DBMSs by database model [3], with a staggering 71.6% in August 2022. SQL is loved by 64.25% of the respondents in Stack Overflow's

© The Author(s), under exclusive license to Springer Nature Switzerland AG 2023
D. Simian and L. F. Stoica (Eds.): MDIS 2022, CCIS 1761, pp. 204–217, 2023.
https://doi.org/10.1007/978-3-031-27034-5_14

2022 Developer Survey [12], it comes third in [13]'s chart of skills an employer looks for in a data science engineer, and is increasingly being used by less or even non-technical people, like sales staff and marketers [14]. Relational repositories are currently opening up to a great variety of users.

Data mining is another key player in our understanding of the world today, as it allows us to uncover interesting patterns in large data sets using techniques from domains like statistics, machine learning, information retrieval, etc. [5]

We propose a unified approach to data exploration, based on SQL queries and machine learning techniques. We focus on relational databases and supervised learning. The main goal is to provide the user with a richer experience when interacting with a database, to enable them to further explore their data once they received a query's result.

In the traditional querying paradigm, the user knows what they want from the database and formulates a query that corresponds to their needs. Upon the query's execution by the database system, the user is presented with the desired answer set.

We extend this paradigm to enable the user to explore a database through SQL queries, while also benefitting from machine learning techniques. The user formulates an initial query on a database and receives not only an answer set, but also several machine learning models that differentiate among the tuples that are wanted in the answer set and those that are not initially desired, along with metrics evaluating the models, and sets of features that are important for the models in question. We don't ask the user to label data or to be a machine learning expert, we just ask them to write SQL queries to explore their data in more depth.

Such a proposition would benefit scientists who are working with SQL queryable datasets, where tables have a large number of attributes storing data obtained from different instruments. Selection conditions cannot always be easily expressed in this context, e.g., one mustn't omit a predicate involving a particular attribute, the numerical value that appears in a predicate on a certain attribute must be correct. People using such datasets can end up spending more time writing a correct query than doing the actual science.

Scientific datasets aside, a user may simply want more than a basic query-answer interaction with the database. Take Jane, a 20-year-old sociology student who analyzes the implementation of dropout prevention policies for schools in underprivileged areas. She works on the *Students* relation shown in Table 1. The *ParentsAbroad* column has value T if both the student's parents have been working abroad for more than 5 years, and F otherwise. Initially, she's interested in students with a low Grade Point Average, knowing they are at a higher dropout risk, so she writes the following SQL query $Q1$:

```
SELECT *
FROM Students
WHERE GPA < 5
```

The answer set for query $Q1$ contains tuples corresponding to *Popescu Andrei, Ionescu Ana, Andrei Dana* and *Dumitru Dan*. Behind the scenes, the

Table 1. The Students relation

SID	Name	DateOfBirth	Grade	School	GPA	ParentsAbroad
1	Popescu Andrei	01.01.2010	VI C	Liceul IC	4.5	T
23	Ion Alexandru	02.03.2009	VII A	Generala AI	6	F
15	Ionescu Ana	02.08.2009	VII B	Liceul IC	4.1	T
2	Andrei Dana	15.07.2010	VI C	Gimnaziala MD	3.2	T
76	Vasilescu Anton	01.02.2011	V B	Generala AI	9.9	F
3	Vasile Alexandra	07.03.2009	VII B	Generala AI	8	F
7	Popescu Mihaela	19.11.2010	VI A	Liceul IC	10	F
5	Vlad Andreea	12.07.2010	VI B	Gimnaziala MD	8.5	F
9	Dumitru Dan	05.08.2011	V A	Generala AI	2.9	T
11	Dan Costin	03.02.2011	V D	Liceul IC	7.75	F
12	Popa Andrei	17.11.2009	VII A	Generala AI	NULL	T

learning set in Table 2 is built. Tuples that Jane wants in her answer set (students with a low GPA) are labeled with a "plus" sign (these are considered as *positive tuples*), and those that are not desired - with a "minus" sign (these are the *negative tuples*). The *GPA* column is excluded from this set.

Table 2. Learning set for query Q1 on the Students relation

SID	Name	DateOfBirth	Grade	School	ParentsAbroad	Label
1	Popescu Andrei	01.01.2010	VI C	Liceul IC	T	+
15	Ionescu Ana	02.08.2009	VII B	Liceul IC	T	+
2	Andrei Dana	15.07.2010	VI C	Gimnaziala MD	T	+
9	Dumitru Dan	05.08.2011	V A	Generala AI	T	+
23	Ion Alexandru	02.03.2009	VII A	Generala AI	F	−
76	Vasilescu Anton	01.02.2011	V B	Generala AI	F	−
3	Vasile Alexandra	07.03.2009	VII B	Generala AI	F	−
7	Popescu Mihaela	19.11.2010	VI A	Liceul IC	F	−
5	Vlad Andreea	12.07.2010	VI B	Gimnaziala MD	F	−
11	Dan Costin	03.02.2011	V D	Liceul IC	F	−

A machine learning model m is trained and evaluated on this learning set. m should distinguish between the positive tuples and the negative tuples. Performance metrics that evaluate the quality of m as well as a set of *the most important features* for m are returned. The user can further explore the database using these features in SQL queries.

While not a machine learning expert, Jane has a fair enough understanding of metrics like accuracy, precision and recall, and is satisfied with the metrics values for model m (e.g., high accuracy, precision, recall). The set of *the most important features* contains feature *ParentsAbroad*. Jane can pose another query Q2 requesting all students whose parents have been working abroad for more than five years:

```
SELECT *
FROM Students
WHERE ParentsAbroad = T
```

Apart from the tuples in the answer set of $Q1$, a new tuple corresponding to *Popa Andrei* is returned in the answer set of query $Q2$. According to the model, Jane might also be interested in this tuple, as it's likely that *Popa Andrei* also has a low GPA. Nevertheless, this tuple was revealed by $Q2$, whose formulation was supported by the machine learning model.

Jane can now look into social policies encouraging parents to stay in the country with their children.

We build on our previous research in [2] and [1] by tapping into new machine learning models, providing additional performance metrics and sets of important features that can guide the user to write subsequent queries. In our previous work we used the C4.5 decision tree algorithm [10] and supplied the user with a new, reformulated query. We now conducted a new experimental study with three different algorithms available in ML.NET [8]: LightGbm, FastTree and GAM. Moreover, the user now has the freedom to build their own subsequent queries, using the new metrics and sets of important features that are returned along with the models.

The rest of the paper is organized as follows. In Sect. 2 we give an overview of our approach. Section 3 presents the experimental study we conducted and the results we obtained. In Sect. 4 we discuss related propositions. Finally, Sect. 5 concludes this paper and presents future research directions.

2   Data Exploration with SQL and Machine Learning

Let U be a user who formulates an initial query Q on a relational database D. The system is to return:

- the answer set A of Q;
- a set of models denoted by M, where each model $m \in M$ differentiates among the tuples in A and those not wanted by U;
- a set of *performance metrics results* denoted by R; an element r in this set corresponds to a specific model $m \in M$ and holds m's values for 8 performance metrics;
- a set of *the most important features sets* denoted by F; an element f in this set corresponds to a specific model $m \in M$ and contains the most important features for m computed with Permutation Feature Importance.

2.1   Chosen Set of Queries

We focus on queries of the form: $Q = \pi_\alpha(\sigma_C(T_1 \bowtie \ldots \bowtie T_m))$, where $T_1 \ldots T_m$ are tables in D, and C is a selection condition built from atomic predicates connected by $\wedge$, $\vee$, $\neg$. We denote by T the result of the join(s) $T_1 \bowtie \ldots \bowtie T_m$. α is a subset of attributes from T.

A predicate in C has the form *attr op v*, where *attr* is an attribute in T, *op* is a comparison operator in $\{<, <=, =, >=, >, <>\}$, and v a numerical or categorical value.

We denote by Q_E the set of attributes that appear in Q's selection condition C.

2.2   Machine Learning Models and Metrics

Building The Training and Test Sets. The answer set A of Q contains exactly the tuples user U is interested in when the query is posed. The system automatically generates a so-called *negation query* $\overline{Q}$ for Q. $\overline{Q}$ is built by negating the selection condition C. We'll denote its answer set by $\overline{A}$.

A set Q_S is built as the union of A and $\overline{A}$. The schema of Q_S contains an additional attribute called *Label*, which takes on value 1 for tuples in A and value 0 for tuples in $\overline{A}$.

Q_S is split into a training set and a test set. All models are trained on the training set and evaluated on the test set.

Machine Learning Algorithms and Performance Metrics. The training set is then fed to a set of machine learning algorithms, each generating a particular model that should allow the classification of new tuples into one of two classes: with *Label* value 1 or 0, i.e., tuples U is interested in versus tuples U is not interested in at the moment. When training a model for query Q we don't consider as features the attributes in Q_E.

We chose the following three trainers in ML.NET:

- LightGbm - "an open source implementation of gradient boosting decision tree" [8];
- FastTree - an "implementation of the MART gradient boosting algorithm" [8];
- GAM (Generalized Additive Models) - "implemented using shallow gradient boosted trees" [8].

Every trained model $m \in M$ comes with a set of *performance metrics results* $r \in R$ that evaluate the quality of m: accuracy, positive precision, positive recall, negative precision, negative recall, area under the ROC curve, area under the precision-recall curve, F1 score.

Based on the values of these metrics, user U could focus on a particular model and use one or several of its most important features to write additional SQL queries and further explore the database.

The Most Important Features. Every trained model $m \in M$ comes with the set of its *most important features* $f \in F$. f is computed with the Permutation Feature Importance technique, which gives the relative contribution of each feature for a prediction [9].

Features are ordered descendingly by the impact they have on the accuracy metric. U is presented with the Top$K$ features with an accuracy mean absolute value greater than 0, where K is a parameter.

3   Experimental Study

We conducted an experimental study on a machine running Windows 11 Home, with a 1.30 GHz Intel Core i7-1065G7 CPU (4 Cores, 8 Logical Processors) and 16 GB RAM.

We used C♯ and ML.NET, a machine learning framework that allows training and evaluating models in the .NET ecosystem [8].

We used the *Wine Data Set* [15], which has 13 attributes: *Alcohol, Malic acid, Ash, Alcalinity of ash, Magnesium, Total phenols, Flavanoids, Nonflavanoid phenols, Proanthocyanins, Color intensity, Hue, OD280/OD315 of diluted wines,* and *Proline*. We eliminated the *class id* class attribute from the dataset. All attributes are continuous; there are no missing values. The dataset has 178 instances.

For simplicity, in the experimental study we focused on numerical attributes and queries of the form $Q = \pi_\alpha(\sigma_C(T))$, where $C = p_1 \wedge \ldots \wedge p_n$, i.e., we generated queries that operate on a single data table T and have only conjunctions in the selection condition. Extending the implementation to queries with joins and disjunctions is straightforward. Table T holds the wine data.

We computed the minimum and maximum value for each attribute in T, denoted by $Min(attr)$ and $Max(attr)$ for attribute $attr$.

We proceeded to generate 10 *query sets*. A *query set* is defined by the number of predicates it contains: query set QS_n has n predicates, where $n \in \{1, \ldots, 10\}$. We set the cardinality of each query set to 15, i.e., for each number of predicates n, we generated 15 queries with n predicates.

For each query Q we generated its predicates as follows: for each predicate p, we randomly chose an attribute $attr$ from T, a comparison operator op, and a value in the range $[Min(attr), Max(attr)]$. For a given attribute $attr$, there is at most one predicate in Q.

We defined a parameter s to control the size of the answer set for all queries, to avoid extremely imbalanced learning sets (e.g., a query that returns a couple of rows, while its negation returns the rest of the table). In the experiments we set the value of s to 20, which means we generate queries that return between 20% and $100 - s = 80\%$ of the rows in the table.

For each query Q, we generated its negation $\overline{Q}$ and the Q_S set. We split Q_S into a training set holding 80% of the tuples in Q_S, and a test set with the remaining tuples. The 150 training sets thus obtained were used to train LightGbm, FastTree, and GAM models, each evaluated on the corresponding test set.

When computing the most important features for model m with Permutation Feature Importance, we set the *permutation count* parameter value to 30 and the value of the K parameter to the smaller of the following values: *number of available features to learn from* and $((number\ of\ columns\ in\ the\ datatable)/2)$; for the Wine dataset, a set f holding the most important features for model m can have at most 6 features.

We looked at the impact each feature has on accuracy. We considered only the features for which the absolute value of the accuracy's mean is greater than 0. We returned the Top$K$ such features ordered by $|mean\ value\ of\ accuracy|$.

3.1 Results

We present the results we obtained in the experimental study for the 450 trained models.

Figure 1 shows the accuracy for the models produced using the three algorithms - the values' distribution into quartiles based on the number of predicates. The mean value for this metric stays around 0.8 for all models (always greater than 0.7), which is quite good. Nevertheless, most of the learning sets are not balanced, so other metrics are investigated as well.

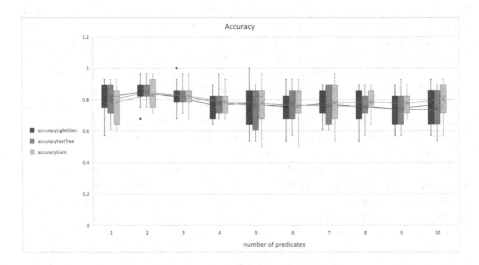

Fig. 1. Accuracy - LightGbm, FastTree, GAM

The F1 Score, positive precision and positive recall for the same models can be seen in Fig. 2. All tend to decrease as the number of predicates increases. With up to 4 predicates, the mean for these metrics remains above 0.6. A possible explanation of this decrease in the predictive performance of the models could be attributed to the fact that as the number of predicates increases, there are less available features to use in the training process. However, this doesn't seem to have the same impact on the negative precision and negative recall, as seen in Fig. 3. The negative precision mean remains above 0.75 and the negative recall mean above 0.7. These aspects can be further investigated.

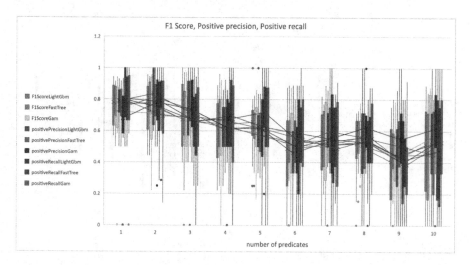

Fig. 2. F1 Score, Positive precision, Positive recall - LightGbm, FastTree, GAM

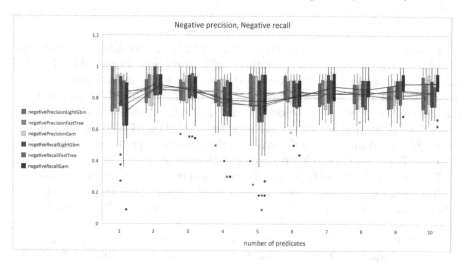

Fig. 3. Negative precision, Negative recall - LightGbm, FastTree, GAM

We examine both the Area under the precision-recall curve and the Area under ROC curve in Fig. 4, as most learning sets are not balanced. The mean for the latter doesn't go below 0.7, meaning the obtained models are pretty good. However, the mean for the former metric tends to decrease as the number of predicates increases.

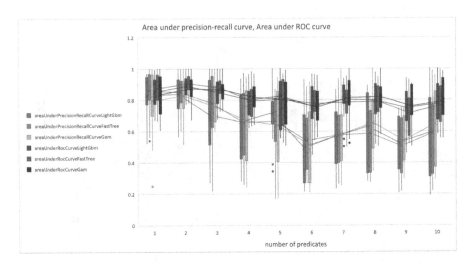

Fig. 4. Area under precision-recall curve, Area under ROC curve - LightGbm, FastTree, GAM

Figures 5, 6 and 7 show the time in ms to train the model, to evaluate the model, and to compute the set of important features with Permutation Feature Importance, respectively. The mean value for the GAM models training time doesn't exceed 2.5 s, whereas the models based on LightGbm and FastTree have a mean training time value below 250 ms. The time to evaluate the models is below 20 ms, with a few exceptions, for all algorithms. The time to compute the set of important features decreases as the number of predicates increases, as there are fewer features to learn from (less than 3.6 s).

The time to split the Q_S set into training and test sets and eliminate features we don't want our model to learn from as they appear in the query's predicates is below one millisecond for all queries with a few exceptions (for which we got 51 ms, 2 ms, and 1 ms, respectively).

We illustrate our entire approach on one of the generated queries, query 17 with 2 predicates: $[OD280/OD315\,of\,diluted\,wines] > 1.7967287491911303$ AND $[Alcalinity\,of\,ash] < 17.527212581876018$. The metrics values obtained for this query can be seen in Fig. 8.

The following sets of important features are returned (ordered by the accuracy impact, as previously explained):

- LightGbm: *Ash, Alcohol, Malic acid, Proanthocyanins, Flavanoids, Color intensity*;
- FastTree: *Ash, Alcohol, Proline, Malic acid, Flavanoids, Proanthocyanins*;

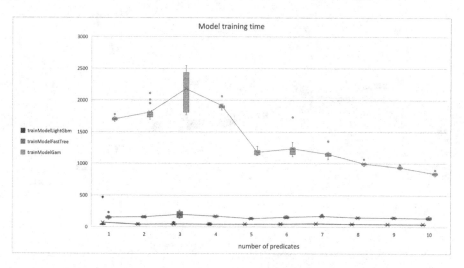

Fig. 5. Model training time (ms) - LightGbm, FastTree, GAM

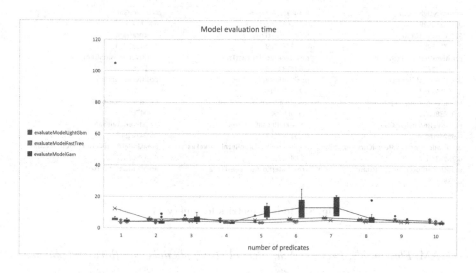

Fig. 6. Model evaluation time (ms) - LightGbm, FastTree, GAM

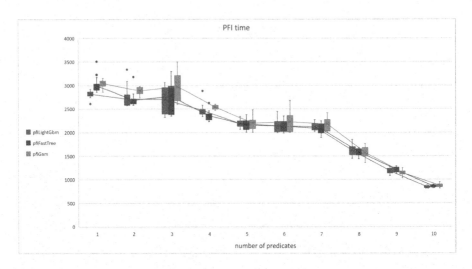

Fig. 7. Permutation Feature Importance time (ms) - LightGbm, FastTree, GAM

accuracyLightGbm	accuracyFastTree	accuracyGam
0.928571429	0.892857143	0.857142857
F1ScoreLightGbm	**F1ScoreFastTree**	**F1ScoreGam**
0.857142857	0.769230769	0.714285714
positivePrecisionLightGbm	**positivePrecisionFastTree**	**positivePrecisionGam**
0.857142857	0.833333333	0.714285714
positiveRecallLightGbm	**positiveRecallFastTree**	**positiveRecallGam**
0.857142857	0.714285714	0.714285714
negativePrecisionLightGbm	**negativePrecisionFastTree**	**negativePrecisionGam**
0.952380952	0.909090909	0.904761905
negativeRecallLightGbm	**negativeRecallFastTree**	**negativeRecallGam**
0.952380952	0.952380952	0.904761905
areaUnderPrecisionRecallCurveLightGbm	**areaUnderPrecisionRecallCurveFastTree**	**areaUnderPrecisionRecallCurveGam**
0.763821892	0.784637188	0.74766543
areaUnderRocCurveLightGbm	**areaUnderRocCurveFastTree**	**areaUnderRocCurveGam**
0.93877551	0.952380952	0.931972789

Fig. 8. Models' metrics for query 17

– GAM: *Flavanoids, Proline, Ash, Alcohol, Total phenols, Nonflavanoid phenols.*

These sets of features can help the user pose subsequent queries on the dataset.

4   Related Work

The realm of data exploration is evolving rapidly. [6] discusses *"Modern Exploration-driven Applications"* with queries' results influencing the writing of subsequent queries. We use machine learning models, along with performance metrics and important features to guide the user to write subsequent queries. [11] also discusses *"ambiguous query intent"*, when users don't know exactly what they want from the data set.

In a number of propositions like [7] and [4], users' feedback is required, sometimes tuples being manually labeled and fed to machine learning tools. We only ask the user to write SQL queries to explore their data.

In our previous work we focused on reformulating the initial query provided by the user. We employed only the C4.5 algorithm for this purpose and built the reformulated query as a disjunction of the positive branches in the obtained tree [1,2]. In our current proposition we are providing the user with several models trained using LightGbm, FastTree, and GAM, accompanied by a rich new set of metrics assessing the quality of the obtained models. Moreover, each model comes with a set of important features, computed using Permutation Feature Importance. The user now has the freedom to choose a particular model and one or several features to build additional queries and thus further explore the data.

In [1] we asked the user to express the selection condition as a conjunction of two formulas: $F_1 \wedge F_2$, where F_1 is a discrimination condition, and F_2 a classical one, possibly empty. In [2] we only tackled conjunctive queries, as we proposed a Knapsack-based heuristic to generate the *balanced* negation query, i.e., the query whose answer size is as close as possible to the size of the initial answer set. We use no such restrictions in the current proposition.

We previously defined metrics assessing the result of the reformulated query in terms of diversity and representativeness with respect to the initial query's answer set A and its negation's answer set N: the reformulated query should return as many tuples as possible from A, as few tuples as possible from N, and new, extra tuples. We also defined the distance between the true *balanced* negation query and the Knapsack-based one. We now return 8 metrics for the model: accuracy, positive precision, positive recall, negative precision, negative recall, area under the ROC curve, area under the precision-recall curve, and F1 score.

5 Conclusion and Future Directions

In this paper we discussed an approach to data exploration using SQL and machine learning. Upon executing an SQL query on a relational dataset, the user is presented with an answer set, and several different classification models, along with performance metrics and a set of features deemed important for each model. Instead of just examining the answer set, the user can now further explore the data by choosing one or several models, guided by the metrics' values, and formulating new SQL queries using the returned features.

Our contribution can be summarized as follows:

- we use three new classification models to capture the difference between the query's answer set and the tuples the user is not interested in: LightGbm, GAM, FastTree;
- the user is presented with a rich new set of metrics assessing the quality of the models; they can now choose the model(s) they regard as the best for their needs;

- the user is presented with sets containing *the most important features*, one set per model, computed with Permutation Feature Importance, which can serve as the starting points for subsequent queries posed by the user;
- to illustrate our approach, we conducted a new experimental study, generating 150 queries - each with its own training and test sets, training and evaluating 450 models, and summarizing the performance metrics and the results related to the most important features.

We plan to extend our work in the following directions:

- we plotted the metrics' values against the number of predicates in the queries, which went from 1 to 10; we are currently working on a new analysis assessing the same metrics by taking into account the degree of balancedness of the learning set;
- the experiments can be conducted on a larger dataset, and the set of models can be further expanded;
- the most important features can be computed by considering the impact they have on metrics other than accuracy (precision, recall, etc.); the actual impact a feature has on a particular metric can be another parameter.

References

1. Cumin, J., Petit, J.-M., Rouge, F., Scuturici, V.-M., Surace, C., Surdu, S.: Requêtes discriminantes pour l'exploration des données. In: de Runz, C., Crémilleux, B. (eds.) 16ème Journées Francophones Extraction et Gestion des Connaissances, EGC 2016, 18–22 Janvier 2016, Reims, France, Revue des Nouvelles Technologies de l'Information, vol. E-30, pp. 195–206. Éditions RNTI (2016)
2. Cumin, J., Petit, J.-M., Scuturici, V.-M., Surdu, S.: Data exploration with SQL using machine learning techniques. In: Markl, V., Orlando, S., Mitschang, B., Andritsos, P., Sattler, K.-U., Breß, S. (eds.) Proceedings of the 20th International Conference on Extending Database Technology, EDBT 2017, Venice, Italy, 21–24 March 2017, pp. 96–107. OpenProceedings.org (2017). https://doi.org/10.5441/002/edbt.2017.10
3. DB-Engines Ranking. https://db-engines.com/en/ranking_categories. Accessed 29 Aug 2022
4. Dimitriadou, K., Papaemmanouil, O., Diao, Y.: Explore-by-example: an automatic query steering framework for interactive data exploration. In: Dyreson, C.E., Li, F., Özsu, M. T. (eds.) International Conference on Management of Data, SIGMOD 2014, Snowbird, UT, USA, 22–27 June 2014, pp. 517–528. ACM (2014). https://doi.org/10.1145/2588555.2610523
5. Han, J., Kamber, M., Pei, J.: Data Mining: Concepts and Techniques, 3rd edn. Morgan Kaufmann, Burlington (2011)
6. Idreos, S., Papaemmanouil, O., Chaudhuri, S.: Overview of data exploration techniques. In: Sellis, T.K., Davidson, S.B., Ives, Z.G. (eds.) Proceedings of the 2015 ACM SIGMOD International Conference on Management of Data, Melbourne, Victoria, Australia, 31 May–4 June 2015, pp. 277–281. ACM (2015). https://doi.org/10.1145/2723372.2731084
7. Li, H., Chan, C.-Y., Maier, D.: Query from examples: an iterative, data-driven approach to query construction. Proc. VLDB Endow. 8(13), 2158–2169 (2015)

8. ML.NET.   https://dotnet.microsoft.com/en-us/apps/machinelearning-ai/ml-dotnet. Accessed 15 Aug 2022

9. ML.NET Documentation. https://docs.microsoft.com/en-us/dotnet/machine-learning/. Accessed 15 Aug 2022

10. Quinlan, J.R.: C4.5: Programs for Machine Learning. Morgan Kaufmann, Burlington (1993)

11. Rahman, P., Jiang, L., Nandi, A.: Evaluating interactive data systems. VLDB J. **29**(1), 119–146 (2020)

12. Stack Overflow 2022 Developer Survey. https://survey.stackoverflow.co/2022/# section-most-loved-dreaded-and-wanted-programming-scripting-and-markup-languages. Accessed 29 Aug 2022

13. Ste, A.: How to Become More Marketable as a Data Scientist. https://www. kdnuggets.com/2019/08/marketable-data-scientist.html. Accessed 29 Aug 2022

14. Tun, L.L.: Why Non-Programmers Should Learn SQL. https://www.udacity.com/ blog/2020/06/why-non-programmers-should-learn-sql.html. Accessed 29 Aug 2022

15. UCI Machine Learning Repository, Wine Data Set. https://archive.ics.uci.edu/ml/ datasets/wine. Accessed 15 Aug 2022

Mathematical Models for Development of Intelligent Systems

Analytical Solution of the Simplest Entropiece Inversion Problem

Jean Dezert[1], Florentin Smarandache[2(✉)], and Albena Tchamova[3]

[1] The French Aerospace Lab, Palaiseau, France
jean.dezert@onera.fr
[2] Department of Mathematics, University of New Mexico, Gallup, NM, USA
smarand@unm.edu
[3] IICT, Bulgarian Academy of Sciences, Sofia, Bulgaria
tchamova@bas.bg

Abstract. In this paper, we present a method to solve analytically the simplest Entropiece Inversion Problem (EIP). This theoretical problem consists in finding a method to calculate a Basic Belief Assignment (BBA) from the knowledge of a given entropiece vector which quantifies effectively the measure of uncertainty of a BBA in the framework of the theory of belief functions. We give an example of the calculation of EIP solution for a simple EIP case, and we show the difficulty to establish the explicit general solution of this theoretical problem that involves transcendental Lambert's functions.

Keywords: Belief functions · Entropy · Measure of uncertainty

1 Introduction

In this paper, we suppose the reader to be familiar with the theory of Belief Functions (BF) introduced by Shafer in [1], and we do not present in details the basics of BF. We just recall that a frame of discernement (FoD) $\Theta = \{\theta_1, \theta_2, \ldots, \theta_N\}$ is a finite exhaustive set of $N > 1$ mutually exclusive elements θ_i ($i = 1, \ldots, N$), and its power set (i.e. the set of all subsets) is denoted by 2^Θ. A FoD represents a set of potential solutions of a decision-making problem under consideration. A Basic Belief Assignment (BBA)[1] is a mapping $m : 2^\Theta \to [0, 1]$ with $m(\emptyset) = 0$, and $\sum_{X \in 2^\Theta} m(X) = 1$.

A new effective entropy measure $U(m)$ for any BBA $m(\cdot)$ defined on a FoD Θ has been defined as follows [2]:

$$U(m) = \sum_{X \in 2^\Theta} s(X) \tag{1}$$

---

[1] For notation convenience, we denote by m or $m(\cdot)$ any BBA defined implicitly on the FoD Θ, and we also denote it as m^Θ to explicitly refer to the FoD when necessary.

© The Author(s), under exclusive license to Springer Nature Switzerland AG 2023
D. Simian and L. F. Stoica (Eds.): MDIS 2022, CCIS 1761, pp. 221–233, 2023.
https://doi.org/10.1007/978-3-031-27034-5_15

where $s(X)$ is named the *entropiece* of X, which is defined by

$$s(X) = -m(X)(1 - u(X))\log(m(X)) + u(X)(1 - m(X)) \tag{2}$$

with

$$u(X) = Pl(X) - Bel(X) = \sum_{Y \in 2^{\Theta} | X \cap Y \neq \emptyset} m(Y) - \sum_{Y \in 2^{\Theta} | Y \subseteq X} m(Y). \tag{3}$$

$Pl(X)$ and $Bel(X)$ are respectively the plausibility and the belief of the element X of the power set of Θ, see [1] for details. $u(X)$ quantifies the imprecision of the unknown probability of X. The vacuous BBA characterizing the total ignorant source of evidence is denoted by m_v, and it is such that $m_v(\Theta) = 1$ and $m_v(X) = 0$ for any $X \subset \Theta$.

This measure of uncertainty $U(m)$ (i.e. entropy measure) is effective because it satisfies the following four essential properties [2]:

1. $U(m) = 0$ for any BBA $m(\cdot)$ focused on a singleton X of 2^{Θ}.
2. $U(m_v^{\Theta}) < U(m_v^{\Theta'})$ if $|\Theta| < |\Theta'|$.
3. $U(m) = -\sum_{X \in \Theta} m(X)\log(m(X))$ if $m(\cdot)$ is a Bayesian[2] BBA. Hence, $U(m)$ reduces to Shannon entropy [7] in this case.
4. $U(m) < U(m_v)$ for any non-vacuous BBA $m(\cdot)$ and for the vacuous BBA $m_v(\cdot)$ defined with respect to the same FoD.

The proof of the three first properties is quite simple to make, whereas the proof of $U(m) < U(m_v)$ is much more difficult, see [2] for proofs and examples. A detailed analysis of other (non-effective) entropy measures proposed in the literature during the last four decades is done in [3].

The entropiece $s(X)$ given by (2) corresponds to the contribution of X to the whole uncertainty measure $U(m)$. The entropiece $s(X)$ involves $m(X)$ and the imprecision $u(X) = Pl(X) - Bel(X)$ about the unknown probability of X in a subtle interwoven manner named *epistemic entanglement*. The cardinality of X is indirectly taken into account in the derivation of $s(X)$ thanks to $u(X)$ which requires the derivation of $Pl(X)$ and $Bel(X)$ functions that depend on the cardinality of X. Because $u(X) \in [0,1]$ and $m(X) \in [0,1]$ one has $s(X) \geq 0$, and $U(m) \geq 0$. The quantity $U(m)$ is expressed in *nats* because we use the natural logarithm. $U(m)$ can be expressed in *bits* by dividing the $U(m)$ value in *nats* by $\log(2) = 0.69314718....$ This measure of uncertainty $U(m)$ is a continuous function in its basic belief mass arguments because it is a summation of continuous functions. In formula (2), we always take $m(X)\log(m(X)) = 0$ when $m(X) = 0$ because $\lim_{m(X)\to 0^+} m(X)\log(m(X)) = 0$ which can be proved using L'Hôpital rule [4]. Note that for any BBA m, one has always $s(\emptyset) = 0$ because $m(\emptyset) = 0$ and $u(\emptyset) = Pl(\emptyset) - Bel(\emptyset) = 0 - 0 = 0$. For the vacuous BBA, one has $s(\Theta) = 0$ because $m_v(\Theta) = 1$ and $u(\Theta) = Pl(\Theta) - Bel(\Theta) = 1 - 1 = 0$.

---

[2] m is Bayesian BBA if it has only singletons as focal elements, i.e. $m(\theta_i) > 0$ for some $\theta_i \in \Theta$ and $m(X) = 0$ for all non-singletons X of 2^{Θ}.

As proved in [2], the entropy of the vacuous BBA on the FoD Θ is equal to

$$U(m_v) = 2^{|\Theta|} - 2 \tag{4}$$

This maximum entropy value $2^{|\Theta|} - 2$ makes perfectly sense because for the vacuous BBA there is no information at all about the conflicts between the elements of the FoD. Actually for all $X \in 2^{\Theta} \setminus \{\emptyset, \Theta\}$ one has $u(X) = 1$ because $[Bel(X), Pl(X)] = [0, 1]$, and one has $u(\emptyset) = 0$ and $u(\Theta) = 0$. Hence, the sum of all imprecisions of $P(X)$ for all $X \in 2^{\Theta}$ is exactly equal to $2^{|\Theta|} - 2$ which corresponds to $U(m_v)$ as expected. Moreover, one has always $U(m_v) > \log(|\Theta|)$ which means that the vacuous BBA has always an entropy greater than the maximum of Shannon entropy $\log(|\Theta|)$ obtained with the uniform probability mass function distributed on Θ. As a dual concept of this entropy measure $U(m)$, we have defined in [8] the measure of information content of any BBA by

$$IC(m) = U(m_v) - U(m) = (2^{|\Theta|} - 2) - \sum_{X \in 2^{\Theta}} s(X) \tag{5}$$

From the definition (5), one sees that for $m \neq m_v^{\Theta}$ one has $IC(m) > 0$ because $U(m) < U(m_v)$, and for $m = m_v$ one has $IC(m_v) = 0$ (i.e. the vacuous BBA carries no information), which is what we naturally expect.

Note that the information content $IC(m^{\Theta})$ of a BBA depends not only of the BBA $m(\cdot)$ itself but also on the cardinality of the frame of discernment Θ because $IC(m)$ requires the knowledge of $|\Theta| = N$ to calculate the max entropy value $U(m_v) = 2^{|\Theta|} - 2$ entering in (5). This remark is important to understand that even if two BBAs (defined on different FoDs) focus entirely on a same focal element, their information contents are necessarily different. This means that the information content depends on the context of the problem, i.e. the FoD. The notions of information gain and information loss between two BBAs are also mathematically defined in [8] for readers interested in this topic.

This paper is organized as follows. Section 2 defines the general entropiece inversion problem (EIP). Section 3 describes the simplest entropiece inversion problem (SEIP). An analytical solution of SEIP is proposed and it is applied on a simple example also in Sect. 3. The conclusion is made in Sect. 4.

2   The General Entropiece Inversion Problem (EIP)

The set $\{s(X), X \in 2^{\Theta}\}$ of the entropieces values $s(X)$ given by (2) can be represented by an entropiece vector $\mathbf{s}(m) = [s(X), X \in 2^{\Theta}]^T$, where any order of elements X of the power set 2^{Θ} can be chosen. For simplicity, we suggest to use the classical N-bits representation if $|\Theta| = N$, with the increasing order (see example in Sect. 3). The general Entropiece Inversion Problem, or EIP for short, is an interesting theoretical problem which can be easily stated as follows:

Suppose that if the entropiece vector $\mathbf{s}(m)$ known (estimated or given), is it possible to calculate a BBA $m(\cdot)$ corresponding to this entropiece vector $\mathbf{s}(m)$? and how?

Also we would like to know if the derivation of $m(\cdot)$ from $\mathbf{s}(m)$ provides a unique BBA solution, or not?

This general entropiece inversion problem is a challenging mathematical problem, and we do not know if a general analytical solution of EIP is possible, or not. We leave it as an open mathematical question for future research. However, we present in this paper the analytical solution for the simplest case where the FoD Θ has only two elements, i.e. when $|\Theta| = N = 2$. Even in this simplest case, the EIP solution is no so easy to calculate as it will be shown in the next section. This is the main contribution of this paper.

The mathematical EIP addressed in this paper is not related (for now) to any problem for the natural world and it cannot be confirmed experimentally using data from nature because the entropy concept is not directly measurable, but only computable from the estimation of probability $p(\cdot)$ or belief mass functions $m(\cdot)$. So, why do we address this entropiece inversion problem? Because in advanced information fusion systems we can imagine to have potentially access to this type of information and it makes sense to assess the underlying BBA provided by a source of evidence to eventually modify it in some fusion systems for some aims. We could also imagine to make adjustments of entropieces values to voluntarily improve (or degrade) $IC(m)$, and to generate the proper modified BBA for some tasks. At this early stage of research work it is difficult to anticipate the practical interests of the calculation of solutions of the general EIP, but to present its mathematical interest for now.

3   The Simplest Entropiece Inversion Problem (SEIP)

3.1   Example

We consider a FoD Θ with only two elements, say $\Theta = \{A, B\}$, where A and B are mutually exclusive and exhaustive, and the following BBA

$$m(A) = 0.5, \quad m(B) = 0.3, \quad m(A \cup B) = 0.2$$

Because $[Bel(\emptyset), Pl(\emptyset)] = [0, 1]$ one has $u(\emptyset) = 0$. Because $[Bel(A), Pl(A)] = [0.5, 0.7]$, $[Bel(B), Pl(B)] = [0.3, 0.5]$, $[Bel(\Theta), Pl(\Theta)] = [1, 1]$, one has $u(A) = 0.2$, $u(B) = 0.2$, and $u(\Theta) = 0$. Applying (2), one gets $s(\emptyset) = 0$, $s(A) \approx 0.377258$, $s(B) \approx 0.428953$ and $s(\Theta) \approx 0.321887$. Using the 2-bits representation with increasing ordering[3], we encode the elements of the power set as $\emptyset = 00$, $A = 01$, $B = 10$ and $A \cup B = 11$. The entropiece vector is

$$\mathbf{s}(m^\Theta) = \begin{bmatrix} s(\emptyset) \\ s(A) \\ s(B) \\ s(A \cup B) \end{bmatrix} \approx \begin{bmatrix} 0 \\ 0.3773 \\ 0.4290 \\ 0.3219 \end{bmatrix} \tag{6}$$

---

[3] Once the binary values are converted into their digit value with the most significant bit on the left (i.e. the least significant bit on the right).

If we use the classical 2-bits (here $|\Theta| = 2$) representation with increasing ordering (as we recommend) the first component of entropiece vector $\mathbf{s}(m)$ will be $s(\emptyset)$ which is always equal to zero for any BBA m, hence the first component of $\mathbf{s}(m)$ is always zero and it can be dropped (i.e. removed of the vector representation actually). By summing all the components of the entropiece vector $\mathbf{s}(m)$ we obtain the entropy $U(m) \approx 1.128098$ nats of the BBA $m(\cdot)$. Note that the components $s(X)$ (for $X \neq \emptyset$) of the entropieces vector $\mathbf{s}(m)$ are not independent because they are linked to each other through the calculation of $Bel(X)$ and $Pl(X)$ values entering in $u(X)$.

3.2 Analytical Solution of SEIP

Because we suppose $\Theta = \{A, B\}$, the expression of three last components[4] of the entropiece vector $\mathbf{s}(m)$ are given by (2), and we have

$$s(A) = -m(A)(1 - u(A)) \log(m(A)) + u(A)(1 - m(A))$$
$$s(B) = -m(B)(1 - u(B)) \log(m(B)) + u(B)(1 - m(B))$$
$$s(A \cup B) = -m(A \cup B)(1 - u(A \cup B)) \log(m(A \cup B)) + u(A \cup B)(1 - m(A \cup B))$$

Because $u(A) = Pl(A) - Bel(A) = (m(A) + m(A \cup B)) - m(A) = m(A \cup B)$, $u(B) = Pl(B) - Bel(B) = (m(B) + m(A \cup B)) - m(B) = m(A \cup B)$ and $u(A \cup B) = Pl(A \cup B) - Bel(A \cup B) = 1 - 1 = 0$, one gets the following system of equations to solve

$$s(A) = -m(A)(1 - m(A \cup B)) \log(m(A)) + m(A \cup B)(1 - m(A)) \quad (7)$$
$$s(B) = -m(B)(1 - m(A \cup B)) \log(m(B)) + m(A \cup B)(1 - m(B)) \quad (8)$$
$$s(A \cup B) = -m(A \cup B) \log(m(A \cup B)) \quad (9)$$

The set of Eqs. (7), (8) and (9) is called the EIP *transcendental equation system* for the case $|\Theta| = 2$.

The plot of function $s(A \cup B) = -m(A \cup B) \log(m(A \cup B))$ is given in Fig. 1 for convenience. By derivating the function $-m(A \cup B) \log(m(A \cup B))$ we see that its maximum value is obtained for $m(A \cup B) = 1/e \approx 0.3679$ for which

$$s(A \cup B) = -\frac{1}{e} \log(1/e) = \frac{1}{e} \log(e) = \frac{1}{e}$$

Therefore, the numerical value of $s(A \cup B)$ always belongs to the interval $[0, 1/e]$.

Without loss of generality, we assume $0 < s(A \cup B) \le 1/e$ because if $s(A \cup B) = 0$ then one deduces directly without ambiguity that either $m(A \cup B) = 1$ (which means that the BBA $m(\cdot)$ is the vacuous BBA) if $s(A) = s(B) = 1$, or $m(A \cup B) = 0$ otherwise. With the assumption $0 < s(A \cup B) \le 1/e$, the Eq. (9) is of the general transcendental form

$$ye^y = a \Leftrightarrow \log(m(A \cup B))m(A \cup B) = -s(A \cup B) \quad (10)$$

---

[4] We always omit the 1st component $s(\emptyset)$ of entropiece vector $\mathbf{s}(m)$ which is always equal to zero and not necessary in our analysis.

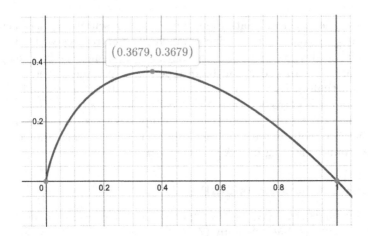

Fig. 1. Plot of $s(A \cup B) = -m(A \cup B) \log(m(A \cup B))$ (in red) with x-axis equals $m(A \cup B) \in [0, 1]$, and y-axis equals $s(A \cup B)$ in nats. (Color figure online)

by considering the known value as $a = -s(A \cup B)$ in $[-\frac{1}{e}, 0)$, and the unknown as $y = \log(m(A \cup B))$.

Unfortunately the solution of the transcendental Eq. (10) does not have an explicit expression involving simple functions. Actually, the solution of this equation is actually given by the Lambert's W-function which is a multivalued function (called also the omega function or product logarithm in mathematics) [6]. It can however be calculated[5] with a good precision by some numerical methods - see [5] for details. The equation $ye^y = a$ admits real solution(s) only if $a \geq -\frac{1}{e}$. For $a \geq 0$, the solution of $ye^y = a$ is $y = W_0(a)$, and for $-\frac{1}{e} \leq a < 0$ there are two possible real values of $W(a)$ - see Fig. 1 of [5] which are denoted respectively $y_1 = W_0(a)$ and $y_2 = W_{-1}(a)$. The principal branch of the Lambert's function $W(x)$ satisfying $-1 \leq W(x)$ is denoted $W_0(x)$, and the branch satisfying $W(x) \leq -1$ is denoted by $W_{-1}(x)$ by Corless et al. in [5]. In our context because we have $a \in [-\frac{1}{e}, 0)$, the solutions of $ye^y = a$ are given by

$$y_1 = W_0(a) = W_0(-s(A \cup B))$$
$$y_2 = W_{-1}(a) = W_{-1}(-s(A \cup B))$$

Hence we get two possible solutions for the value of $m(A \cup B)$, which are

$$m_1(A \cup B) = e^{y_1} = e^{W_0(-s(A \cup B))} \tag{11}$$
$$m_2(A \cup B) = e^{y_2} = e^{W_{-1}(-s(A \cup B))} \tag{12}$$

Of course, at least one of these solutions is necessarily correct but we do not know which one. So, at this current stage, we must consider[6] and the two

---

[5] Lambert's W-function is implemented in Matlab^{TM} as *lambertw* function.

[6] If the two masses values are admissible, that is if $m_1(A \cup B) \in [0, 1]$ and if $m_2(A \cup B) \in [0, 1]$. If one of them is non-admissible it is eliminated.

solutions $m_1(A \cup B)$ and $m_1(A \cup B)$ for $m(A \cup B)$ as acceptable, and we must continue to solve Eqs. (7) and (8) to determine the mass values $m(A)$ and $m(B)$.

Let's now determine $m(A)$ at first by solving (7). Suppose we set the value of $m(A \cup B)$ is known and taken either as $m_1(A \cup B)$, or as $m_2(A \cup B)$, then we can rearrange the Eq. (7) as

$$-\frac{s(A) - m(A \cup B)}{1 - m(A \cup B)} = m(A)[\log(m(A)) + \frac{m(A \cup B)}{1 - m(A \cup B)}]$$

which can be rewritten as the general equation of the form

$$(y + a)e^y = b \tag{13}$$

by taking

$$y = \log(m(A)) \tag{14}$$

$$a = \frac{m(A \cup B)}{1 - m(A \cup B)} \tag{15}$$

$$b = -\frac{s(A) - m(A \cup B)}{1 - m(A \cup B)} \tag{16}$$

The solution of (13) are given by [5]

$$y = W(be^a) - a \tag{17}$$

Once y is calculated by formula (17) and since $y = \log(m(A))$ we obtain the solution for $m(A)$ given by

$$m(A) = e^y = e^{W(be^a)-a} \tag{18}$$

Similarly, the solution for $m(B)$ will be given by

$$m(B) = e^y = e^{W(be^a)-a} \tag{19}$$

by solving the equation $(y + a)e^y = b$ with

$$y = \log(m(B)) \tag{20}$$

$$a = \frac{m(A \cup B)}{1 - m(A \cup B)} \tag{21}$$

$$b = -\frac{s(B) - m(A \cup B)}{1 - m(A \cup B)} \tag{22}$$

We must however check if there is one solution only $m(A) = e^{W_0(be^a)-a}$, or in fact two solutions $m_1(A) = e^{W_0(be^a)-a}$ and $m_2(A) = e^{W_{-1}(be^a)-a}$, and similarly for the solution for $m(B)$. This depends on the parameters a and b with respect to $[-1/e, 0)$ interval and $[0, \infty)$.

We illustrate in the next subsection how to calculate the SEIP solution from these analytical formulas for the previous example.

3.3   SEIP Solution of the Previous Example

We recall that we have for this example $s(\emptyset) = 0$, $s(A) \approx 0.3773$, $s(B) \approx 0.4290$ and $s(\Theta) \approx 0.3219$. If we apply formulas (11)–(12) for this example, we have $a = -s(A \cup B) = -0.3219$ and therefore

$$y_1 = W_0(-0.3219) = -0.5681$$
$$y_2 = W_{-1}(-0.3219) = -1.6094$$

Hence the two potential solutions for the mass $m(A \cup B)$ are

$$m_1(A \cup B) = e^{y_1} \approx 0.5666$$
$$m_2(A \cup B) = e^{y_2} = 0.2000$$

It can be easily verified that

$$- m_1(A \cup B) \log(m_1(A \cup B)) = 0.3219 = s(A \cup B)$$
$$- m_2(A \cup B) \log(m_2(A \cup B)) = 0.3219 = s(A \cup B)$$

We see that the second potential solution $m_2(A \cup B) = 0.2000$ is the solution that corresponds to the original mass of $A \cup B$ of the BBA $m(A \cup B)$ of our example.

Now, we examine what would be the values of $m(A)$ and $m(B)$ given respectively by (18) and (19) by taking either $m(A \cup B) = m_1(A \cup B) = 0.5666$ or $m(A \cup B) = m_2(A \cup B) = 0.20$.

– Let's examine the 1st possibility with the potential solution

$$m(A \cup B) = m_1(A \cup B) = 0.5666$$

For determining $m(A)$, we have to solve $(y + a)e^y = b$ with the unknown $y = \log(m(A))$ and with

$$a = \frac{m(A \cup B)}{1 - m(A \cup B)} \approx \frac{0.5666}{1 - 0.5666} = 1.3073$$

$$b = -\frac{s(A) - m(A \cup B)}{1 - m(A \cup B)} \approx -\frac{0.3773 - 0.5666}{1 - 0.5666} = 0.4369$$

Hence, $be^a = 0.4368 \cdot e^{1.3073} \approx 1.6148$.
Applying formula (18), one gets[7]

$$m_1(A) = e^{W_0(be^a) - a} = 0.5769$$
$$m_2(A) = e^{W_{-1}(be^a) - a} = -0.0216 + 0.0924i$$

---

[7] Using *lambertw* Matlab™ function.

For determining $m(B)$ we have to solve $(y + a)e^y = b$ with the unknown $y = \log(m(B))$ and with

$$a = \frac{m(A \cup B)}{1 - m(A \cup B)} \approx \frac{0.5666}{1 - 0.5666} = 1.3073$$

$$b = -\frac{s(B) - m(A \cup B)}{1 - m(A \cup B)} \approx -\frac{0.4290 - 0.5666}{1 - 0.5666} = 0.3176$$

Hence, $be^a = 0.3176 \cdot e^{1.3073} \approx 1.1739$.
Applying formula (19), one gets

$$m_1(B) = e^{W_0(be^a) - a} = 0.5065$$

$$m_2(B) = e^{W_{-1}(be^a) - a} = -0.0204 + 0.0657i$$

One sees that there is no effective choice for the values of $m(A)$ and $m(B)$ if we suppose $m(A \cup B) = m_1(A \cup B) = 0.5666$ because if one takes as real values solutions $m(A) = m_1(A) = 0.5769$ and $m(B) = m_1(B) = 0.5065$ one would get

$$m(A) + m(B) + m(A \cup B) = 0.5769 + 0.5065 + 0.5666 = 1.65$$

which is obviously greater than one. This generates an improper BBA.

– Let's consider the 2nd possibility with the potential solution

$$m(A \cup B) = m_2(A \cup B) = 0.20$$

For determining $m(A)$, we have to solve $(y + a)e^y = b$ with the unknown $y = \log(m(A))$ and with

$$a = \frac{m(A \cup B)}{1 - m(A \cup B)} = \frac{0.20}{1 - 0.20} = 0.25$$

$$b = -\frac{s(A) - m(A \cup B)}{1 - m(A \cup B)} \approx -\frac{0.3773 - 0.20}{1 - 0.20} = -0.2216$$

Hence, $be^a = -0.2216 \cdot e^{0.25} \approx -0.2845$.

$$m_1(A) = e^{W_0(be^a) - a} = 0.5000$$

$$m_2(A) = e^{W_{-1}(be^a) - a} = 0.1168$$

For determining $m(B)$ we have to solve $(y + a)e^y = b$ with the unknown $y = \log(m(B))$ and with

$$a = \frac{m(A \cup B)}{1 - m(A \cup B)} \approx \frac{0.20}{1 - 0.20} = 0.25$$

$$b = -\frac{s(B) - m(A \cup B)}{1 - m(A \cup B)} \approx -\frac{0.4290 - 0.20}{1 - 0.20} = -0.2862$$

Hence, $be^a = -0.2862 \cdot e^{0.25} \approx -0.3675$.

Applying formula (19), one gets

$$m_1(B) = e^{W_0(be^a)-a} = 0.3000$$
$$m_2(B) = e^{W_{-1}(be^a)-a} = 0.2732$$

Based on this 2nd possibility for potential solution $m(A \cup B) = 0.20$, one sees that the only possible effective choice of mass values $m(A)$ and $m(B)$ is to take $m(A) = m_1(A) = 0.50$ and $m(B) = m_1(B) = 0.30$ which gives the proper sought BBA such that $m(A) + m(B) + m(A \cup B) = 1$ which exactly corresponds to the original BBA that has been used to generate the entropiece vector $s(m)$ for this example.

In summary, for the case $|\Theta| = 2$ it is always possible to calculate the BBA $m(\cdot)$ from the knowledge of the entropiece vector, and the solution of SEIP is obtained by analytical formulas.

3.4 Remark

In the very particular case where $s(A \cup B) = 0$ the Eq. (9) reduces to

$$- m(A \cup B) \log(m(A \cup B)) = 0 \qquad (23)$$

which has two possible solutions $m(A \cup B) = m_1(A \cup B) = 1$, and $m(A \cup B) = m_2(A \cup B) = 0$.

If $m(A \cup B) = 1$, then it means that necessarily the BBA is the vacuous BBA, and so $m(A) = m(B) = 0$, $u(A) = Pl(A) - Bel(A) = 1$, $u(B) = Pl(B) - Bel(B) = 1$. Therefore[8]

$$
\begin{aligned}
s(A) &= -m(A)(1 - u(A)) \log(m(A)) + u(A)(1 - m(A)) \\
&= -m(A)(1 - m(A \cup B)) \log(m(A)) + m(A \cup B)(1 - m(A)) \\
&= 0(1 - 1) \log(0) + 1(1 - 0) = 1 \\
s(B) &= -m(B)(1 - u(B)) \log(m(B)) + u(B)(1 - m(B)) \\
&= -m(B)(1 - m(A \cup B)) \log(m(B)) + m(A \cup B)(1 - m(B)) \\
&= 0(1 - 1) \log(0) + 1(1 - 0) = 1
\end{aligned}
$$

So the choice of $m(A \cup B) = m_1(A \cup B) = 1$ is the only possible if the entropiece vector is $s(m) = [110]^T$.

---

[8] We use the formal notation $\log(0)$ even if $\log(0)$ is $-\infty$ because in our derivations we have always a $0 \log(0)$ product which is equal to zero due to L'Hôpital's rule [4].

If $s(A) < 1$, or if $s(B) < 1$ (or both) then we must choose $m(A \cup B) = m_2(A \cup B) = 0$, and in this case we have to solve the equations

$$
\begin{aligned}
s(A) &= -m(A)(1 - u(A)) \log(m(A)) + u(A)(1 - m(A)) \\
&= -m(A)(1 - m(A \cup B)) \log(m(A)) + m(A \cup B)(1 - m(A)) \\
&= -m(A) \log(m(A)) \\
s(B) &= -m(B)(1 - u(B)) \log(m(B)) + u(B)(1 - m(B)) \\
&= -m(B)(1 - m(A \cup B)) \log(m(B)) + m(A \cup B)(1 - m(B)) \\
&= -m(B) \log(m(B))
\end{aligned}
$$

The possible solutions of equation $s(A) = -m(A) \log(m(A))$ are given by

$$m_1(A) = e^{W_0(-s(A))} \tag{24}$$
$$m_2(A) = e^{W_{-1}(-s(A))} \tag{25}$$

and the possible solutions of equation $s(B) = -m(B) \log(m(B))$ are given by

$$m_1(B) = e^{W_0(-s(B))} \tag{26}$$
$$m_2(B) = e^{W_{-1}(-s(B))} \tag{27}$$

In this particular case where $s(A \cup B) = 0$, and $s(A) < 1$ or $s(B) < 1$, we have to select the pair of possible solutions among the four possible choices

$$
\begin{aligned}
(m(A), m(B)) &= (m_1(A), m_1(B)), \\
(m(A), m(B)) &= (m_1(A), m_2(B)), \\
(m(A), m(B)) &= (m_2(A), m_1(B)), \\
(m(A), m(B)) &= (m_2(A), m_2(B)).
\end{aligned}
$$

The judicious choice of pair $(m(A), m(B))$ must satisfy the proper BBA constraint $m(A) + m(B) + m(A \cup B) = 1$, where $m(A \cup B) = 0$ because $s(A \cup B) = 0$ in this particular case.

For instance, if we consider $\Theta = \{A, B\}$ and the following (bayesian) BBA

$$m(A) = 0.6, m(B) = 0.4, m(A \cup B) = 0$$

The entropiece vector $\mathbf{s}(m)$ is

$$
\mathbf{s}(m) = \begin{bmatrix} s(A) \\ s(B) \\ s(A \cup B) \end{bmatrix} \approx \begin{bmatrix} 0.3065 \\ 0.3665 \\ 0 \end{bmatrix} \tag{28}
$$

Hence from $s(m)$ we can deduce $m(A \cup B) = 0$ because we cannot consider $m(A \cup B) = 1$ as a valid solution because $s(A) < 1$ and $s(B) < 1$. The possible solutions of equation $s(A) = -m(A)\log(m(A))$ are

$$m_1(A) = e^{W_0(-s(A))} = e^{W_0(-0.3065)} = 0.6000$$
$$m_2(A) = e^{W_{-1}(-s(A))} = e^{W_{-1}(-0.3065)} = 0.1770$$

and the possible solutions of equation $s(B) = -m(B)\log(m(B))$ are

$$m_1(B) = e^{W_0(-s(B))} = e^{W_0(-0.3665)} = 0.4000$$
$$m_2(B) = e^{W_{-1}(-s(B))} = e^{W_{-1}(-0.3665)} = 0.3367$$

One sees that the only effective (or judicious) choice for $m(A)$ and $m(B)$ is to take $m(A) = m_1(A) = 0.60$ and $m(B) = m_1(B) = 0.40$, which coincides with the original bayesian BBA that has been used to generate the entropiece vector $s(m) = [0.3065, 0.3665, 0]^T$.

4   Conclusion

In this paper we have introduced for the first time the entropiece inversion problem (EIP) which consists in calculating a basic belief assignment from the knowledge of a given entropiece vector which quantifies effectively the measure of uncertainty of a BBA in the framework of the theory of belief functions. The general analytical solution of this mathematical problem is a very challenging open problem because it involves transcendental equations. We have shown however how it is possible to obtain an analytical solution for the simplest EIP involving only two elements in the frame of discernment. Even in this simplest case the analytical solution of EIP is not easy to obtain because it requires a calculation of values of the transcendental Lambert's functions. Even if no general analytical formulas are found for the solution of general EIP, it would be interesting to develop numerical methods to approximate the general EIP solution, and to exploit it in future advanced information fusion systems.

References

1. Shafer, G.: A Mathematical Theory of Evidence. Princeton University Press, Princeton (1976)
2. Dezert, J.: An effective measure of uncertainty of basic belief assignments. In: Fusion 2022 International Conference on Proceedings, ISIF Editor, Linköping, Sweden (2022)
3. Dezert, J., Tchamova, A.: On effectiveness of measures of uncertainty of basic belief assignments, information & security journal. Int. J. (ISIJ) **52**, 9–36 (2022)
4. Bradley, R.E., Petrilli, S.J., Sandifer, C.E.: L'Hôpital's analyse des infiniments petits (An annotated translation with source material by Johann Bernoulli), Birkhäuser, p 311 (2015)

5. Corless, R.M., Gonnet, G.H., Hare, D.E.G., Jeffrey, D.J., Knuth, D.E.: On the Lambert W function. Adv. Comput. Math. **5**, 329–359 (1996)
6. Lambert W function (2022). https://en.wikipedia.org/wiki/Lambert_W_function. Accessed 1 Dec 2022
7. Shannon, C.E.: A mathematical theory of communication, in [9] and in The Bell System Technical Journal, **27**, 379–423 & 623–656 (1948)
8. Dezert, J., Tchamova, A., Han, D.: Measure of information content of basic belief assignments. In: Le Hégarat-Mascle, S., Bloch, I., Aldea, E. (eds.) BELIEF 2022. LNCS, vol. 13506, pp. 119–128. Springer, Cham (2022). https://doi.org/10.1007/978-3-031-17801-6_12

Latent Semantic Structure in Malicious Programs

John Musgrave[1(✉)], Temesguen Messay-Kebede[2], David Kapp[2],
and Anca Ralescu[1]

[1] University of Cincinnati, Cincinnati, OH, USA
`musgrajw@mail.uc.edu, ralescal@ucmail.uc.edu`
[2] Air Force Research Lab, Wright-Patt Air Force Base, Dayton, OH, USA
`{temesgen.kebede.1,david.kapp}@us.af.mil`

Abstract. Latent Semantic Analysis is a method of matrix decomposition used for discovering topics and topic weights in natural language documents. This study uses Latent Semantic Analysis to analyze the composition binaries of malicious programs. The semantic representation of the term frequency vector representation yields a set of topics, each topic being a composition of terms. The vectors and topics were evaluated quantitatively using a spatial representation. This semantic analysis provides a more abstract representation of the program derived from its term frequency analysis. We use a metric space to represent a program as a collection of vectors, and a distance metric to evaluate their similarity within a topic. The segmentation of the vectors in this dataset provides increased resolution into the program structure.

Keywords: Malware analysis · Latent semantic analysis · Security

1 Introduction

Analysis methods for malicious programs face challenges to the interpretation of results. This is due to the loss of semantic information when languages are translated between abstraction layers. A "semantic gap" exists, which is a gap between higher level language specification or theorem proving and the lower level instructions obtained from reverse engineering binary programs into their instruction sets. Proofs about the operational semantics of higher level languages are not easily translated to other layers in the abstraction hierarchy, such as binary or hex representations [15,21].

A representation of operational semantics without abstraction cannot easily bridge the gap between architectural layers. If program semantics are to be correlated across architectural layers, a greater level of abstraction is required. Once a more abstract representation is found, then it is possible for programs to be classified by the identification of patterns in their structural properties.

© The Author(s), under exclusive license to Springer Nature Switzerland AG 2023
D. Simian and L. F. Stoica (Eds.): MDIS 2022, CCIS 1761, pp. 234–246, 2023.
https://doi.org/10.1007/978-3-031-27034-5_16

```
mov      ecx , rbp − 44
mov      eax , ecx
and      eax , 400
or       eax , 140
or       ecx , 1
cmp      rip + 170, 0
cmovne   ecx , eax
mov      rbp − 44, ecx
mov      rip + 180, 0
jmp      0x100000000
```

Fig. 1. Basic block segment of assembly instructions

Abstraction enables the potential identification of a process generating patterns of syntax to be uncovered. Greater levels of abstraction allow for the ability to predict features of syntax. These syntactic features can be predicted based on the structure of syntax provided by the abstraction. In the case of Griffiths, Steyvers, and Tenenbaum, this was accomplished by the addition of a bi-partite graph with nodes representing concepts and words [6,20].

The strength of a semantic representation depends on the ability of the representation to express structural properties, to express syntactic patterns at a sufficient level of abstraction, and the degree of accuracy. For greater accuracy we use a segmentation approach, which provides a more fine grained resolution.

One approach to providing both abstraction and structure is to discover the latent or hidden structure of a program by the analysis of terms in the document, called $tf - idf$ methods. One advantage this offers is a representation not based on specific patterns of syntax. A term frequency representation of a program contains latent structure with a greater level of abstraction, and datasets using term frequency representations can be analyzed with natural language analysis methods. The abstraction of a latent structure would be a more descriptive feature of program semantic structure, as the correlation of syntactic features would be easily explained by the structure, and would allow for prediction.

Latent Semantic Analysis is already a widely used method of analysis for the discovery of topics in documents of natural language, and uses matrix decomposition to derive a more abstract representation with respect to a document's semantic content [6,11,17,20].

We use Latent Semantic Analysis to provide a greater level of abstraction on a segmented term-frequency dataset. We construct a metric space to evaluate the structure of a program through a spatial representation. In order to have a measure of correlation in the metric space, we use Cosine Similarity as a similarity metric.

1.1 Background

Malware analysis has used Machine Learning methods to offer new solutions to program analysis. Patterns of both syntax and semantics have been successfully classified using machine learning methods, but semantic program representation is still a difficult problem [19].

Latent Semantic Analysis has shown to be an effective method of analyzing natural language [5,6].

The process of Singular Value Decomposition (SVD) obtained from a linear algebraic approach allows the decomposition of matrices into orthonormal bases, along with weights [2,11,17].

1.2 Related Work

Machine Learning has been applied in many contexts to successfully identify malicious programs based on a variety of features. Several different classification methods have been used for supervised learning including artificial neural networks and support vector machines. Many datasets have been collected with several different kinds of features, including assembly instructions, n-gram sequences of instructions and system calls, and program metadata, [3,4,10,14,19].

A number of studies have explored the use of static features at the level of file format, and their impact on the recognition of malicious programs. A decision tree was used as a method of classification for Windows PE files. Subsequent studies have focused on malware classification using ensemble methods which include random forest with support vector machines and principal component analysis that was focused on file header features of Trojans [16,18,22].

While many studies of supervised learning rely on on $tf - idf$ representations for labeled data, n-gram sequences are typically computed linearly based on the placement of terms in the document as a whole, and are not segmented into blocks. The instruction sequence is determined by the structural properties of the control flow graph, and not the placement in the document as in natural language documents. This structure is often evaluated in isolation, and not included in the term frequency representation. The evaluation of term frequencies in the document without segmentation assumes a linear sequence of terms with respect to their spatial placement, and this is not the case for executable binaries. While edges in a control flow graph define the sequence order, sequential statements in the program are represented by nodes in the control flow graph. This leads us to use a representation that is segmented in order to provide a suitable level of resolution.

The absence of accurate segmentation of a term frequency representation in program analysis would introduce noise from the existence of n-gram sequences that are not representative of deterministic instruction sequences.

1.3   Outline

Section 2 covers the experiments performed. Section 3 covers the Results and Discussion. Section 4 is a Summary and Conclusion.

2   Experiment

This section details the data collection and matrix decomposition methods that were performed.

2.1   Data Collection

An executable program is a sequence of instructions, each instruction being composed of opcodes and data operands. The dataset used in this study was composed by assembling a series of program binaries. Each binary program was reverse engineered using Linux decompilation tools such as GNU objdump to obtain an assembly representation of opcodes and data operands. This assembly representation was separated into segments called basic blocks, which are blocks of sequential instructions separated by a jump instruction. Figure 1 shows a basic block segment of contiguous assembly instructions from a binary program that was obtained after the decompilation process. From a program segment we can obtain a sequence of terms and their term frequencies [8,9,13].

Additional structure was obtain from the program control flow graph in an adjacency matrix format was recovered using a concolic testing and symbolic execution tool *radare2* for static analysis.

The following section outlines the data collection method and the data obtained.

Define Term Dictionary. An initial term dictionary was composed by selecting all possible opcodes present in the x86/64 opcode instruction set architecture. If we were to plot vectors in the space defined by all potential terms, each vector in the dataset would have a dimensionality of 527 [1].

Stemmed Dictionary. The term dictionary was then simplified by grouping terms that expressed the same arithmetic and logic or CPU control operation in terms of their operational behavior. These terms may differ slightly in operating on different data types, e.g. floating point operations. After grouping like terms in the dictionary the dimensionality of each vector in the term frequency vector space was reduced to 32. This eliminates unnecessary redundancy, and reduces the dimensionality of the metric space being considered for each vector [1,11,17].

Reverse Engineering Binary. A set of binaries able to be executed within the x86/64 instruction set architecture were collected. Each binary was analyzed by using the *objdump* tool to recover the assembly code representations of the opcodes present in the binary executable.

Convert Program Document into Segments. Each program sample was split into its component basic block segments by segmenting on each jump instruction or similar control state transition. A jump instruction represents a transition of program control flow.

A basic block is a representation of a deterministic portion of the program. Each basic block was then represented as a vector by counting the frequencies of terms in the dictionary.

Term Distribution. The distribution of opcode term frequencies is positively skewed towards data movement was measured. The distribution is positively skewed and follows a power law distribution.

2.2 Dataset

The dataset resulting from the data collection contains two representations. The first is a matrix vectors with one dimension per term in the term dictionary, and the numeric value representing the term frequency. Each vector corresponds to a basic block segment of the program. From a perspective of $tf - idf$, the program's basic blocks would be considered to be documents. The document corpus is a collection of programs which comprise the dataset.

The second is an adjacency matrix representation of the program structure in terms of it's control flow graph edges. This captures the inter-segment structural composition in a network.

Figure 1 shows a basic block program segment of assembly instructions from a binary program that was obtained after the decompilation process [8, 9].

2.3 Matrix Decomposition

Singular Value Decomposition (SVD) was then performed on each of these datasets in order to decompose the matrices into their component parts. Figure 2 shows the top weights of topic dimensions obtained from Singular Value Decomposition of term-document co-occurrence matrix. These weights correspond to topic abstractions which are composed of multiple terms. The control flow graph represents the structural relationships of the segments of the document, and for this reason this adjacency matrix was also decomposed into component matrices. In natural language documents, the syntactic structure is a linear sequence, but the sequence of program instructions is instead determined by program control flow edges in the adjacency matrix. For the matrices representing control flow networks, the SVD was performed on the adjacency matrix representation.

$$
\begin{pmatrix}
109.85 & 0 & 0 & 0 & 0 \\
0 & 35.10 & 0 & 0 & 0 \\
0 & 0 & 21.34 & 0 & 0 \\
0 & 0 & 0 & 19.55 & 0 \\
0 & 0 & 0 & 0 & 15.35
\end{pmatrix}
$$

Fig. 2. Top 5 weights of dimensions obtained from Singular Value Decomposition of term-document co-occurrence matrix.

Singular Value Decomposition. The matrix of term frequencies decomposes into:

U is an orthonormal basis for instructions in a segment.
D is a matrix of weights per dimension (topic).
V is an orthonormal basis for programs.

The othonormal basis for instructions in a segment shows the relationship between a document and its abstraction. The topic weight shows the strength of the individual abstract topic representation. The orthonormal basis for the program shows the degree to which a program is associated with an abstract representation or topic. These components were evaluated using a spatial representation.

Construction of Metric Space. Constructing a metric space allows us to evaluate vectors for relative similarity based on quantitative measurements. We can quantitatively evaluate documents for the degree to which they correspond to a topic, and which datasets are have clear exemplars of topics with high strength based on their similarity.

By multiplying the matrix D of weights per dimension with the matrix U of terms, and selecting the dimensions with the highest weight, we are able to plot vectors in a dimension that captures a majority of the variance for each term frequency vector. This dimension represents the degree to which an opcode relates to a higher level abstraction of program structure.

If we construct a metric space over U obtained from SVD with a distance metric selected being Cosine similarity, then we have a basis for comparing the similarity of blocks by their terms.

Figures 4 and 5 show a metric space constructed from the highest weighted topics obtained from SVD. Each dimension in the metric space represents one of the two highest weighted dimensions. This gives us a spatial representation of the topic abstraction, and its relation to the documents. We discuss the applications of this representation in Sect. 3.

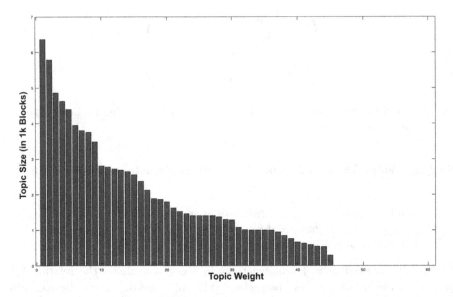

Fig. 3. The distribution of discovered topic representations obtained from matrix decomposition of a program's control flow graph, which defines the structural relations between documents in the program.

3 Results

This section discusses the results of the experiment performed in Sect. 2.

3.1 Discussion

The goal of this work is to provide an additional level of abstraction in order to determine the patterns generating the syntactic elements of malicious program binaries. The features used for this are a vector representation using $tf - idf$ term frequency methods and the program's control flow adjacency matrix representation. These features are extracted based upon segmentation of the program, which provides additional feature resolution. The abstract representation is obtained from matrix decomposition using Singular Value Decomposition (SVD). Metric spaces are defined for the program vectors as well as their topic abstractions in order to evaluate program features for similarity. Once abstract topic representations are obtained from matrix decomposition, these topics are used to evaluate the original static artifacts of the program by projecting the vector for each program segment into the metric space defined by the topic. In this section we explore a single program in metric spaces of topics selected from the component matrices.

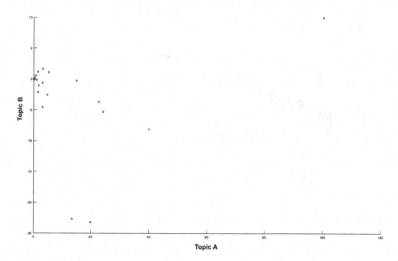

Fig. 4. Metric space constructed from highest weighted topics. Individual vectors correspond to term frequency vectors. This shows the contribution of individual terms in the dictionary to the topics discovered. The dimensionality of each vector is 32. Using the topic projection we can evaluate each vector in a 2 dimensional Euclidean space. Both topics have a high degree of correlation with data movement.

Three resulting matrices are obtained from each matrix decomposition. The matrix U is an orthonormal basis for instructions in a segment which shows the relationship between a document and its abstraction. The matrix D is a matrix of topic weights per dimension which shows the strength of the individual abstract topic representation. The matrix V is an orthonormal basis for programs which shows the degree to which a program is associated with an abstract representation or topic. These components were evaluated individually using a spatial representation.

The goal of the matrix decomposition is to obtain a set of latent variables from each of the matrices. These variables would represent a conceptual abstraction from the previous resolution of the data.

The square matrix D obtained from performing SVD on an adjacency matrix representation of the control flow graph for a given program shows the abstract representations of the network. The distribution of weights for each dimension is shown in Fig. 3. This figure shows that there are a large number, a number of unique topics approximately 30% of the number of nodes in the network, each with a low weight (less than 10). The space of nodes and the network were then reconstructed using the topics to create a scatter plot of the metric space obtained from decomposition of the control flow network [7].

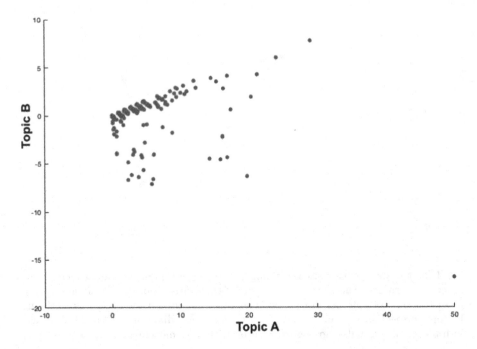

Fig. 5. Metric space constructed from highest weighted topics for the same program and using the same topics. This shows the complete representation of the program as a set of segment vectors. Each vector represents a basic block segment, and can be viewed as a node in the program's control flow graph. A spatial representation allows for an increase in resolution in comparison to analysis and hashing of whole document feature representations. A single program is a collection of segments, and the program's impact on the topic is distributed across the set of segments. We can see that the two topics differ in their representation of the program, as specific vectors have a larger magnitude in single dimensions.

One view of this representation is that a program is composed of a corpus of "documents", which are basic block segments. Each basic block is a segment of the complete program. The term frequencies in each vector represent the document term frequency of instructions in a block.

The decomposed matrix U from the opcode frequency matrix is representative of how an opcode relates to a given abstraction. The matrix D gives the weight of each abstract representation. The matrix V represents how a basic block relates to a given abstraction. This makes the assumption that term co-occurrence is significant.

Figure 4 shows a projection of the individual terms into the metric space defined by the highest weighted topics. Individual vectors correspond to term frequency vectors selected from matrix U using topics from matrix D. This shows the contribution of individual terms in the dictionary to the topics discovered.

The dimensionality of each vector is 32. Using the topic projection we can evaluate each vector in a 2 dimensional Euclidean space. If we view correlation as the vector cosine, we can see that both topics A and B have a low degree of correlation with a high number of terms and a very high correlation with a single term. Both topics have a high degree of correlation with data movement. This finding is validated by the distribution of term frequencies across the corpus [12].

Figure 5 shows a scatter plot of term frequency vectors. Vectors are selected from the matrix V using the topics from matrix D. This projection shows a plot of the entire program represented as a set of vectors in the metric space. This figure shows us how a program contributes to a given topic. Each vector is plotted in a spatial representation, and corresponds directly to a basic block segment. The segments were selected from a document corpus, which composes a program. The program shown in this figure is a collection of vectors in the metric space. Vectors in the space are plotted in the highest weighted dimensions obtained from SVD on each axis in two dimensions. The dimensions were selected from the topic matrix, and used to construct the vector space. This vector space provides a greater level of abstraction is suitable for further analysis into which documents are similar, and which documents are representative of specific concepts. For example, the value of the cosine similarity in outliers has a value that is drastically different from that of a majority of the data points in the dataset. This also shows a correlation that would be subject to regression analysis, as a linear trend is visible at several points in the space. We hope to explore clustering in the topic space in future work.

In this figure we can see the contribution of individual vectors in the topic space from the vector magnitude. It is important to note in this example that topics A and B differ in that the vector with the largest contribution to topic A is not present in topic B. This is an outlier, and would be a good candidate for a differentiating feature of this abstract topic representation. This syntactic feature has a correspondence from the topic abstraction. This represents an increase in the feature resolution that would not have been possible without the segmentation of the program document.

The adjacency matrix of a program's control flow decomposes into matrices of weights and space for nodes in control flow network. As previously discussed, the abstraction shows a significantly large distribution of relatively small weights. This implies that we are given a large number of very small features to detect, and their abstractions are not correlated. This can be seen again when plotting node and network level embeddings of the topics, which are close to their respective axes and do not show a strong correlation via cosine similarity. Program control flow networks appear to have properties that do not easily allow for decomposition, due to the low weighting of a large number of dimensional weights obtained in the square matrix after decomposition. This indicates that there are not clear abstractions for this structure.

Figure 4 shows a metric space for each of the dimensions with the highest weights as obtained from SVD. The vectors in this metric space are the right-singular vectors, which represent the degree to which a document corresponds to a more abstract representation, such as a topic as a collection of terms for natural language processing. In this context, a "document" is a deterministic segment of a program, as a collection of operations. This figure shows each of the program segments plotted in the dimensions that capture the highest variance.

We can see a clear cluster of similar segments of the program that have high to moderate magnitude in the y-dimension. We can also see at least three outliers, two with low magnitude in each dimension, one with high magnitude in both dimensions. While the magnitude of these vectors differs, they would be similar in terms of a cosine similarity metric.

4 Conclusion

In this study we have collected a dataset of term frequencies based on a dictionary. We have reduced the dimensionality of the dataset by using a stemming approach from over 500 terms to 32 terms. We have segmented the dataset into sequential blocks and an adjacency matrix of control flow representing the structure of sequences to provide an increase in feature resolution. Segmentation was done for increased accuracy and a more fine grained resolution. The control flow adjacency matrix was collected to evaluate the structural properties of the document. When these matrices were decomposed to obtain an abstract representation of the structure, low weights of a large number of dimensions obtained in the square matrix after decomposition, indicating that there are not clear abstractions for this structure. We have decomposed the datasets collected by using Singular Value Decomposition. This was done to provide an increased level of abstraction to the term frequency representation. This was measured by the strength of the topics in the matrix representing the topic weights. The value in the topic matrix provides a measurement of the topic weight. The weight of this topic is the strength of the abstraction discovered. We have evaluated the data using a spatial representation by constructing a metric space for the highest weighted abstract representations. We have evaluated the similarity of vectors in the metric space by measuring their cosine similarity. By providing structure, abstraction, and accuracy, we can analyze structural patterns, find generative processes for syntactic patterns through abstraction, and show the correlation between representations.

Acknowledgement. This research was supported in part by Air Force Research Lab grant #FA8650 to the University of Cincinnati.

References

1. X86 opcode and instruction reference. http://ref.x86asm.net/#HTML-Editions
2. Axler, S.: Linear Algebra Done Right. Springer, Heidelberg (1997)
3. Chandrasekaran, M., Ralescu, A., Kapp, D., Kebede, T.M.: Context for API calls in malware vs benign programs. In: Simian, D., Stoica, L.F. (eds.) MDIS 2020. CCIS, vol. 1341, pp. 222–234. Springer, Cham (2021). https://doi.org/10.1007/978-3-030-68527-0_14

4. Djaneye-Boundjou, O., Messay-Kebede, T., Kapp, D., Greer, J., Ralescu, A.: Static analysis through topic modeling and its application to malware programs classification. In: 2019 IEEE National Aerospace and Electronics Conference (NAECON), pp. 226–231. IEEE (2019)

5. Dumais, S.T.: Latent semantic analysis. Ann. Rev. Inf. Sci. Technol. (ARIST) **38**, 189–230 (2004)

6. Griffiths, T.L., Steyvers, M., Tenenbaum, J.B.: Topics in semantic representation. Psychol. Rev. **114**(2), 211 (2007)

7. Hagberg, A., Swart, P., Chult, D.S.: Exploring network structure, dynamics, and function using NetworkX. Technical report, Los Alamos National Lab. (LANL), Los Alamos, NM (United States) (2008)

8. Hennessy, J.L., Patterson, D.A.: Computer Architecture: A Quantitative Approach. Elsevier (2011)

9. Hopcroft, J.E., Motwani, R., Ullman, J.D.: Introduction to automata theory, languages, and computation. ACM SIGACT News **32**(1), 60–65 (2001)

10. Kebede, T.M., Djaneye-Boundjou, O., Narayanan, B.N., Ralescu, A., Kapp, D.: Classification of malware programs using autoencoders based deep learning architecture and its application to the Microsoft malware classification challenge (Big 2015) dataset. In: 2017 IEEE National Aerospace and Electronics Conference (NAECON), pp. 70–75. IEEE (2017)

11. Manning, C., Schutze, H.: Foundations of Statistical Natural Language Processing. MIT Press, Cambridge (1999)

12. Musgrave, J., Purdy, C., Ralescu, A.L., Kapp, D., Kebede, T.: Semantic feature discovery of trojan malware using vector space kernels. In: 2020 IEEE 63rd International Midwest Symposium on Circuits and Systems (MWSCAS), pp. 494–499. IEEE (2020)

13. Nar, M., Kakisim, A.G., Yavuz, M.N., Soğukpinar, İ.: Analysis and comparison of disassemblers for opcode based malware analysis. In: 2019 4th International Conference on Computer Science and Engineering (UBMK), pp. 17–22. IEEE (2019)

14. Rawashdeh, O., Ralescu, A., Kapp, D., Kebede, T.: Single property feature selection applied to malware detection. In: IEEE National Aerospace and Electronics Conference, NAECON 2021, pp. 98–105. IEEE (2021)

15. Sebesta, R.W., Mon, T., day Mon, R., Class, L., Fri, N.: Programming languages (1999)

16. Shafiq, M.Z., Tabish, S.M., Mirza, F., Farooq, M.: PE-miner: mining structural information to detect malicious executables in realtime. In: Kirda, E., Jha, S., Balzarotti, D. (eds.) RAID 2009. LNCS, vol. 5758, pp. 121–141. Springer, Heidelberg (2009). https://doi.org/10.1007/978-3-642-04342-0_7

17. Shawe-Taylor, J., Cristianini, N., et al.: Kernel Methods for Pattern Analysis. Cambridge University Press, Cambridge (2004)

18. Siddiqui, M., Wang, M.C., Lee, J.: Detecting trojans using data mining techniques. In: Hussain, D.M.A., Rajput, A.Q.K., Chowdhry, B.S., Gee, Q. (eds.) IMTIC 2008. CCIS, vol. 20, pp. 400–411. Springer, Heidelberg (2008). https://doi.org/10.1007/978-3-540-89853-5_43

19. Souri, A., Hosseini, R.: A state-of-the-art survey of malware detection approaches using data mining techniques. HCIS **8**(1), 1–22 (2018). https://doi.org/10.1186/s13673-018-0125-x

20. Steyvers, M., Tenenbaum, J.B.: The large-scale structure of semantic networks: Statistical analyses and a model of semantic growth. Cogn. Sci. **29**(1), 41–78 (2005)

21. Turi, D., Plotkin, G.: Towards a mathematical operational semantics. In: Proceedings of Twelfth Annual IEEE Symposium on Logic in Computer Science, pp. 280–291. IEEE (1997)
22. Witten, I.H., Frank, E., Trigg, L.E., Hall, M.A., Holmes, G., Cunningham, S.J.: Weka: practical machine learning tools and techniques with Java implementations (1999)

Innovative Lattice Sequences Based on Component by Component Construction Method for Multidimensional Sensitivity Analysis

Venelin Todorov[1,2](✉) 🆔 and Slavi Georgiev[1,3] 🆔

[1] Department of Information Modeling, Institute of Mathematics
and Informatics, Bulgarian Academy of Sciences, Acad. Georgi Bonchev Str.,
Block 8, 1113 Sofia, Bulgaria
{vtodorov,sggeorgiev}@math.bas.bg
[2] Department of Parallel Algorithms, Institute of Information and Communication
Technologies, Bulgarian Academy of Sciences, Acad. Georgi Bonchev Str.,
Block 25A, 1113 Sofia, Bulgaria
venelin@parallel.bas.bg
[3] Department of Applied Mathematics and Statistics, Faculty of Natural Sciences
and Education, Angel Kanchev University of Ruse, 8 Studentska Str.,
7004 Ruse, Bulgaria
sggeorgiev@uni-ruse.bg

Abstract. Many challenges in the environmental protection exist since
this is one of the leading priorities worldwide. Sensitivity analysis plays
a foundational role in the validating process of the large-scale computational air pollution models to guarantee their efficiency and reliability.
The mathematical problem for the sensitivity analysis leads to computation of multidimensional integrals. In this paper for the first time we
develop three new highly accurate lattice sequences based on component
by component construction methods: construction of rank-1 lattice rules
with prime number of points and with product weights; construction of
rank-1 lattice rules with prime number of points and with order dependent weights; construction of polynomial rank-1 lattice sequences in base
2 and with product weights. Our methods show significantly optimized
results when compared with plain Monte Carlo algorithm and the most
widely used lattice sequence.

Keywords: Air pollution modelling · Lattice sequences · Sensitivity
analysis

1 Introduction

The Sensitivity Analysis (SA), according to many papers [15,35,36,41,42], studies the level of uncertainty at which the model input data impacts the model
output in terms of accuracy. The perturbation in the input might be caused

© The Author(s), under exclusive license to Springer Nature Switzerland AG 2023
D. Simian and L. F. Stoica (Eds.): MDIS 2022, CCIS 1761, pp. 247–263, 2023.
https://doi.org/10.1007/978-3-031-27034-5_17

by various factors as instrumental error, data approximation and compression, and many others. The real-world experiments are very costly, time-consuming and they require highly trained crew to conduct them. Due to this fact, the mathematical modeling is very useful and important when describing natural phenomena [18–20, 45]. So is the SA.

Furthermore, the calculation of total sensitivity indices (SIs) is the main ingredient in SA [10, 11, 28–30]. In mathematical terms, the estimation of SIs comes down to calculation of multidimensional integrals (MIs) [5, 6, 12, 40]. Amongst the existing scientific instruments, Monte Carlo (MC) methods are most suitable for calculating MIs [1, 16, 22, 23, 25] as well as being applied in other areas as transportation [34], finance [17], etc. The most basic MC method, namely the Crude Monte Carlo (CRU) [7–9, 24, 31–33], will be the benchmark method in the paper. The other benchmark method will be the most widely used lattice sequence – Fibonacci based lattice rule (FIBO) [2, 21, 44].

The large-scale model under consideration for remote transport of air pollutants – **U**nified **D**anish **E**ulerian **M**odel (UNI-DEM) [14, 49] is described [46–48] by the following PDEs:

$$\frac{\partial c_s}{\partial t} = -\frac{\partial (u c_s)}{\partial x} - \frac{\partial (v c_s)}{\partial y} - \frac{\partial (w c_s)}{\partial z}$$
$$+ \frac{\partial}{\partial x}\left(K_x \frac{\partial c_s}{\partial x}\right) + \frac{\partial}{\partial y}\left(K_y \frac{\partial c_s}{\partial y}\right) + \frac{\partial}{\partial z}\left(K_z \frac{\partial c_s}{\partial z}\right)$$
$$+ E_s + Q_s(c_1, c_2, \ldots, c_q) - (k_{1s} + k_{2s})c_s, \quad s = 1, 2, \ldots, q.$$

We have the following parameters in the model:

c_s - pollutant concentrations,
u, v, w - wind components along the coordinate axes,
K_x, K_y, K_z - diffusion coefficients,
E_s - space emissions,
k_{1s}, k_{2s} - dry and wet deposit coefficients, respectively ($s = 1, \ldots, q$),
$Q_s(c_1, c_2, \ldots, c_q)$ - nonlinear functions describing chemical reactions between pollutants.

The considered model accounts for a big geographical region (more than 23 million km^2), which include the continents Europe in full and Africa and Asia in part, and the Mediterranean Sea. In particular, it regards the main physical, chemical and photochemical processes between the studied species as well as the dynamical meteorological conditions and the emissions in an extremely precise manner. This is of vital importance for the environmental protection thus for both health care and economics.

Apparently, the model is to a great extent non-linear and stiff. This is due to the involved chemistry. The chemical scheme employed by the model is the condensed CBM-IV, which stands for Carbon Bond Mechanism. This scheme is both computationally inexpensive and highly accurate.

The outputs of UNI-DEM, which are most widely used, are the average monthly concentrations of a number of hazardous chemical species according to the concrete chemical scheme. These concentrations are numerically derived at the grid nodes of the computational domain. The particular studies are explained further in the text.

The paper is organized as follows. In the next section, the evaluated quantities, i.e. the sensitivity indices are explained in detail as well as the stochastic approaches used to compute them, in particular lattice sequences. Section 3 is devoted to the extended presentation of the results, and it is divided into two parts. In the first one, a sensitivity analysis with respect to the emission levels is conducted, while in the second one a sensitivity analysis with respect to the chemical reaction rates is performed. The paper is closed with some conclusions and remarks.

2 Methods

Variance-based methods are employed for providing SA. In this paper Sobol Variance-Based Approach (SVBA) has been applied to evaluate SIs. The concept of SVBA is founded on a decomposition of an integrable model function f [41]:

$$f(\mathbf{x}) = f_0 + \sum_{\nu=1}^{s} \sum_{l_1 < ... < l_\nu} f_{l_1...l_\nu}(x_{l_1}, x_{l_2}, \ldots, x_{l_\nu}), \tag{1}$$

where f_0 is a constant. The expression (1) is called the ANOVA-representation of $f(\mathbf{x})$ if for every term is satisfied [41]:

$$\int_0^1 f_{l_1...l_\nu}(x_{l_1}, x_{l_2}, \ldots, x_{l_\nu}) dx_{l_k} = 0, \quad 1 \leq k \leq \nu, \quad \nu = 1, \ldots, s.$$

The functions on the right-hand side of (1) are determined in a unique way, where $f_0 = \int_{[0,1]^s} f(\mathbf{x}) dx$. The quantities

$$\mathbf{D} = \int_{[0,1]^s} f^2(\mathbf{x}) dx - f_0^2, \quad \mathbf{D}_{l_1 \, ... \, l_\nu} = \int f_{l_1 \, ... \, l_\nu}^2 dx_{l_1} \ldots dx_{l_\nu} \tag{2}$$

are respectively referred to total and partial variances. Considering the total variance, we have $\mathbf{D} = \sum_{\nu=1}^{s} \sum_{l_1 < ... < l_\nu} \mathbf{D}_{l_1...l_\nu}$. The main SIs following the SVBA are referred to Sobol global SIs [36, 41] determined by

$$S_{l_1 \, ... \, l_\nu} = \frac{\mathbf{D}_{l_1 \, ... \, l_\nu}}{\mathbf{D}}, \quad \nu \in \{1, \ldots, s\}. \tag{3}$$

and the **Total Sensitivity Index** (TSI) of an input parameter $x_i, i \in \{1, \ldots, s\}$ defined by [36,41]:

$$S_i^{\text{tot}} = S_i + \sum_{l_1 \neq i} S_{il_1} + \sum_{l_1, l_2 \neq i, l_1 < l_2} S_{il_1 l_2} + \ldots + S_{il_1...l_{s-1}}, \tag{4}$$

where S_i is called the *first-order sensitivity index (or main effect)* of x_i and $S_{il_1...l_{j-1}}$ is the jth-order sensitivity index. Then the problem of evaluating global SA consists in estimating (4), based on the formulas (2)–(3), and this task transforms to evaluating MIs.

Let $\mathbf{z} \in \mathbb{N}^s$, and N is a natural number. According to the definition [37], the point set $P = \mathbf{x_1}, \ldots, \mathbf{x_N}$ in $[0,1]^s$ with $\mathbf{x}_k = k\mathbf{z}/N$ for all $1 \le k \le N$ is called a lattice point set and $\mathbf{z}$ is called its generating vector. The algorithm with point set $P = \mathbf{x_1}, \ldots, \mathbf{x_N}$ is called a lattice sequence (LS) [38].

We will use the rank-1 LS defined by [44]:

$$\mathbf{x}_k = \left\{ \frac{k}{N}\mathbf{z} \right\}, \ k = 1, \ldots, N,$$

where $N \ge 2$ is an integer, $\mathbf{z} = (z_1, z_2, \ldots z_s)$ is the generating vector.

The performance of the LS can be optimized by the Component by Component (CBC) Fast Construction method [3,27,39]. Now we generate several special LS following the idea in [4,26]. The first LS will be the CBC construction of rank-1 lattice rules with prime number of points and with product weights, we will call it **1PT**. The second LS will be the CBC construction of rank-1 lattice rules with prime number of points and with order dependent weights, we will call it **1OD**. The third LS will be the CBC construction of polynomial rank-1 lattice sequences in base 2 and with product weights, we will call it **2POLY**.

The derivation of optimal generating vectors in higher dimension s for fixed N has great computational difficulty. Sampling a good generator vector that leads to small integration errors is not an easy task. It is used the advanced recently developed technique CBC construction method, mentioned earlier, by means of which different generating vectors are found.

On the first stage of LS,

$$\mathbf{z} = (z_1, z_2, \ldots z_s)$$

is obtained by the CBC. In particular, in the beginning $z_1 := 1$. Then, z_1 is hold fixed, and $z_2 \in U^N := \{z \in \mathbb{N} : 1 \le z \le N - 1, \ \gcd(z, N) = 1\}$ is chosen in such a way that the predefined error criterion, such as worst function errors or Zaremba index, is minimized in two dimensions. Then, iteratively for $i = 3, \ldots, s$, z_i is chosen from U^N in such a way to minimize the predefined error criterion in i dimensions.

The corresponding generating vector, which is built in this way, is extensible in s. So, the lattice points are generated following

$$\mathbf{x}_k = \left\{ \frac{k}{N}\mathbf{z} \right\}, \ k = 1, \ldots, N.$$

Finally, the approximate value I_N of the integral is obtained:

$$I_N = \frac{1}{N} \sum_{k=1}^{N} f\left(\left\{ \frac{k}{N}\mathbf{z} \right\} \right).$$

The model UNI-DEM has been already studied. For instance, in the paper [13] the Sobol quasi-Monte Carlo algorithm, Latin hypercube sampling algorithm, quasi-Monte Carlo Fibonacci lattice rule are applied in the estimation of Sobol sensitivity indices. Further, the experimental methods are supplemented by modified Sobol sequences and digital sequences, based on Sobol and Niederreiter-Xing sequences in [43]. The present study is a continuation of the mentioned, as new algorithms are tested for their superiority over the already applied ones. The results are given in the next section.

3   Results and Discussion

This section is dedicated to the detailed presentation of the numerical simulations, investigating the described algorithms in the previous section.

Initially, the model is considered in detail in the book [46], where data for the past century is provided. The actual data, used for the simulations, is given in [10, 11].

The computational experiments are conducted using the MATLAB® environment. The tests described below are performed on a user laptop with 6-core processor and 16 GB RAM.

3.1   Sensitivity Studies with Respect to Emission Levels

First we will conduct the SA of the mean monthly concentrations of ammonia in Milan. The anthropogenic emissions input is composed by four different components $\mathbf{E} = (\mathbf{E^A}, \mathbf{E^N}, \mathbf{E^S}, \mathbf{E^C})$ as follows:

$$\mathbf{E^A} - \text{ammonia } (NH_3);$$
$$\mathbf{E^S} - \text{sulphur dioxide } (SO_2);$$
$$\mathbf{E^N} - \text{nitrogen oxides } (NO + NO_2);$$
$$\mathbf{E^C} - \text{anthropogenic hydrocarbons.}$$

The output of the model is the average monthly concentration of the following three pollutants:

s_1 – ozone (O_3);
s_2 – ammonia (NH_3);
s_3 – ammonium sulphate and ammonium nitrate $(NH_4SO_4 + NH_4NO_3)$.

The latter are given together since the model produces joint results about them and they are further regarded as a single pollutant.

Since the grid used for emission levels is coarse (96×96, with a step of 50 km), an interpolation is used to obtain values closest to Milan. The concentrations for the aforementioned three pollutants are measured in January 1997, which represents a typical winter scenario. It is not a problem that the data is old since the total sensitivity indices do not depend on the changing meteorological conditions within years, when the month is fixed.

Henceforward, we will use the notation REL for the relative error and EQ for the estimated quantity. The number of samples is denoted by NS. As so far, CRU, 1PT, 1OD, 2POLY and FIBO will stand for the crude Monte Carlo method, the three lattice sequences, based on CBC construction, and the Fibonacci-based sequence. RV will denote the reference value.

RELs for the approximate evaluation of the quantities f_0, **D** and first and total SIs are presented in Tables 1, 2, 3, 4, 5, 6 respectively. The quantity f_0 is given by four-dimensional integral while the rest of quantities are given by eight-dimensional integrals.

For f_0 it follows directly from Table 1 that for the largest number of samples the best algorithm is 1PT, followed by the 1OD and 2POLY. However, for number of samples between 2^{20} and 2^{22} FIBO is better.

For **D** form Table 2 it is visible that for the largest number of samples the best algorithm is 1OD, followed by FIBO and 1PT, and for number of samples 2^{22} 1PT is the best approach.

Table 1. Relative errors for $f_0 \approx 0.048$.

NS n	CRU	1PT	1OD	2POLY	FIBO
	Relative error	Relative error	Relative error	Relative error	Relative error
2^{10}	1.0203e−02	2.3606e−04	**1.9175e−04**	2.2367e−03	2.0933e−04
2^{12}	3.4246e−03	1.3544e−04	6.7981e−05	5.4661e−04	**4.3178e−05**
2^{14}	2.5052e−03	**1.8217e−05**	3.0088e−05	1.2375e−04	2.2526e−05
2^{16}	1.7268e−03	**4.6572e−06**	8.5625e−06	3.5660e−05	8.7031e−06
2^{18}	4.3151e−04	3.2560e−06	**1.7842e−06**	8.5348e−06	1.7861e−06
2^{20}	6.7226e−05	8.4365e−07	5.7304e−07	2.3296e−06	**4.2130e−07**
2^{22}	6.4635e−05	7.6756e−08	1.0806e−07	3.1698e−07	**5.4420e−08**
2^{24}	1.6251e−05	**1.8688e−08**	3.1692e−08	1.0916e−07	1.5098e−07

Table 2. Relative errors for **D** ≈ 0.0002.

NS n	CRU	1PT	1OD	2POLY	FIBO
	Relative error	Relative error	Relative error	Relative error	Relative error
2^{10}	1.1512e−01	9.5910e−02	3.3770e−02	**1.7981e−02**	1.6298e−01
2^{12}	2.8713e−02	1.1282e−02	**5.0718e−03**	1.7996e−02	2.3878e−02
2^{14}	4.3025e−02	5.4575e−03	1.1414e−02	2.7210e−02	**2.8985e−03**
2^{16}	1.7631e−02	6.1309e−04	6.3908e−04	4.2120e−03	**2.6458e−04**
2^{18}	1.1619e−02	2.3935e−04	**9.4319e−05**	9.9209e−05	3.0109e−04
2^{20}	5.7971e−03	5.7851e−05	1.3522e−05	**6.0946e−06**	1.1919e−04
2^{22}	7.3641e−04	**2.5767e−05**	3.2597e−05	3.4462e−05	2.5938e−05
2^{24}	1.9965e−03	6.7076e−06	**1.7255e−06**	2.8621e−05	4.9071e−06

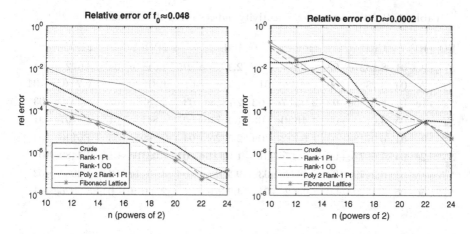

Fig. 1. Relative errors for the calculation of $f_0 \approx 0.048$ (left) and $D \approx 0.0002$ (right).

Figure 1 shows that for the four-dimensional integral f_0 all of the methods except CRU have similar performance, while for the eight-dimensional integral D for smaller number of samples all the methods have similar performance, while for larger number of samples all LSs are better and 1OD is the best.

Table 3. Relative errors for sensitivity indices using the methods ($n = 2^{12}$).

EQ	RV	CRU	1PT	1OD	2POLY	FIBO
S_1	9e−01	2.6015e−02	**5.4483e−03**	6.6155e−03	1.0321e−02	1.9214e−02
S_2	2e−04	8.1368e+00	**1.6094e−02**	6.0510e−01	9.2859e−01	1.5752e+01
S_3	1e−01	1.5309e−01	2.7681e−02	6.2924e−02	**3.4366e−03**	1.0267e−01
S_4	4e−05	1.5072e+01	4.0561e+00	2.0488e+00	**1.5116e+00**	1.3980e+01
S_1^{tot}	9e−01	2.1584e−02	3.9344e−03	7.8691e−03	**9.9561e−05**	1.7723e−02
S_2^{tot}	2e−04	9.5698e+00	1.2220e+00	1.2083e−01	**4.5788e−02**	1.7151e+01
S_3^{tot}	1e−01	1.8180e−01	**4.1154e−02**	5.6745e−02	8.4492e−02	1.4168e−01
S_4^{tot}	5e−05	1.9228e+01	6.7052e+00	4.8032e+00	**9.4174e−02**	4.1143e+01

From the Table 3 it could be observed that for most of the SIs the best REL is produced by 2POLY, and for the others the best approach is 1PT. However, for small in value SIs like S_4 and S_4^{tot} which are most important for the reliability of the model results, the best method is 2POLY.

Table 4. Relative errors for sensitivity indices using the methods ($n = 2^{16}$).

EQ	RV	CRU	1PT	1OD	2POLY	FIBO
S_1	9e$-$01	8.2327e$-$03	1.2828e$-$04	**2.1750e$-$05**	4.6117e$-$03	3.6222e$-$04
S_2	2e$-$04	2.2865e+00	1.9932e$-$01	4.4597e$-$01	1.8153e$-$01	**1.7365e$-$01**
S_3	1e$-$01	5.3236e$-$02	**2.0517e$-$03**	2.7858e$-$03	2.2825e$-$02	3.2236e$-$03
S_4	4e$-$05	3.0153e+00	**4.4630e$-$03**	1.4074e+00	1.8413e$-$01	4.8707e$-$01
S_1^{tot}	9e$-$01	6.9713e$-$03	**1.9432e$-$04**	2.5726e$-$04	2.8095e$-$03	4.6146e$-$04
S_2^{tot}	2e$-$04	1.6677e+00	4.6117e$-$01	**1.0819e$-$01**	1.4310e$-$01	3.4489e$-$01
S_3^{tot}	1e$-$01	6.2479e$-$02	2.2677e$-$03	**9.2985e$-$04**	3.7371e$-$02	1.9639e$-$03
S_4^{tot}	5e$-$05	1.1430e+00	7.0522e$-$01	4.2961e$-$01	**2.4159e$-$01**	5.0626e$-$01

From Table 4 it could be observed that for three of the SIs the best approach is 1PT, and for other three the best approach is 1OD. Only for S_2 the best approach is FIBO. For small in value S_4^{tot} the best method is again 2POLY.

From the Table 5 it can be seen that for most of the SIs the best REL is produced by the FIBO algorithm, and for most important small in value SIs S_4, S_2^{tot} and S_4^{tot} the best approach is 1OD.

From Table 6 it can be seen that only for S_2^{tot} the best approach is FIBO. For the other SIs 1PT gives better results for four of the quantities, and for the other three the best REL is produced by the 1OD algorithm. Generally, the increased number of samples improve the results produced by 1PT and 1OD. For small in value SI S_4 the best approach is 1PT and for small in value S_4^{tot} the best approach is 1OD.

Table 5. Relative errors for sensitivity indices using the methods ($n = 2^{20}$).

EQ	RV	CRU	1PT	1OD	2POLY	FIBO
S_1	9e$-$01	7.8667e$-$04	2.1331e$-$06	4.7294e$-$05	1.2366e$-$05	**5.2856e$-$08**
S_2	2e$-$04	5.5512e$-$01	5.9804e$-$02	2.4606e$-$02	1.5653e$-$02	**3.1685e$-$03**
S_3	1e$-$01	3.4181e$-$03	2.0816e$-$04	1.9248e$-$04	1.0666e$-$04	**6.8845e$-$05**
S_4	4e$-$05	2.1558e$-$01	3.0313e$-$01	**8.0971e$-$02**	3.3948e$-$01	1.8759e$-$01
S_1^{tot}	9e$-$01	5.2742e$-$04	2.8432e$-$05	2.8374e$-$05	3.2882e$-$05	**2.1379e$-$05**
S_2^{tot}	2e$-$04	4.4329e$-$01	8.7193e$-$02	**5.6201e$-$04**	1.4556e$-$03	4.5626e$-$03
S_3^{tot}	1e$-$01	5.3285e$-$03	2.4503e$-$04	3.4734e$-$04	9.8445e$-$05	**4.6919e$-$05**
S_4^{tot}	5e$-$05	1.4729e$-$01	2.4701e$-$02	**6.8235e$-$03**	8.6469e$-$03	6.0848e$-$02

Table 6. Relative errors for sensitivity indices using the methods ($n = 2^{24}$).

EQ	RV	CRU	1PT	1OD	2POLY	FIBO
S_1	9e−01	2.0736e−04	2.1781e−06	**2.1669e−06**	5.3531e−06	3.4282e−06
S_2	2e−04	3.3428e−03	**1.2707e−04**	1.0530e−03	7.4211e−03	1.5679e−03
S_3	1e−01	1.3562e−03	**3.8944e−06**	1.6627e−05	2.9098e−05	1.1510e−05
S_4	4e−05	1.6781e−01	**4.1349e−04**	5.8271e−03	4.3866e−02	1.2408e−02
S_1^{tot}	9e−01	1.6667e−04	**5.1453e−07**	2.4255e−06	6.5536e−06	2.0793e−06
S_2^{tot}	2e−04	6.3335e−02	3.3297e−04	5.8544e−04	2.3172e−04	**4.1606e−05**
S_3^{tot}	1e−01	1.7633e−03	1.7565e−05	**1.7294e−05**	3.3825e−05	2.4602e−05
S_4^{tot}	5e−05	1.5280e−02	1.4029e−03	**9.6576e−04**	6.2087e−03	1.5739e−02

From Tables 1, 2, 3, 4, 5 and 6 it can be summarized that 1OD gives the most reliable results for the considered problem.

3.2 Sensitivity Studies with Respect to Chemical Reaction Rates

In this section we will make SA of the ozone concentration in Genova according to the rate variation of some chemical reactions: # 1, 3, 7, 22 (time-varying) and # 27, 28 (time unvarying) reactions of the condensed CBM-IV scheme [46]. The respective simplified chemical equations of these reactions looks:

[#1] $NO_2 + h\nu \Longrightarrow NO + O$;   [#22] $HO_2 + NO \Longrightarrow OH + NO_2$;
[#3] $O_3 + NO \Longrightarrow NO_2$;   [#27] $HO_2 + HO_2 \Longrightarrow H_2O_2$;
[#7] $NO_2 + O_3 \Longrightarrow NO_3$;   [#28] $OH + CO \Longrightarrow HO_2$.

It is not the ozone that obligatory takes part in the reactions. Some of its popular precursors also participate.

Now, the chemical reaction rates are the input parameters and the concentration of pollutants are the output parameters. The data is related to the concentrations near Genova in July 1998. The age of the data is not a problem, as discussed earlier.

RELs for the approximate evaluation of the quantities f_0, **D** and first, second order and total SIs are presented in Tables 7, 8, 10, 11, 12 respectively. The quantity f_0 is given by six-dimensional integral whereas the rest of quantities are given by twelve-dimensional integrals.

Table 7. Relative errors for $f_0 \approx 0.27$.

NS n	CRU	1PT	1OD	2POLY	FIBO
	Relative error	Relative error	Relative error	Relative error	Relative error
2^{10}	6.9655e−04	5.4367e−03	**3.6828e−04**	8.6668e−04	2.0775e−03
2^{12}	8.0553e−04	1.5507e−03	**7.2191e−06**	3.7151e−04	1.4018e−04
2^{14}	2.8035e−03	3.4648e−04	**4.8522e−05**	1.1677e−04	3.9808e−04
2^{16}	5.9493e−04	4.3231e−05	**1.2827e−05**	2.9617e−05	2.6116e−04
2^{18}	7.6591e−04	2.6326e−05	**1.7568e−06**	2.9189e−06	7.2863e−06
2^{20}	3.4494e−04	6.6963e−06	6.5648e−07	1.5606e−06	**4.5748e−07**
2^{22}	4.6980e−05	7.0237e−07	**1.5057e−07**	2.6543e−06	5.6709e−07
2^{24}	8.6192e−06	2.7795e−07	**1.1488e−07**	3.0362e−07	1.1892e−06

Table 7, with $f_0 \approx 0.27$ shows that for all of the number of samples except 2^{20} the best algorithm is 1OD, followed by the 2POLY and 1PT. Only for 2^{20} FIBO is better.

Table 8. Relative errors for $\mathbf{D} \approx 0.0025$.

NS n	CRU	1PT	1OD	2POLY	FIBO
	Relative error	Relative error	Relative error	Relative error	Relative error
2^{10}	7.2185e−02	3.9665e−01	**5.0978e−03**	8.7639e−02	6.7333e+00
2^{12}	9.5413e−02	9.3069e−02	**8.0257e−03**	3.8016e−02	5.2657e−01
2^{14}	6.3987e−02	1.9043e−02	**4.3188e−03**	2.6355e−02	1.0198e−01
2^{16}	2.9741e−02	8.1587e−04	**1.4004e−04**	6.7750e−03	1.9703e−03
2^{18}	7.4173e−03	2.4942e−03	**3.6638e−04**	7.4647e−04	4.5271e−03
2^{20}	8.9182e−03	8.4610e−04	2.0600e−04	**5.5513e−06**	9.3318e−03
2^{22}	2.2089e−03	**4.3003e−05**	5.9544e−05	9.0933e−05	2.2052e−02
2^{24}	1.2915e−03	**3.2604e−06**	3.4328e−05	2.1879e−05	5.0299e−04

Table 8 for $\mathbf{D} \approx 0.0025$ shows that for smaller number of samples the best algorithm is 1OD, and for larger number of samples 2POLY and 1OPT give better RELs.

On Fig. 2 it can be seen that for six-dimensional integral f_0 overall 1OD is the best algorithm and for twelve-dimensional integral $\mathbf{D}$ 2POLY and 1PT give better RELs then the 1OD.

From the Table 9 it can be seen that for most of the SIs the best RELs is produced by the proposed 1OD and for three of the quantities the best algorithm is 2POLY. None of the proposed methods obtain good results for S_5^{tot} and S_{45} and CRU is the best for S_{12}.

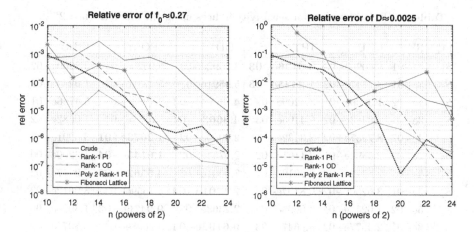

Fig. 2. Relative errors for the calculation of $f_0 \approx 0.27$ (left) and $\mathbf{D} \approx 0.0025$ (right).

Table 9. Relative errors for sensitivity indices using the methods ($n \approx 2^{12}$).

EQ	RV	CRU	1PT	1OD	2POLY	FIBO
S_1	4e−01	1.2279e−01	1.1831e−01	8.7056e−03	**9.9830e−04**	1.5500e−01
S_2	3e−01	5.3654e−01	1.0020e−01	**1.6724e−02**	4.0915e−02	2.9180e−02
S_3	5e−02	9.9530e−01	8.0227e−01	**5.5664e−02**	1.6734e−01	2.5132e−01
S_4	3e−01	7.6622e−02	6.8514e−02	**1.8398e−02**	2.3556e−02	4.1440e−01
S_5	4e−07	8.9638e+03	4.7829e+03	**1.2738e+03**	1.4878e+03	4.2522e+03
S_6	2e−02	9.8081e−02	2.4238e−01	**2.0436e−03**	6.3677e−03	1.9069e+00
S_1^{tot}	4e−01	2.0417e−01	1.2368e−01	**1.4252e−02**	2.6843e−02	2.0755e−01
S_2^{tot}	3e−01	4.7382e−01	8.7713e−02	1.5857e−02	**1.3992e−02**	4.0030e−01
S_3^{tot}	5e−02	1.1102e+00	1.0325e+00	**1.0483e−01**	1.3184e−01	1.0300e+00
S_4^{tot}	3e−01	2.1265e−01	9.8093e−02	**3.3088e−03**	1.5530e−02	7.2211e−01
S_5^{tot}	2e−04	6.9894e+00	**3.7308e+00**	4.1131e+00	7.9208e+00	2.6991e+01
S_6^{tot}	2e−02	3.9265e−01	3.2392e−01	1.3625e−01	**6.0369e−02**	3.9152e+00
S_{12}	6e−03	**1.6117e−02**	8.2744e−02	2.1173e+00	1.4394e+00	2.3716e+01
S_{14}	5e−03	5.1347e+00	9.4221e−01	**2.4440e−01**	4.7961e−01	9.5928e+00
S_{24}	3e−03	3.4837e+00	1.0753e+00	2.0816e−01	**1.8671e−02**	9.1121e+00
S_{45}	1e−05	1.1317e+02	**6.5409e+00**	9.8030e+01	1.6769e+01	2.3222e+01

From the Table 10 it can be seen that for six of the SIs the best RELs is produced by the proposed 1OD, for six of the quantities the best algorithm is 2POLY and for four of the quantities the best approach is 1PT. None of the methods is good for the most important small in value SI S_5 but 2POLY gives the best result for it.

From the Table 11 it can be seen that for seven of the quantitites the best RELs is produced by the proposed 1OD, for five of the quantities the best

Table 10. Relative errors for sensitivity indices using the methods ($n \approx 2^{16}$).

EQ	RV	CRU	1PT	1OD	2POLY	FIBO
S_1	4e−01	5.6096e−03	**3.0126e−03**	3.3592e−03	5.9928e−03	3.8200e−02
S_2	3e−01	4.7347e−02	**1.2425e−03**	5.9591e−03	5.4738e−03	1.0263e−02
S_3	5e−02	1.3830e−01	3.3200e−02	**4.0658e−03**	9.1963e−03	5.4766e−01
S_4	3e−01	1.0483e−02	6.4105e−03	**4.8663e−04**	5.6857e−04	1.0682e−02
S_5	4e−07	8.7257e+02	2.3232e+01	2.0080e+02	**1.2673e+01**	3.3991e+03
S_6	2e−02	2.6892e−01	2.1161e−02	2.5754e−02	**1.6145e−04**	1.3200e+00
S_1^{tot}	4e−01	2.8312e−02	8.2919e−03	7.0001e−03	**4.5562e−03**	7.9247e−02
S_2^{tot}	3e−01	3.6260e−02	6.8394e−03	**2.1701e−03**	4.1559e−03	3.0612e−02
S_3^{tot}	5e−02	1.6244e−01	3.0698e−02	2.2363e−02	**9.9476e−03**	1.3090e+00
S_4^{tot}	3e−01	2.4677e−02	8.0647e−03	**9.6192e−04**	3.2008e−03	3.8375e−01
S_5^{tot}	2e−04	2.4371e+00	**1.7889e−01**	1.2750e+00	1.5567e+00	8.8515e+01
S_6^{tot}	2e−02	2.7228e−01	4.7099e−03	**3.0394e−03**	3.0474e−02	2.1534e+00
S_{12}	6e−03	6.3125e−01	3.6604e−01	3.1937e−01	**4.5936e−02**	3.2088e+00
S_{14}	5e−03	8.2786e−01	**1.4026e−02**	2.8073e−02	8.7938e−02	8.6366e+00
S_{24}	3e−03	2.2332e−01	4.6330e−02	5.2742e−02	**4.4146e−02**	1.3661e+01
S_{45}	1e−05	1.8853e+01	3.3732e+00	**1.2941e+00**	2.2953e+00	4.2516e+01

Table 11. Relative errors for sensitivity indices using the methods ($n \approx 2^{20}$).

EQ	RV	CRU	1PT	1OD	2POLY	FIBO
S_1	4e−01	2.5476e−03	**5.5429e−05**	2.0160e−04	1.3385e−04	9.2073e−03
S_2	3e−01	1.9128e−02	3.6138e−04	**2.6997e−04**	3.5548e−04	1.4697e−02
S_3	5e−02	3.4996e−03	**5.4389e−04**	1.2984e−03	1.1577e−03	6.5013e−01
S_4	3e−01	1.7948e−02	2.1778e−03	4.5678e−04	**1.6977e−04**	1.5332e−01
S_5	4e−07	1.8681e+02	2.4054e+02	6.6954e+01	**2.8167e+00**	2.6805e+03
S_6	2e−02	5.3782e−02	**1.7669e−04**	2.1234e−03	7.4131e−04	1.1299e+00
S_1^{tot}	4e−01	7.9477e−03	8.4170e−04	**2.3031e−05**	8.5798e−05	9.6889e−03
S_2^{tot}	3e−01	1.9734e−02	7.8061e−04	2.3010e−04	**5.0400e−05**	3.0059e−02
S_3^{tot}	5e−02	3.0996e−03	5.3788e−03	**2.2004e−03**	2.4749e−03	1.3651e+00
S_4^{tot}	3e−01	2.0145e−02	1.6670e−03	9.2995e−05	**4.8531e−05**	3.6701e−01
S_5^{tot}	2e−04	5.8023e−01	1.4925e−01	**1.0619e−01**	4.7780e−01	3.8981e+01
S_6^{tot}	2e−02	7.2660e−02	6.1494e−04	**7.3465e−05**	5.5207e−04	1.7630e+00
S_{12}	6e−03	1.3345e−01	5.9771e−02	**7.2983e−03**	9.4106e−03	8.3981e−02
S_{14}	5e−03	2.7148e−01	3.9964e−03	**1.1830e−03**	4.3435e−03	1.8503e−01
S_{24}	3e−03	3.0686e−01	2.2941e−02	3.0215e−02	**3.3275e−03**	1.4056e+01
S_{45}	1e−05	3.1523e+00	**1.4279e−01**	2.9606e−01	4.0888e+00	2.5998e+01

Table 12. Relative errors for sensitivity indices using the methods ($n \approx 2^{24}$).

EQ	RV	CRU	1PT	1OD	2POLY	FIBO
S_1	4e−01	2.4330e−03	**1.0316e−05**	1.1644e−05	1.3608e−04	1.3278e−03
S_2	3e−01	1.8600e−03	2.4055e−05	1.5499e−05	**3.4644e−06**	6.0920e−04
S_3	5e−02	3.9300e−03	6.9747e−04	4.5117e−04	**1.0049e−04**	3.4739e−03
S_4	3e−01	1.7259e−03	6.6590e−05	**3.9532e−05**	4.2325e−05	6.2489e−04
S_5	4e−07	6.2991e+01	**1.0123e+00**	2.2922e+00	1.0216e+00	8.9491e+00
S_6	2e−02	1.3933e−02	4.3252e−05	**7.9079e−06**	6.1158e−05	2.8759e−03
S_1^{tot}	4e−01	1.9688e−03	**1.1852e−05**	1.3781e−05	2.2831e−05	3.1442e−04
S_2^{tot}	3e−01	2.6026e−03	**3.9920e−06**	9.3048e−06	2.9469e−05	2.2658e−04
S_3^{tot}	5e−02	2.0711e−03	3.1878e−04	**4.9894e−05**	1.3466e−04	3.0986e−03
S_4^{tot}	3e−01	1.7469e−03	1.1292e−04	**1.3869e−05**	5.7511e−05	4.2628e−04
S_5^{tot}	2e−04	4.2987e−02	**3.7190e−03**	6.1847e−03	6.4344e−03	2.2012e−02
S_6^{tot}	2e−02	5.3593e−03	**3.9144e−05**	1.1329e−04	4.8143e−04	1.9901e−02
S_{12}	6e−03	2.4931e−02	**6.2827e−05**	7.7267e−04	4.2029e−03	3.9075e−02
S_{14}	5e−03	1.2130e−02	**4.0231e−05**	7.5581e−04	2.2898e−03	5.4462e−03
S_{24}	3e−03	2.6639e−02	1.4740e−03	**6.4054e−04**	4.3661e−03	9.2478e−03
S_{45}	1e−05	1.5179e−01	**1.2320e−02**	3.9643e−02	1.5846e−02	5.8353e−02

algorithm is 2POLY and for four of the quantities the best approach is 1PT. None of the methods is good for the most important small in value SI S_5 but 2POLY gives result that is one order better than the others.

From the Table 12 it can be seen that for nine of the quantities the best RELs is produced by the proposed 1PT, for five of the quantities the best algorithm is 1OD and for two of the quantities the best approach is 2POLY. None of the methods is good for the most important small in value SI S_5 but 1PT and 2POLY give the best result. For the small in value SI S_{45} the best approach is again 1PT followed by 2POLY. Generally the increase number of samples improve the results produced by 1PT.

The overall conclusion is that 1OD and 1PT give the best results.

The aim of the paper was to discover the most effective and reliable approaches to study the global sensitivity of the calculated concentration levels of major pollutants due to variation of emission levels and chemical reactions in a real-world framework of air pollution transport. These practical results in terms of superior efficiency, especially for small in value sensitivity indices, give insights which quasi-Monte Carlo approaches are most suitable to conduct sound, robust and fast sensitivity analysis. Such analyses are important in deciding how to split and/or simplify the respective large-scale model, since its computation remains challenging even for the contemporary supercomputers. Moreover, the conducted sensitivity analysis allows studying the performance of the reasonable simplifications; identifying the "bottleneck" processes, which must be considered with greater accuracy and efficiency; increasing the results reliability; and improving the model.

4 Conclusion

A complex novel experimental study of Monte Carlo methods based on lattice sequences with specific constructions has been performed for the UNI-DEM model. Studies of sensitivity indices and especially for small in value ones are very important for the reliability of the UNI-DEM. Variance-based methods are employed for conducting sensitivity analysis. Sobol variance-based approach for global sensitivity indices has been applied to compute the corresponding sensitivity indices. In this paper for the first time for the UNI-DEM three different lattice sequences based on fast component-by-component construction has been used: construction of rank-1 lattice rules with prime number of points and with product weights; construction of rank-1 lattice rules with prime number of points and with order dependent weights; construction of polynomial rank-1 lattice sequences in base 2 and with product weights. The three methods are significant optimization techniques over the results produced by the plain Monte Carlo approach and the Fibonacci-based lattice rule. The obtained results will help to validate and enhance the mathematical models through our sensitivity analysis. The mathematical model will contribute to a more accurate estimation of sensitivity indices and enable an evaluation of the harmful emissions effects on human health.

Acknowledgement. Slavi Georgiev is supported by the Bulgarian National Science Fund under Project KP-06-M32/2 - 17.12.2019 "Advanced Stochastic and Deterministic Approaches for Large-Scale Problems of Computational Mathematics" and Scientific Research Fund of University of Ruse under FNSE-04.

Venelin Todorov is supported by the Bulgarian National Science Fund under Projects KP-06-N52/5 "Efficient methods for modeling, optimization and decision making" and KP-06-N62/6 "Machine learning through physics-informed neural networks". The work is also supported by the Project KP-06-Russia/17 "New Highly Efficient Stochastic Simulation Methods and Applications", funded by the National Science Fund – Bulgaria.

The authors thank the anonymous referees whose invaluable comments improved significantly the quality of the paper.

References

1. Antonov, I., Saleev, V.: An economic method of computing LP_τ-sequences. USSR Comput. Math. Phy. **19**, 252–256 (1979)
2. Bahvalov, N.: On the approximate computation of multiple integrals. Vestn. Mosc. State Univ. **4**, 3–18 (1959)
3. Baldeaux, J., Dick, J., Leobacher, G., Nuyens, D., Pillichshammer, F.: Efficient calculation of the worst-case error and (fast) component-by-component construction of higher order polynomial lattice rules. Numer. Algor. **59**, 403–431 (2012)
4. Cools, R., Kuo, F., Nuyens, D.: Constructing embedded lattice rules for multivariate integration. SIAM J. Sci. Comput. **28**, 2162–2188 (2006)
5. Dimov, I.: Monte Carlo Methods for Applied Scientists. World Scientific, Singapore (2008)

6. Dimov, I., Atanassov, E.: What Monte Carlo models can do and cannot do efficiently? Appl. Math. Model. **32**, 1477–1500 (2007)
7. Dimov, I., Georgieva, R.: Monte Carlo method for numerical integration based on Sobol's sequences. In: Dimov, I., Dimova, S., Kolkovska, N. (eds.) NMA 2010. LNCS, vol. 6046, pp. 50–59. Springer, Heidelberg (2011). https://doi.org/10.1007/978-3-642-18466-6_5
8. Dimov, I.T., Georgieva, R.: Multidimensional sensitivity analysis of large-scale mathematical models. In: Iliev, O.P., et al. (eds.) Numerical Solution of Partial Differential Equations: Theory, Algorithms, and Their Applications, Springer Proceedings in Mathematics & Statistics. PROMS, vol. 45, pp. 137–156. Springer, New York (2013). https://doi.org/10.1007/978-1-4614-7172-1_8
9. Dimov, I., Georgieva, R.: Monte Carlo algorithms for evaluating Sobol's sensitivity indices. Math. Comput. Simul. **81**(3), 506–514 (2010)
10. Dimov, I., Georgieva, R., Ostromsky, T., Zlatev, Z.: Variance-based sensitivity analysis of the unified Danish Eulerian model according to variations of chemical rates. In: Dimov, I., Faragó, I., Vulkov, L. (eds.) NAA 2012. LNCS, vol. 8236, pp. 247–254. Springer, Heidelberg (2013). https://doi.org/10.1007/978-3-642-41515-9_26
11. Dimov, I.T., Georgieva, R., Ostromsky, Tz., Zlatev, Z.: Sensitivity studies of pollutant concentrations calculated by UNI-DEM with respect to the input emissions. Central Eur. J. Math. Numer. Methods Large Scale Sci. Comput. **11**(8), 1531–1545 (2013)
12. Dimov, I.T., Georgieva, R., Ostromsky, Tz., Zlatev, Z.: Advanced algorithms for multidimensional sensitivity studies of large-scale air pollution models based on Sobol sequences. Special Issue Comput. Math. Appl. **65**(3), 338–351 (2013)
13. Dimov, I.T., Georgieva, R., Todorov, V., Ostromsky, Tz.: Efficient stochastic approaches for sensitivity studies of an Eulerian large-scale air pollution model. In: AIP Conference Proceedings, vol. 1895, no. 1, 050004 (2017). https://doi.org/10.1063/1.5007376
14. Dimov, I., Zlatev, Z.: Testing the sensitivity of air pollution levels to variations of some chemical rate constants. Notes Numer. Fluid Mech. **62**, 167–175 (1997)
15. Ferretti, F., Saltelli, A., Tarantola, S.: Trends in sensitivity analysis practice in the last decade. J. Sci. Total Environ. **568**, 666–670 (2016)
16. Fidanova, S.: Simulated annealing: a Monte Carlo method for GPS surveying. In: Alexandrov, V.N., van Albada, G.D., Sloot, P.M.A., Dongarra, J. (eds.) ICCS 2006. LNCS, vol. 3991, pp. 1009–1012. Springer, Heidelberg (2006). https://doi.org/10.1007/11758501_160
17. Georgiev, I., Centeno, V., Mihova, V., Pavlov, V.: A modified ordinary differential equation approach in price forecasting. In: AIP Conference on Proceedings, vol. 2459, p. 030008 (2022)
18. Gery, M., Whitten, G., Killus, J., Dodge, M.: A photochemical kinetics mechanism for urban and regional scale computer modelling. J. Geophys. Res. **94**, 12925–12956 (1989)
19. Grozev, D., Milchev, M., Georgiev, I.: Analysis of the load on the taxi system in a medium-sized city. In: IOP Conference Series: Materials Science and Engineering, vol. 664, no. 1, p. 012035 (2019)
20. Homma, T., Saltelli, A.: Importance measures in global sensitivity analysis of nonlinear models. Reliab. Eng. Syst. Saf. **52**, 1–17 (1996)
21. Hua, L.K., Wang, Y.: Applications of Number Theory to Numerical Analysis. Springer, New York (1981)

22. Joe, S., Kuo, F.: Remark on algorithm 659: implementing Sobol's quasirandom sequence generator. ACM Trans. Math. Softw. **29**(1), 49–57 (2003)

23. Karaivanova, A., Atanassov, E., Gurov, T., Stevanovic, R., Skala, K.: Variance reduction MCMs with application in environmental studies: sensitivity analysis. In: AIP Conference on Proceedings, vol. 1067, no. 1, pp. 549–558 (2008)

24. Karaivanova, A., Dimov, I.: Error analysis of an adaptive Monte Carlo method for numerical integration. Math. Comput. Simulatio **47**, 201–213 (1998)

25. Karaivanova, A., Dimov, I., Ivanovska, S.: A quasi-Monte Carlo method for integration with improved convergence. In: Margenov, S., Waśniewski, J., Yalamov, P. (eds.) LSSC 2001. LNCS, vol. 2179, pp. 158–165. Springer, Heidelberg (2001). https://doi.org/10.1007/3-540-45346-6_15

26. Kuo, F.Y., Nuyens, D.: Application of quasi-Monte Carlo methods to elliptic PDEs with random diffusion coefficients - a survey of analysis and implementation. Found. Comput. Math. **16**, 1631–1696 (2016)

27. Nuyens, D., Cools, R.: Fast algorithms for component-by-component construction of rank-1 lattice rules in shift-invariant reproducing kernel Hilbert spaces. Math. Comput. **75**, 903–920 (2006)

28. Ostromsky, Ts., Dimov, I.T., Georgieva, R., Zlatev, Z.: Air pollution modelling, sensitivity analysis and parallel implementation. Int. J. Environ. Pollut. **46**, 83–96 (2011)

29. Ostromsky, T., Dimov, I., Georgieva, R., Zlatev, Z.: Parallel computation of sensitivity analysis data for the Danish Eulerian model. In: Lirkov, I., Margenov, S., Waśniewski, J. (eds.) LSSC 2011. LNCS, vol. 7116, pp. 307–315. Springer, Heidelberg (2012). https://doi.org/10.1007/978-3-642-29843-1_35

30. Ostromsky, Ts., Dimov, I.T., Marinov, P., Georgieva, R., Zlatev, Z.: Advanced sensitivity analysis of the Danish Eulerian Model in parallel and grid environment. In: AIP Conference on Proceedings, vol. 1404, pp. 225–232 (2011)

31. Owen, A.: Randomly permuted (t, m, s)-nets and (t, s)-sequences. In: Niederreiter, H., Shiue, P.J.S. (eds.) Monte Carlo and Quasi-Monte Carlo Methods in Scientific Computing. LNS, vol. 106, pp. 299–317. Springer, Heidelberg (1995). https://doi.org/10.1007/978-1-4612-2552-2_19

32. Owen, A.: Scrambled net variance for integrals of smooth functions. Ann. Stat. **25**, 1541–1562 (1997)

33. Owen, A.: Variance and discrepancy with alternative scramblings. ACM Trans. Comput. Logic. **V**, 1–16 (2002)

34. Pencheva, V., Georgiev, I., Asenov, A.: Evaluation of passenger waiting time in public transport by using the Monte Carlo method. In: AIP Conference on Proceedings, vol. 2321, p. 030028 (2021)

35. Saltelli, A.: Making best use of model valuations to compute sensitivity indices. Comput. Phys. Commun. **145**, 280–297 (2002)

36. Saltelli, A., Tarantola, S., Campolongo, F., Ratto, M.: Sensitivity Analysis in Practice: A Guide to Assessing Scientific Models. Halsted Press, New York (2004)

37. Sloan, I.H., Joe, S.: Lattice Methods for Multiple Integration. Oxford University Press, Oxford (1994)

38. Sloan, I.H., Kachoyan, P.J.: Lattice methods for multiple integration: theory, error analysis and examples. SIAM J. Numer. Anal. **24**, 116–128 (1987)

39. Sloan, I.H., Reztsov, A.V.: Component-by-component construction of good lattice rules. Math. Comput. **71**, 263–273 (2002)

40. Sobol, I.: Numerical Methods Monte Carlo. Nauka, Moscow (1973)

41. Sobol, I.M.: Sensitivity estimates for nonlinear mathematical models. Math. Model. Comput. Exp. **1**(4), 407–414 (1993)

42. Sobol, I.M., Tarantola, S., Gatelli, D., Kucherenko, S., Mauntz, W.: Estimating the approximation error when fixing unessential factors in global sensitivity analysis. Reliab. Eng. Syst. Saf. **92**, 957–960 (2007)
43. Todorov, V., Dimov, I.: Innovative digital stochastic methods for multidimensional sensitivity analysis in air pollution modelling. Mathematics **10**, 2146 (2022)
44. Wang, Y., Hickernell, F.J.: An historical overview of lattice point sets. In: Fang, K.T., Niederreiter, H., Hickernell, F.J. (eds.) Monte Carlo and Quasi-Monte Carlo Methods 2000, pp. 158–167. Springer, Heidelberg (2002). https://doi.org/10.1007/978-3-642-56046-0_10
45. Zaharieva, S.L., Georgiev, I.R., Mutkov, V.A., Neikov, Y.B.: Arima approach For forecasting temperature in a residential premises part 2. In: 20th International Symposium INFOTEH-JAHORINA (INFOTEH), pp. 1–5. IEEE (2021)
46. Zlatev, Z.: Computer Treatment of Large Air Pollution Models. KLUWER Academic Publishers, Dorsrecht (1995)
47. Zlatev, Z., Dimov, I.T., Georgiev, K.: Three-dimensional version of the Danish Eulerian model. Z. Angew. Math. Mech. **76**(S4), 473–476 (1996)
48. Zlatev, Z., Dimov, I.T.: Computational and Numerical Challenges in Environmental Modelling. Elsevier, Amsterdam (2006)
49. The Danish Eulerian Model. https://www2.dmu.dk/AtmosphericEnvironment/DEM/. Accessed 2 Oct 2022

On an Optimization of the Lattice Sequence for the Multidimensional Integrals Connected with Bayesian Statistics

Venelin Todorov[1,2] and Slavi Georgiev[1,3(✉)]

[1] Department of Information Modeling, Institute of Mathematics and Informatics, Bulgarian Academy of Sciences, Acad. Georgi Bonchev Str., Block 8, 1113 Sofia, Bulgaria
{vtodorov,sggeorgiev}@math.bas.bg
[2] Department of Parallel Algorithms, Institute of Information and Communication Technologies, Bulgarian Academy of Sciences, Acad. Georgi Bonchev Str., Block 25A, 1113 Sofia, Bulgaria
venelin@parallel.bas.bg
[3] Department of Applied Mathematics and Statistics, Faculty of Natural Sciences and Education, Angel Kanchev University of Ruse, 8 Studentska Str., 7004 Ruse, Bulgaria
sggeorgiev@uni-ruse.bg

Abstract. Lots of challenges in the Bayesian statistic exist since this is one of the fundamental discipline for neural network and machine learning. The latters being extensively developed, they impose a number of problems, which require contemporary solutions. During training of the artificial neural networks, eventually multidimensional integrals arise, which have to be computed extremely fast and accurately. In this paper, for the purpose of multidimensional integration, for the first time we develop three new highly accurate lattice sequences based on component by component construction methods: construction of rank-1 lattice rules with prime number of points and with product weights; construction of rank-1 lattice sequences with prime number of points and with product weights; construction of polynomial rank-1 lattice sequences in base 2 and with product weights. Our methods show significant optimization compared to the results produced by the plain Monte Carlo algorithm and the most widely used lattice sequence. The outcomes could play a major multi-sided role.

Keywords: Multidimensional integrals · Optimization of the lattice sequences · Bayesian statistics

1 Introduction

There are two paradigms in modern statistics. The one that is most familiar and most used is called *frequentist* statistics. The other, which is to a great extent

D. Simian and L. F. Stoica (Eds.): MDIS 2022, CCIS 1761, pp. 264–275, 2023.
https://doi.org/10.1007/978-3-031-27034-5_18

unknown and is used only by a limited number of scientists and researchers, is called *Bayesian*. But even where Bayesian statistics is known, it is not interpreted as a separate paradigm. This is equally valid for the representatives of the frequentist paradigm as for representatives of Bayesian statistics.

In the framework of Bayesian statistics we have the following simplified setup – at each study we only have the survey data and we want to check some hypotheses. In such a case, the probability that a particular hypothesis is true and at the same time to have exactly the data we have, could be represented in two ways. First, it is the probability that the hypothesis is correct multiplied by the probability that we have exactly that data, given conditional on the hypothesis being true; or second, the probability of having exactly these data multiplied by the probability that the hypothesis is true, given that this is the data. In practice, apart from the survey data and the testable hypotheses, in any study we also have *a priori* information. This may be theoretical knowledge about the object under study, or it may be information from some past research on the same topic. A priori probability is the probability that the hypothesis is true only in the light of the a priori information about the object under study. The sampling distribution is the probability of exactly the data that are obtained if the hypothesis is true. Full probability is the probability of obtaining exactly the data that are obtained, whether or not the hypothesis is true.

One of the fundamental differences between frequentist and Bayesian statistics is in the understanding of what probability is. According to requentist statistics, probability is an objective characteristic of the object under study that occurs in an infinite number of trials. According to Bayesian statistics, probability is a measure of knowledge (state of knowledge) about the object under study. In this sense, from the perspective of frequentist statistics, probability is an intrinsic characteristic of the object under study, and from the perspective of Bayesian statistics, probability does not characterize the object, but the knowledge of it.

A very important problem in Bayesian statistics considered by Lin in [10,11] is the problem of calculating multidimensional integrals used in machine learning, deep learning and many other areas [6,14]. For example, the training of a network is performed by minimization of an error function, based on the maximum likelihood principle, which in turn often involves computation of multidimensional integrals of a specific form. The first multidimensional Lin integrals are of the form

$$\int_\Omega p_1^{u_1}(x)\dots p_s^{u_s}(x)\mathrm{d}x, \tag{1}$$

and the second Lin integrals are of the form

$$\int_\Omega e^{-Nf(x)}\phi(x)\mathrm{d}x, \tag{2}$$

where $\Omega \in \mathbb{R}^s$, $x = (x_1,\dots,x_s)$, $p_i(x)$ are polynomials, u_i are integers, $i = 1,\dots,s$, $f(x)$ and $\phi(x)$ are multidimensional polynomials and N is an integer. Lin integrals (1) and (2) are often computed with unsatisfactory precision with deterministic methods [22,23], and it is well known that the Monte Carlo (MC) methods [3, 13, 18] perform better than the deterministic methods, because MC does not suffer from the so called "curse of dimensionality" [3, 5], which exhibits especially when the number of dimensions becomes large.

The original Monte Carlo idea was a major breakthrough decades ago, but it has become clear it has a couple of drawbacks and it is still far from completely studied area. The greatest disadvantage of the classical Monte Carlo methods is their slow order of convergence, which is $\mathcal{O}(N^{1/2})$. A large amount of effort has been put to investigate possible approaches to improve this estimate. It is found that this poor convergence is due to the non-uniform population of the multidimensional computational domain by the sampled random points. This phenomenon is not surprising, though, because of the random nature of the points. So, if a deterministic rather than stochastic sequences are used, they aim to minimize the clustering of points and 'empty regions' of the domain. Such sequences are named to feature low discrepancy and are well dispersed along the integration domain. Obviously, these sequences are expected to be superior (in probabilistic sense) to the ordinary Monte Carlo sequences and experience faster convergence than the latter. This idea is continued in Sect. 2. Section 3 approaches the computational investigation of the Lin multidimensional integrals. The results are discussed and a comparison with other studies is performed in Sect. 4. To conclude, some implications are presented in Sect. 5.

2   Methods

In this section we will describe the concept behind the quasi-Monte Carlo lattice sequences and their improved convergence properties.

Let $\mathbf{z} \in \mathbb{N}^s$, and $N \in \mathbb{N}$. According to [15], the set $P = \mathbf{x_1}, \ldots, \mathbf{x_N}$ in $[0, 1]^s$ with $\mathbf{x}_k = k\mathbf{z}/N$ for all $1 \leq k \leq N$ is named a *lattice point set* and $\mathbf{z}$ is named its *generating vector*. The algorithm with point set $P = \mathbf{x_1}, \ldots, \mathbf{x_N}$ is named a *lattice sequence (LS)* [16].

We will use this rank-1 LS defined by [21]:

$$\mathbf{x}_k = \left\{ \frac{k}{N} \mathbf{z} \right\}, \ k = 1, \ldots, N,$$

where $N \geq 2$ is an integer, $\mathbf{z} = (z_1, z_2, \ldots z_s)$ is the generating vector.

The performance of the LS can be optimized through the Component-by-Component (CBC) Fast Construction method [1, 12, 17]. Now we construct several special LS following the idea in [2, 9]. The first LS will be the CBC construction of rank-1 lattice rules with prime number of points and with product weights, we will call **1PT**. The second LS will be the CBC construction of rank-1 lattice sequences with prime power of points and with product weights, this can also be used for fast construction of a rank-1 lattice rule with a prime power of points, we will call it **1EXPT**. The third LS will be the CBC construction of polynomial rank-1 lattice sequences in base 2 and with product weights, we will call **2POLY**.

The search for optimal generating vectors in high dimensions s at fixed N has high computational difficulty. Choosing a good generator vector that leads to small integration errors is not trivial. It is advisable that the single-dimensional projections of the lattice rule to have N distinct values, so each component of $\mathbf{z}$ has to be in turn restricted to the set

$$U_N := \{z \in \mathbb{N} : 1 \le z \le N - 1, \ \gcd(z, N) = 1\}.$$

In fact, the size of U_N is $|U_N| = \varphi(N)$, which is the Euler totient function. If the prime factorization of N is defined as $N = p_1^{\alpha_1} p_2^{\alpha_2} \cdots p_k^{\alpha_k}$, it is true that

$$\varphi(N) = (p_1^{\alpha_1} - p_1^{\alpha_1 - 1})(p_2^{\alpha_2} - p_2^{\alpha_2 - 1}) \cdots (p_k^{\alpha_k} - p_k^{\alpha_k - 1}).$$

$\varphi(N)$ grows asymptotically at a rate close to N, more precisely $1/\varphi(N) = \mathcal{O}(\log \log N / N)$. In case N is a prime, then $\varphi(N) = N - 1$, and further there exist $N - 1$ possible choices for every component of $\mathbf{z}$, and $(N - 1)^s$ possible choices for the generating vector $\mathbf{z}$.

If N and s are relatively big, a bruteforcing search to obtain the generating vector which fulfills a predefined error criterion, is practically infeasible. There exists sophisticated number theory methods based on various criteria such as the Zaremba index and worst function errors. A simple construction algorithm, named after Korobov [8], is as follows.

If an integer a is chosen in such a way that $1 \le a \le N - 1$ and $\gcd(a, N) = 1$, then

$$\mathbf{z} = \mathbf{z}(a) := (1, a, a^2, a^3, \dots, a^{s-1}) \bmod N.$$

Obviously, there are maximum $N - 1$ choices for the parameter a, which leads to maximum $N - 1$ choices for the generating vector $\mathbf{z}$. In this case sometimes it is technically possible to check all $N - 1$ choices for $\mathbf{z}$ and pick up the one with best properties.

Here we use another advanced recently developed technique known as component by component (CBC) construction method. Using this component by component construction method, different generating vectors are obtained.

We construct the lattice rule following the three steps. At the beginning of the algorithm the input is the number of dimensionality s and the number of samples N. At the first step of the algorithm the s dimensional optimal generating vector

$$\mathbf{z} = (z_1, z_2, \dots, z_s) \tag{3}$$

is generated by the component by component construction method shortly described below. For a given N, it is possible to construct a generating vector $\mathbf{z} = (z_1, z_2, \dots, z_s)$ as follows.

o Fix $z_1 = 1$.
o With z_1 being known, obtain $z_2 \in U_N$ to satisfy the chosen error criterion in two dimensions.
o With z_1, z_2 being known, obtain $z_3 \in U_N$ to satisfy the chosen error criterion in three dimensions.
o With z_1, z_2, z_3 being known, obtain $z_4 \in U_N$ to satisfy the chosen error criterion in four dimensions.

o Similarly, the other generating vector components are obtained.

The obtained generating vector by the CBC construction method is then extensible in s. The second step of the algorithm is generating the points of lattice rule by formula

$$\mathbf{x}_k = \left\{ \frac{k}{N} \mathbf{z} \right\}, \ k = 1, \ldots, N. \tag{4}$$

And at the third step of the algorithm an approximate value I_N of the multidimensional integral is evaluated by the formula:

$$I_N = \frac{1}{N} \sum_{k=1}^{N} f\left(\left\{ \frac{k}{N} \mathbf{z} \right\} \right). \tag{5}$$

3   Numerical Results

The computational experiments are conducted using the MATLAB® environment. The tests described below are performed on a user laptop with 6-core processor and 16 GB RAM.

In this section the results produced by the most simple and widely used crude Monte Carlo CRU [4] and standard lattice LAT [9] will be compared with our optimizations 1PT, 1EXPT and 2POLY. Tables hereinafter give the relative errors (REER) obtained with the five approaches for the corresponding multidimensional integral (MI), together with the required computational time t, measured in seconds [s].

We explore the following Lin MIs (1) and (2):

Example 1. $s = 3$:

$$\int_{[0,1]^3} \exp(x_1 x_2 x_3) \mathrm{d}\mathbf{x} \approx 1.14649913323497. \tag{6}$$

Example 2. $s = 4$:

$$\int_{[0,1]^4} x_1 x_2^2 e^{x_1 x_2} \sin(x_3) \cos(x_4) \mathrm{d}\mathbf{x} \approx 0.10897491798381. \tag{7}$$

Example 3. $s = 5$:

$$\int_{[0,1]^5} \exp(-100 x_1 x_2 x_3)\big(\sin(x_4) + \cos(x_5)\big) \mathrm{d}\mathbf{x} \approx 0.185429894674946. \tag{8}$$

Example 4. $s = 7$:

$$\int_{[0,1]^7} e^{1 - \sum_{i=1}^{3} \sin(\frac{\pi}{2} x_i)} \arcsin\left(\sin(1) + \frac{\sum_{j=1}^{7} x_j}{200} \right) \mathrm{d}\mathbf{x} \approx 0.481088487600829. \tag{9}$$

Example 5. $s = 10$:

$$\int\limits_{[0,1]^{10}} \frac{4x_1 x_3^2 e^{2x_1 x_3}}{(1 + x_2 + x_4)^2} e^{\sum\limits_{j=5}^{10} x_j} d\mathbf{x} \approx 14.8087092095592. \tag{10}$$

Example 6. $s = 15$:

$$\int\limits_{[0,1]^{15}} \left(\sum\limits_{i=1}^{10} x_i^2 \right) \left(x_{11} - x_{12}^2 - x_{13}^3 - x_{14}^4 - x_{15}^5 \right)^2 d\mathbf{x} \approx 1.9644304795203. \tag{11}$$

Example 7. $s = 25$:

$$\int\limits_{[0,1]^{25}} \frac{4x_1 x_3^2 e^{2x_1 x_3}}{(1 + x_2 + x_4)^2} e^{\sum\limits_{j=5}^{20} x_j} \prod\limits_{j=21}^{25} x_j d\mathbf{x} \approx 103.987049568116. \tag{12}$$

Example 8. $s = 30$:

$$\int\limits_{[0,1]^{30}} \frac{4x_1 x_3^2 e^{2x_1 x_3}}{(1 + x_2 + x_4)^2} e^{\sum\limits_{j=5}^{20} x_j} \prod\limits_{j=21}^{30} x_j d\mathbf{x} \approx 3.242993405561. \tag{13}$$

Now, the REERs follow in the Tables 1, 2, 3, 4, 5, 6, 7 and 8.

Table 1. REER for the 3-MI (6).

N	CRU	t,[s]	LAT	t,[s]	1PT	t,[s]	1EXPT	t,[s]	2POLY	t,[s]
2^{10}	4.2662e−03	0.0062	**5.2314e−05**	0.0093	2.6268e−03	0.0041	1.4349e−03	0.0042	8.0558e−04	0.0040
2^{12}	3.8070e−03	0.0158	3.0650e−04	0.0153	2.0163e−03	0.0152	**9.3216e−05**	0.0156	6.2458e−04	0.0157
2^{14}	5.0007e−04	0.0598	1.7529e−04	0.0600	6.2057e−04	0.0610	**9.9337e−06**	0.0608	1.0234e−04	0.0624
2^{16}	2.9793e−04	0.2384	1.2526e−04	0.2405	6.6394e−05	0.2418	**1.4342e−05**	0.2475	2.3206e−05	0.2479
2^{18}	8.5327e−04	0.9539	**6.6274e−07**	0.9636	1.5716e−06	0.9773	4.7892e−06	0.9832	5.4321e−06	0.9737
2^{20}	9.2668e−05	3.8138	8.0126e−07	3.8003	9.7784e−07	3.9026	**8.8419e−09**	3.9049	7.1570e−07	3.9280

The results are discussed in detail in the next section.

The orders of convergence, schematically given, are plotted on Figs. 1 and 2.

4 Discussion

The following observations from Tables 1, 2, 3, 4, 5, 6, 7 and 8 could be done.

For the 3-MI the best approach is 1EXPT – it reaches a relative error 8.8419e−09 (Table 1). For the 4-MI the best method is again 1EXPT - the best error for $N = 2^{20}$ is 7.6816e−06 and for both MIs the computational time for

Table 2. REER for the 4-MI (7).

N	CRU	t,[s]	LAT	t,[s]	1PT	t,[s]	1EXPT	t,[s]	2POLY	t,[s]
2^{10}	3.1234e−02	0.0055	**6.2940e−03**	0.0039	2.9554e−02	0.0042	1.6564e−02	0.0041	1.0733e−02	0.0058
2^{12}	4.4953e−03	0.0160	1.1417e−03	0.0160	2.4898e−02	0.0173	**8.5685e−04**	0.0160	7.3288e−03	0.0158
2^{14}	1.4956e−02	0.0623	1.5498e−03	0.0610	7.5403e−03	0.0625	5.7848e−04	0.0657	**5.0731e−04**	0.0613
2^{16}	2.5006e−03	0.2472	1.6257e−03	0.2474	3.0061e−04	0.2480	**5.1902e−05**	0.2581	2.6567e−04	0.2462
2^{18}	3.5109e−03	0.9811	7.2795e−05	0.9833	9.1357e−05	0.9877	**4.9823e−05**	1.0055	1.0032e−04	0.9880
2^{20}	1.4482e−04	3.8783	9.2342e−05	3.8896	1.1092e−05	3.9818	**7.6816e−06**	3.9899	2.8624e−05	4.0043

Table 3. REER for the 5-MI (8).

N	CRU	t,[s]	LAT	t,[s]	1PT	t,[s]	1EXPT	t,[s]	2POLY	t,[s]
2^{10}	1.4902e−01	0.0053	**2.2806e−03**	0.0040	1.5880e−02	0.0042	2.4388e−03	0.0043	2.5208e−02	0.0041
2^{12}	9.2531e−02	0.0159	3.7245e−03	0.0156	5.5395e−03	0.0160	**3.2610e−03**	0.0160	7.1276e−03	0.0157
2^{14}	1.7645e−02	0.0624	**1.0345e−03**	0.0621	2.1792e−03	0.0631	1.5021e−03	0.0626	2.6700e−03	0.0629
2^{16}	9.5565e−03	0.2445	7.6277e−04	0.2438	5.8889e−04	0.2491	**1.4251e−05**	0.2502	6.0769e−04	0.2486
2^{18}	2.6760e−03	1.0039	6.2269e−05	0.9838	1.5094e−04	0.9961	**1.6224e−05**	1.0089	2.6240e−04	0.9975
2^{20}	1.1653e−03	3.9090	4.9697e−05	3.9262	6.7333e−05	4.0286	**1.8133e−05**	4.0089	7.3474e−05	4.0171

Table 4. REER for the 7-MI (9).

N	CRU	t,[s]	LAT	t,[s]	1PT	t,[s]	1EXPT	t,[s]	2POLY	t,[s]
2^{10}	1.7891e−02	0.0117	4.0172e−03	0.0046	7.6272e−03	0.0049	**3.3134e−03**	0.0048	1.1692e−02	0.0050
2^{12}	1.4043e−02	0.0183	3.4764e−04	0.0182	7.9199e−04	0.0183	**3.0337e−04**	0.0190	2.1256e−03	0.0184
2^{14}	6.8774e−03	0.0710	1.5248e−04	0.0722	4.6962e−04	0.0798	**3.0716e−05**	0.0747	6.6916e−04	0.0742
2^{16}	2.3765e−03	0.2825	2.4174e−04	0.2856	6.4629e−05	0.2949	**3.1352e−05**	0.2949	2.7154e−04	0.2926
2^{18}	9.3172e−04	1.1320	1.7541e−05	1.1444	4.2966e−05	1.1760	**1.0777e−05**	1.1818	4.2813e−05	1.1727
2^{20}	7.0195e−05	4.5071	2.1958e−06	4.5774	1.0120e−05	4.7032	**3.0921e−07**	4.6804	1.3633e−05	4.6761

Table 5. REER for the 10-MI (10).

N	CRU	t,[s]	LAT	t,[s]	1PT	t,[s]	1EXPT	t,[s]	2POLY	t,[s]
2^{10}	1.5310e−01	0.0063	**2.7680e−03**	0.0047	6.7250e−02	0.0054	4.8282e−02	0.0048	6.7397e−02	0.0049
2^{12}	4.6633e−02	0.0175	2.9217e−02	0.0178	5.6807e−02	0.0179	**1.4294e−02**	0.0182	8.1931e−02	0.0178
2^{14}	1.1315e−03	0.0680	2.7959e−03	0.0724	**1.4020e−04**	0.0705	4.7519e−04	0.0719	3.3077e−02	0.0707
2^{16}	3.5038e−03	0.2762	6.2916e−03	0.2817	**7.5167e−04**	0.2819	3.6607e−03	0.2806	9.1079e−04	0.2796
2^{18}	4.2802e−03	1.0807	1.1984e−03	1.1036	1.2669e−03	1.1332	8.7544e−04	1.1237	**4.3233e−04**	1.1407
2^{20}	1.7967e−03	4.3619	2.0709e−03	4.4221	1.1130e−04	4.4889	**8.7057e−05**	4.5203	2.6133e−04	4.4892

Table 6. REER for the 15-MI (11).

N	CRU	t,[s]	LAT	t,[s]	1PT	t,[s]	1EXPT	t,[s]	2POLY	t,[s]
2^{10}	9.6008e−03	0.0066	**2.1065e−04**	0.0051	3.3415e−02	0.0051	2.6363e−02	0.0051	1.5978e−02	0.0053
2^{12}	4.3430e−02	0.0190	**1.1570e−03**	0.0191	1.5391e−02	0.0197	6.3514e−03	0.0199	8.3775e−03	0.0204
2^{14}	**2.1527e−03**	0.0781	4.7587e−03	0.0761	4.7130e−03	0.0781	2.9244e−03	0.0775	3.7451e−03	0.0775
2^{16}	5.2886e−04	0.3021	2.3316e−04	0.3039	4.8827e−04	0.3091	8.7396e−04	0.3094	**6.3174e−07**	0.3114
2^{18}	2.0420e−03	1.1968	**1.2830e−04**	1.2187	1.8603e−04	1.2357	2.1315e−04	1.2389	1.4915e−04	1.2541
2^{20}	1.8109e−03	4.7511	1.7447e−04	4.8115	1.7635e−04	4.9946	**2.7557e−05**	4.9786	1.2229e−04	4.9631

Table 7. REER for the 25-MI (12).

N	CRU	t,[s]	LAT	t,[s]	1PT	t,[s]	1EXPT	t,[s]	2POLY	t,[s]
2^{10}	2.6106e−01	0.0066	3.3046e−01	0.0050	3.2365e−01	0.0055	**1.5479e−01**	0.0054	3.3848e−01	0.0052
2^{12}	1.1141e−01	0.0205	**1.5034e−04**	0.0209	1.7291e−01	0.0201	7.9812e−02	0.0204	1.9826e−01	0.0204
2^{14}	1.6777e−01	0.0751	2.2509e−02	0.0770	7.1751e−02	0.0798	3.7016e−02	0.0793	**2.2011e−02**	0.0778
2^{16}	2.8483e−02	0.2941	**1.5280e−02**	0.3088	3.9921e−02	0.3165	2.4898e−02	0.3205	9.4390e−02	0.3144
2^{18}	2.6824e−03	1.1854	4.7105e−03	1.2317	**2.7591e−04**	1.2711	3.1380e−03	1.2702	4.5965e−02	1.2651
2^{20}	8.0614e−03	4.6930	4.7018e−03	4.8771	**4.0799e−03**	5.1334	5.2808e−03	5.1088	1.8766e−01	5.0451

Table 8. REER for the 30-MI (13).

N	CRU	t,[s]	LAT	t,[s]	1PT	t,[s]	1EXPT	t,[s]	2POLY	t,[s]
2^{10}	3.6722e−01	0.0066	2.0833e−01	0.0057	5.0829e−01	0.0054	**6.8940e−03**	0.0056	9.0823e−02	0.0053
2^{12}	2.9645e−01	0.0192	2.9292e−01	0.0203	3.6716e−01	0.0211	1.3886e−01	0.0208	**4.9670e−02**	0.0206
2^{14}	9.6492e−02	0.0749	1.2486e−01	0.0787	2.3992e−01	0.0813	**4.2109e−02**	0.0815	1.1213e−01	0.0801
2^{16}	7.9793e−02	0.2973	3.4576e−01	0.3145	1.2737e−01	0.3255	**1.2226e−02**	0.3234	1.4197e−01	0.3213
2^{18}	7.5115e−02	1.1819	5.5549e−02	1.2536	2.8080e−02	1.2954	**9.9234e−04**	1.2888	4.9720e−01	1.2940
2^{20}	2.2634e−02	4.7422	9.0972e−03	4.9054	**4.1314e−03**	5.1515	4.5463e−03	5.1749	4.8006e+00	5.1349

the best case is less than 4 s. The same conclusion is true for 5-MI: the most suitable approach is 1EXPT with best relative error 1.8133e−05. So far, in four of six cases (for N) 1EXPT gives the best result, and in two of six cases - LAT. The latter is not true only for 4-MI, where one of the best error is produced by 2POLY for $N = 2^{14}$.

The results for 7-MI (Table 4) stand out since all the best result are given by 1EXPT approach. Only for this method the error drops below $1e − 6$, of course for $N = 2^{20}$.

Considering the 10-MI, it is difficult to state whether there is a best approach, but it must be noted that for $N = 2^{20}$ the lowest error is again reached by 1EXPT. For the 15-MI, LAT gives three of six best result, but as the same as the previous case - for $N = 2^{20}$ the best error is given by 1EXPT.

The case with the 25-MI is similar to 10-MI: there is no method standing out, and the best error for $N = 2^{20}$ is given by 1PT. For 30-MI, though, again 1EXPT produces best results in four of six cases, while the lowest error is again given by 1PT.

All these findings unequivocally demonstrate that the most powerful approach is **1EXPT** (rank-1 lattice sequences with prime number of points and with product weights), which usually outperforms the other methods even in higher dimensions. If this is not the case, it produces low errors for large numbers of points N. Nevertheless, all algorithms work for reasonable amount of time and could be easily integrated in practical applications.

Due to the importance of the integrals studied (Examples 1–8), this is not the first investigation of possible stochastic approachers, aiming to supply accurate algorithms for their computation. We will compare out finding with previously obtained results, which compose far from exhaustive literature review. In the

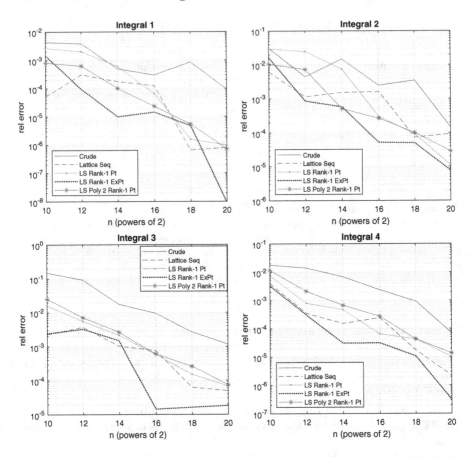

Fig. 1. The convergence of the approaches CRU, LAT, 1PT, 1EXPT, 2POLY for MIs (6)–(9).

study [7] three MIs are considered: a four-dimensional one, which resembles our 10-MI (Example 5), and the high-dimensional 25-MI and 30-MI (Examples 7 and 8). An adaptive Monte Carlo method is proposed and a superconvergent method is presented, combining the ideas of separation of the domain and importance sampling. The tests are performed on a supercomputer (in 1998), so the required time is comparable with the time shown in Tables 1, 2, 3, 4, 5, 6, 7 and 8. However, the relative errors are around 1.5–2 times higher than reported in Table 7, considering 25-MI. For 30-MI, the relative errors are of one order worse that the presented in this paper. For 1EXPT, however, the errors for small N are two orders better in our study.

In [20], 4-MI, 7-MI, 10-MI and 30-MI (Examples 2, 4, 5 and 8) are considered. The methods employed are the Sobol quasi-sequences, augmented by the Matousek linear scrambling, compared with an adaptive Monte Carlo algorithm

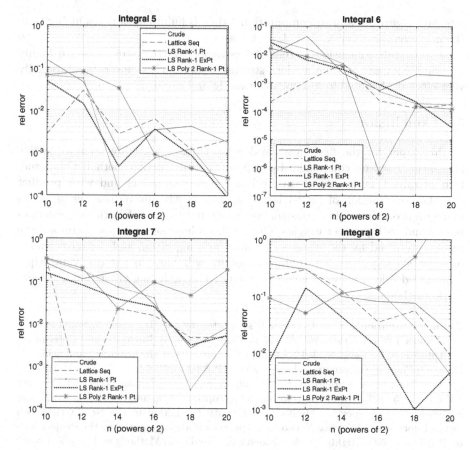

Fig. 2. The convergence of the approaches CRU, LAT, 1PT, 1EXPT, 2POLY for MIs (10)–(13).

and a lattice rule based on generalized Fibonacci numbers. For the lower dimensional case studies, the Fibonacci lattice sets algorithm shows some dominance for lower N, but for higher values of N 1EXPT has comparable or better performance. For the higher dimensional cases, the previously best scrambled Sobol algortihm is outperformed by 1EXPT by one order. It is also noticeable that the routines implemented here are hundreds of times faster that their counterparts from the regarded source.

Similar implications could be drawn when comparing our results with the recent study [19]. There, all MIs are considered without 10-MI (Example 5). The methods investigated there are the Sobol sequence, the Latin hypercube sampling method and an optimal stochastic approach. The latter appears to be the best method among the studied ones. Compared to the lattice rules and sequences, explored in this paper, it has similar performance, even for some MIs it achieves one-order better errors. However, it is extremely computationally intensive algorithm and its computational time is hundreds of times higher than the measured in Tables 1, 2, 3, 4, 5, 6, 7 and 8.

The overall conclusion is that the developed lattice sequences are highly accurate, compared to the modern stochastic approaches, and computationally very efficient, being able to implement easily and rapidly.

5   Conclusion

In this paper for the first time we develop three new highly accurate lattice sequences based on component-by-component construction methods: construction of rank-1 lattice rules with prime number of points and with product weights; construction of rank-1 lattice sequences with prime number of points and with product weights; construction of polynomial rank-1 lattice sequences in base 2 and with product weights. Our methods have significantly optimized the results produced by the standard Monte Carlo algorithm and the most widely used lattice sequence. The obtained results will play an extremely principal multi-sided role. This work could be extended as designing other (quasi-)Monte Carlo methods.

Acknowledgement. Slavi Georgiev is supported by the Bulgarian National Science Fund under Project KP-06-M32/2 - 17.12.2019 "Advanced Stochastic and Deterministic Approaches for Large-Scale Problems of Computational Mathematics" and Scientific Research Fund of University of Ruse under FNSE-04.

Venelin Todorov is supported by the Bulgarian National Science Fund under Project KP-06-N52/5 "Efficient methods for modeling, optimization and decision making" and KP-06-N52/2 "Perspective Methods for Quality Prediction in the Next Generation Smart Informational Service Networks". The work is also supported by the Project KP-06-Russia/17 "New Highly Efficient Stochastic Simulation Methods and Applications", funded by the National Science Fund – Bulgaria.

The authors thank the anonymous referees whose invaluable comments improved significantly the quality of the paper.

References

1. Baldeaux, J., Dick, J., Leobacher, G., Nuyens, D., Pillichshammer, F.: Efficient calculation of the worst-case error and (fast) component-by-component construction of higher order polynomial lattice rules. Numer. Algor. **59**, 403–431 (2012). https://doi.org/10.1007/s11075-011-9497-y
2. Cools, R., Kuo, F., Nuyens, D.: Constructing embedded lattice rules for multivariate integration. SIAM J. Sci. Comput. **28**, 2162–2188 (2006)
3. Dimov, I.: Monte Carlo Methods for Applied Scientists. World Scientific, Singapore (2008)
4. Dimov, I., Atanassov, E.: What Monte Carlo models can do and cannot do efficiently? Appl. Math. Model. **32**, 1477–1500 (2007)
5. Dimov, I.T., Georgieva, R.: Monte Carlo Method for numerical integration based on Sobol' sequences. In: Dimov, I., Dimova, S., Kol'kovska, N. (eds.) Numerical Methods and Applications, LNCS, vol. 6046, pp. 50–59. Springer, Heidelberg (2011). https://doi.org/10.1007/978-3-642-18466-6_5

6. Gery, M., Whitten, G., Killus, J., Dodge, M.: A photochemical kinetics mechanism for urban and regional scale computer modelling. J. Geophys. Res. **94**, 12925–12956 (1989)
7. Karaivanova, A., Dimov, I.: Error analysis of an adaptive Monte Carlo method for numerical integration. Math. Comput. Simul. **47**, 201–213 (1998)
8. Korobov, N.M.: Number-Theoretical Methods in Approximate Analysis. Fizmat-giz, Moscow (1963)
9. Kuo, F.Y., Nuyens, D.: Application of quasi-Monte Carlo methods to elliptic PDEs with random diffusion coefficients - a survey of analysis and implementation. Found. Comput. Math. **16**, 1631–1696 (2016). https://doi.org/10.1007/s10208-016-9329-5
10. Lin, S.: Algebraic Methods for Evaluating Integrals in Bayesian Statistics. PhD dissertation, UC Berkeley, May 2011
11. Lin, S., Sturmfels, B., Xu, Z.: Marginal likelihood integrals for mixtures of independence models. J. Mach. Learn. Res. **10**, 1611–1631 (2009)
12. Nuyens, D., Cools, R.: Fast algorithms for component-by-component construction of rank-1 lattice rules in shift-invariant reproducing kernel Hilbert spaces. Math. Comp. **75**, 903–920 (2006)
13. Paskov, S.H.: Computing High Dimensional Integrals with Applications to finance. Technical report CUCS-023-94, Columbia University, New York (1994)
14. Saltelli, A., Tarantola, S., Campolongo, F., Ratto, M.: Sensitivity Analysis in Practice: A Guide to Assessing Scientific Models. Halsted Press, New York (2004)
15. Sloan, I.H., Joe, S.: Lattice Methods for Multiple Integration. Oxford University Press, Oxford (1994)
16. Sloan, I.H., Kachoyan, P.J.: Lattice methods for multiple integration: theory, error analysis and examples. SIAM J. Numer. Anal. **24**, 116–128 (1987)
17. Sloan, I.H., Reztsov, A.V.: Component-by-component construction of good lattice rules. Math. Comp. **71**, 263–273 (2002)
18. Sobol, I.: Numerical Methods Monte Carlo. Nauka, Moscow (1973)
19. Todorov, V., Dimov, I., Fidanova, S., Georgieva, R., Ostromsky, Tz., Poryazov, S.: An optimized monte Carlo approach for multidimensional integrals related to intelligent systems. Ann. Comput. Sci. Inf. Syst. **32**, 101–104, (2022)
20. Todorov, V., Dimov, I.: Efficient stochastic approaches for multidimensional integrals in Bayesian statistics. In: Lirkov, I., Margenov, S. (eds.) LSSC 2019. LNCS, vol. 11958, pp. 454–462. Springer, Cham (2020). https://doi.org/10.1007/978-3-030-41032-2_52
21. Wang, Y., Hickernell, F.J.: An historical overview of lattice point sets. In: Fang, K.T., Niederreiter, H., Hickernell, F.J. (eds.) Monte Carlo and Quasi-Monte Carlo Methods 2000, pp. 158–167. Springer, Berlin/Heidelberg (2002). https://doi.org/10.1007/978-3-642-56046-0_10
22. Song, J., Zhao, S., Ermon S.: A-nice-mc: adversarial training for MCMC. In: Advances in Neural Information Processing Systems, pp. 5140–5150 (2017)
23. Watanabe, S.: Algebraic analysis for nonidentifiable learning machines. Neural Comput. **13**, 899–933 (2001)

Modelling and Optimization of Dynamic Systems

Numerical Optimization Identification of a Keller-Segel Model for Thermoregulation in Honey Bee Colonies in Winter

Atanas Z. Atanasov, Miglena N. Koleva$^{(\boxtimes)}$, and Lubin Vulkov

University of Ruse, 8 Studentska str., 7017 Ruse, Bulgaria
{aatanasov,mkoleva,lvalkov}@uni-ruse.bg

Abstract. The present work is inspired by laboratory experiments providing measurements in a few places of the hive for a long period of time. Based on Keller-Segel model in form of coupled nonlinear parabolic equations for the local temperature T and the bee density $\rho \geq 0$, using the real data, we investigate numerically the thermoregulation in honey bee colonies in winter. We propose a numerical approach, based on conjugate gradient method into two stages: first, we solve a semilinear parabolic inverse problem to recover the density ρ and the temperature T. On the second stage we solve the strongly nonlinear convection-diffusion equation to recover again the density ρ. The numerical tests show the efficiency of the method at the calibration of thermoregulation model.

Keywords: Keller-segel model · Honeybee thermoregulation · Inverse problem · Optimization · Conjugate gradient method

1 Introduction

Winter mortality is one of the main reason for the decline of bee colonies in recent years. In 2006 a huge wintering losses have been reported in Canada, caused by Varroa mites [14,18]. In many other regions around the world, the reasons of bee colony winter mortality have not been established. A number of possible stressors for the survival of the overwintering population have been suggested, such as temperature, humidity, availability of food, parasitic mites in the colony, intensive agriculture [25]. An important factor for the survival of bees in the winter is the generation and preservation of heat [6,17,29].

During the winter season, the bees remain active and maintain the temperature of the hive within a certain range. They make cells for brood and storage of honey and pollen and bees and form a thermoregulatory cluster, with the highest temperature in the core [6,22]. This temperature varies from 18 °C to 32 °C, when the environment temperature is between −15 °C and 10 °C. This allows them to survive long periods of cold.

© The Author(s), under exclusive license to Springer Nature Switzerland AG 2023
D. Simian and L. F. Stoica (Eds.): MDIS 2022, CCIS 1761, pp. 279–293, 2023.
https://doi.org/10.1007/978-3-031-27034-5_19

Bees produce heat through the activity of their flight muscles - when the temperature become lower than the specific threshold, the bee begins to tremble its flight muscles, otherwise it remains at rest. Additionally, the variable temperature in the honey bee's local surroundings, gives rise for their thermotactic movement [6,12].

In winter, bees do not collect nectar and pollen and relies on its available reserves. Also, since during the winter the brood rearing stops, the survival of the colony depends on a long-lived cohort of bees that is created in the autumn [6,8,29].

One of the efficient approaches for monitoring and understanding important processes and factors for improving the survival rate of colonies is the mathematical modeling. Usually these problems are described by systems of ordinary differential equations, see e.g. [3,4,11,14,18,25], with coefficients that characterize population and disease development. To model colony dynamics more realistically and adequately, heat production and thermoregulation must be taken into account. In this case, the process is described by partial differential equations (PDEs) [20,24]. In [6,29] a Keller–Segel model [7] with a sign-changing chemotactic coefficient - as bees have a preferred temperature, is proposed to describe the self-organized thermoregulation of honeybee clusters. In [6] the authors extend the model proposed in [29], to take into account the mortality of individual bees.

In this article we deal with the model of two PDEs, suggested in [6]. The prior information about the functional form of the model is unknown. But it is possible its estimation, based on transient temperature measurements, taken by the sensors embedded in the hive. They are used in the least squares cost functional for first reconstruction of the bee density ρ. This allows us to develop a two-stage algorithm in order to find numerically the temperature and density. First, concentrated to the one of the equations, we solve an identification problem to recover the temperature and density, fitted to the temperature measurements. Then we refine the density from the numerical solution of the second equation.

The remaining sections are arranged as follows. In Sect. 2, we describe the PDEs system. The Sect. 3 uses the temperature measurements to solve an inverse problem that provides real values for the local temperature T and density ρ. In Sect. 4, using the results of the inverse problem, we develop a positivity preserving for the density, numerical method for the full direct problem. The realization of the numerical method and provide results from numerical simulation are described in Sect. 5. The paper is closed by concluding remarks.

2   The Keller-Segel Honey Bee Model

In this section we introduce the model problem, derived in [6]. In contrast to [29], where the edge of the bee colony is a free boundary, the model in [6] is simplified assuming fixed boundary at length L, but involves nontrivial mortality rate θ, formulated on the base of observations. This model is more relevant for the honey bee colonies with bigger mortality. For the readers convenience we will provide also some details and assumptions of the model derivation.

For $x \in [0, L]$, $t \in [0, t_f]$, the model reads [6]

$$\frac{\partial T}{\partial t} = \frac{\partial^2 T}{\partial x^2} + f(T)\rho, \tag{1}$$

$$\frac{\partial \rho}{\partial t} = \frac{\partial^2 \rho}{\partial x^2} - \frac{\partial}{\partial x}\left(\chi(T)\rho\frac{\partial T}{\partial x}\right) - \theta(\rho, T)\rho, \tag{2}$$

$$\frac{\partial T}{\partial x}(0, t) = 0; \quad T(L, t) = T_a < T_\chi, \tag{3}$$

$$\frac{\partial \rho}{\partial x}(0, t) = 0; \quad \left(\frac{\partial \rho}{\partial x} - \chi(T)\rho\frac{\partial T}{\partial x}\right)(L, t) = 0, \tag{4}$$

$$\rho(x, 0) = \rho^0(x), \quad T(x, 0) = T^0(x). \tag{5}$$

Here T is the local temperature and $\rho \geq 0$ is the bee density, $f(T)\rho$ represents the heat generation by bees and $\chi(T)$ is sign-changing function, describing the movement of the bees towards a higher temperature, when $T < T_\chi$ and moving away to lower temperature, when $T > T_\chi$. Thus the chemotactic coefficient generates very different dynamics compared to the models studied in the literature. The the sign of the coefficient $\chi(T)$ changes at T_χ, and this differ the present model from the known generalized Keller-Segel model, where χ has a fixed positive sign.

In [6] the bee losses is studied by incorporating individual mortality bees function $\theta(\rho, T) > 0$ and the functions f and χ are simplified, based on those in [29]. Since $\chi(T)$ changes its sign from positive, when T is small to negative for bigger values of T, they chose χ to be a step-function. Likewise, in view of the investigations in [29] f is also defined as a step-function

$$f(T) = \begin{cases} f_{\text{low}}, & T < T_f, \\ f_{\text{high}}, & T > T_f, \end{cases} \qquad \chi(T) = \begin{cases} +\chi_1, & T < T_\chi, \\ -\chi_2, & T > T_\chi, \end{cases} \tag{6}$$

Here $f_{\text{low}}, f_{\text{high}}, \chi_1, \chi_2 > 0$, T_f and T_χ are temperature thresholds at which f and $\chi(T)$ change value and sign, respectively ($T_f < T_\chi$). The temperature T_χ can be considered the preferred temperature, since bees choose to go toward to places with such temperature.

The mortality function is constructed as a product of three different effects:

(i) The local temperature (θ_T);
(ii) Effective refresh frequency of bees, that generate heat (θ_D);
(iii) Parasitic mites in the colony (θ_M).

Thus, the mortality rate of the individual bee is represented by

$$\theta(T, \rho) = \theta_0 \theta_T(T)\theta_D(\rho)\theta_M(\rho), \tag{7}$$

where θ_0 is a constant that must fit to the measurements.

The first effect expresses that when the local temperature is above a certain thresholds $T_\theta > T_a$, the mortality does not increase. If the temperature is excessively low ($T(x) < T_\theta$), a bee in that place needs to made great efforts to produce heat, reducing her lifespan. This effect is simplified in [6] and represented by the step-function

$$\theta_T(T) = \begin{cases} 1, \text{ if } T < T_\theta, \\ 0, \text{ if } T \geq T_\theta. \end{cases} \tag{8}$$

The second effect comes from the refresh rate by recovered bees – the ratio between local bee density ρ and the colony size ρ_{tot}.

Each bee is able to warm the colony by trembling its flight muscles, but just for about 30 min, after which she must feed on honey, in order to recuperate and replenish store [27]. The bees rotates from the periphery and inside the colony, at warmer location, so that the exhausted bees rest and their role is taken over by the already recovered bees [26]. Thus, as bigger is the colony as longer is the time for recovering and this influence the mortality. This effect is modeled as

$$\theta_D(\rho) = \frac{\rho}{(\rho_{tot})^\gamma}, \quad \gamma > 0, \quad \rho_{tot} = \int_0^L \rho(x)dx, \tag{9}$$

where the power $\gamma > 0$ is unknown.

The last effect is connected with the occurrence of the parasitic mite Varroa destructor in honey bee colonies, who reduces the body weight and protein content of individual bees and shorten their lifespan [14,25]. Therefore, the increasing of the amount of mites per bee increases the mortality rate. Moreover, this fraction increases with decreasing colony size, because when the host bee dies, the mite is passed on to another bee. Let denote the amount of mites by m. The corresponding effect is modeled as

$$\theta_M(\rho) = 1 + \frac{m}{\rho_{tot}}. \tag{10}$$

In [6], authors study the model (1)–(10). Since the mortality of the bees is a slow process in time, they use a steady-state formulation, where ρ_{tot} decreases, in order to study the evolution of the colony size.

3   The Auxiliary Inverse Problem

In this section, we introduce an inverse coefficient problem that uses a real information from temperature measurements in the hive.

3.1   Formulation of the Problem

We are going to study the problem of identifying $\rho = \rho(x,t)$ as an unknown reaction coefficient in the parabolic Eq. (1). Every inverse problem begins with a well-posed problem called the *direct problem* (DP). In our case it is formed by the Eq. (1) with boundary conditions (3) and initial data (5)

$$T(x,0) = T_0(x), \quad x \in (0,L).$$

Let T be the solution of this direct (forward) problem. Following the possible measurements, we investigate the *inverse problem* IP, about reconstruction of the unknown coefficient $\rho = \rho(x, t)$ such that:

IP: suppose that the initial temperature $T_0(x)$ is known and measurements

$$T(x_m, t_k; \rho) = G_{mk}, \quad m = 1, \ldots, M, \quad k = 1, \ldots, K, \quad \text{are given.} \quad (11)$$

We study the inverse problem IP for seeking $\rho(x, t) \in A = \{\rho : 0 \leq \rho \leq \rho_{max}\}$, using additional information G_{mk}. Here A is the set *admissible* solutions of the inverse problem.

Let $\mathcal{A} : A \rightarrow G$ be an injective operator, $\rho \in A$, $g \in G$ is an Euclidian space of data $g = \{G_{11}, \ldots, G_{MK}\}$. Then, the inverse problem IP can be formulated by $\mathcal{A}(\rho) = g$. This problem is *ill-posed*, namely at least one of the following properties of the solution is not fulfilled: existence, uniqueness, stability to errors in measurements (11), see e.g. [1,9,10,16,19].

The unknowns can be not uniquely identified due to limited amount of measurements, the identification problem is often reduced to a minimization problem [28]. Here, the IP is reformulated as the optimization problem

$$\rho^* = \mathrm{argmin} J(\rho), \quad J(\rho) = \frac{1}{2} < \mathcal{A}(\rho) - g, \ \mathcal{A}(\rho) - g >,$$

where $J(\rho)$ characterize the quadratic error of the model data from the measurements, and we take it as follows:

$$J(\rho) = \frac{1}{KM} \sum_{k=1}^{K} \sum_{m=1}^{M} (T(x_m, t_k; \rho) - G_{mk})^2. \quad (12)$$

We rewrite the functional $J(\rho)$ as follows:

$$J(\rho) = \frac{1}{KM} \sum_{k=1}^{K} \sum_{m=1}^{M} \int_0^{t_f} \int_0^L (T(x, t; \rho) - G_{mk})^2 \delta(x - x_m) \delta(t - t_k) dx dt, \quad (13)$$

where $\delta(\cdot)$ is the Dirac-delta distribution, reformulating the inverse problem as a minimization problem for the functional $J(\rho)$, i.e.,

$$J(\rho) = \min_{a \in A} J(\rho). \quad (14)$$

If $\hat{\rho} \in A$ such that $J(\hat{\rho}) = 0$, then the solution of the minimization problem (14) is called *exact solution* of the inverse problem IP. Otherwise, the solution of minimizing problem is called the *quasi solution*.

We ignore any questions of existence and uniqueness of the minimizer but focus only on the construction of numerical algorithms. For this we derive an explicit formula for the gradient of $J(\rho)$ in the problem IP, which could then be used in construction a conjugate gradient scheme (CGS) for approximation the solution to the IP.

3.2   The Conjugate Gradient Method of Minimization

Next, if $\delta\rho$ is an increment we denote the deviation of solution T by $\delta T(x,t;\rho) = T(x,t;\rho + \delta\rho) - T(x,t;\rho)$. Then, it satisfies with accuracy up to terms of order $O(|\delta\rho|^2)$, the *sensitivity problem*:

$$\frac{\partial \delta T}{\partial t} = \frac{\partial^2 \delta T}{\partial x^2} + f'(T)\rho\delta T + f(T)\delta\rho,$$

$$\frac{\partial \delta T}{\partial x}(0,t) = 0, \quad \delta T(L,t) = 0, \tag{15}$$

$$\delta T(x,0) = 0.$$

Next, we present without derivation the gradient for IP namely, the gradient of the functional (13) has the form

$$J'(\rho) = \frac{1}{KM}f(T)Y(x,t), \tag{16}$$

where the function $Y(x,t)$ satisfies the adjoint problem

$$\frac{\partial Y}{\partial t} = -\frac{\partial^2 Y}{\partial x^2} - f'(T)\rho Y$$

$$+ \frac{2}{KM}\sum_{k=1}^{K}\sum_{m=1}^{M}(T(x,t;\rho) - G_{mk})\delta(x - x_m)\delta(t - t_k)dxdt, \tag{17}$$

$$\frac{\partial Y}{\partial x}(0,t) = 0, \quad Y(l,t) = 0, \quad Y(x,t_f) = 0.$$

Most of the gradient methods (GM) are presented as iterative sequence

$$\rho_{l+1} = \rho_l - \alpha_l J'(\rho_l),$$

where l is the number of iteration, α_l is a descent parameter, $J'(\rho_l)$ is the gradient of the cost functional at point ρ_l. The choice of the initial approximation ρ_0 is very important for the convergence of GM [1, 10, 16, 17]. Very often, the descent parameter α_l define the GM.

We extend the GM proposed in [10, 13, 20, 21]. Consider the iteration process

$$\rho_{l+1} = \rho_l - \alpha_l\beta_l, \tag{18}$$

where β is the search step size and

$$\alpha_0 = J'(\rho_0),$$
$$\alpha_l = J'(\rho_l) + \gamma^l J'(\rho_{l-1}), \quad l = 1, 2, \ldots,$$

$$\gamma_0 = 0, \quad \gamma_l = \frac{\int_0^{t_f}\int_0^L [J(\rho_l)]^2 dtdx}{\int_0^{t_f}\int_0^L [J(\rho_{l-1})]^2 dtdx}, \quad l = 1, 2, \ldots, \tag{19}$$

$$\beta_l = \frac{\sum_{k=1}^{K} \sum_{m=1}^{M} [T(x_m, t_k, \rho_l) - G_{mk}] \, \delta T(x_m, t_k, \rho_l)}{\sum_{k=1}^{K} \sum_{m=1}^{M} [\delta T(x_m, t_k, \rho_l)]^2} \tag{20}$$

The first stage of our numerical method is to solve IP. For this purpose we use the following Algorithm 1.

Algorithm 1. Algorithm for IP

Require: ρ_0, ε
Ensure: ρ_{l+1}, $T(x, t; \rho_{l+1})$
  $l \leftarrow 0$
Calculate $J(\rho_l)$
  while $J(\rho_l) > \varepsilon$ **do**
    $T(x_m, t_k; \rho_l)$, $m = 1, \ldots, M$, $k = 1, \ldots, K$ ← direct problem (1), (3), (5)
    Calculate $J(\rho_l)$
    $Y(x, t; \rho_l)$ ← adjoint problem (17)
    Calculate $J'(\rho_l)$, using (16)
    Calculate α_l, using (19)
    $\delta\rho \leftarrow \alpha_l$
    $\delta T(x, t; \rho_l)$ ← sensitivity problem (15)
    Calculate β_l, using (20)
    $\rho_{l+1} \leftarrow \rho_l - \alpha_l \beta_l$
    $l \leftarrow l + 1$
  end while
  $\rho^* \leftarrow \rho_{l+1}$

We further implement this algorithm numerically and develop a two-stage method for recovering the local temperature and bee density.

4 Finite Difference Method for the Direct and Inverse Problems

We use the equally space and time meshes

$$\overline{\omega}_h = \{x_i : x_i = ih, \ i = 0, 1, \ldots, N_x, \ x_{N_x} = L\},$$
$$\overline{\omega}_\tau = \{t_n : t_n = n\tau \ n = 0, 1, \ldots, N_t, \ t_{N_t} = t_f\}$$

and denote by T_i^n, ρ_i^n the mesh functions at point (x_i, t_n). We will also use the notations [23]

$$u^{\pm} = \max\{0, \pm u\}, \ u_{x,i} = \frac{u_{i+1} - u_i}{h}, \ u_{\overline{x},i} = u_{x,i-1} \ u_{\overline{x}x,i} = \frac{u_{x,i+1/2} - u_{x,i-1/2}}{h}.$$

First, we provide the discretization for solving the direct problem. Equations (1), (5) are approximated by the second-order implicit-explicit finite difference scheme

$$\frac{T_i^{n+1} - T_i^n}{\tau} = T_{\bar{x}x,i}^{n+1} + f(T_i^n)\rho_i^{n+1}, \quad i = 1, 2, \ldots, N_x - 1,$$

$$\frac{T_0^{n+1} - T_0^n}{\tau} = \frac{2}{h}T_{x,0}^{n+1} + f(T_0^n)\rho_0^{n+1}, \quad T_N^{n+1} = T_a. \tag{21}$$

The second Eq. (2) with the corresponding boundary conditions (4) are approximated by upwind finite difference scheme

$$\frac{\rho_i^{n+1} - \rho_i^n}{\tau_n} - \rho_{\bar{x}x,i}^{n+1} + \theta(\rho_i^*, T_i^{n+1})\rho_i^{n+1}$$
$$= -\frac{F_{i+1/2}(\rho_i^*, T_i^{n+1}) - F_{i-1/2}(\rho_i^*, T_i^{n+1})}{h}, \quad i = 1, 2, \ldots, N_x - 1,$$

$$\frac{\rho_N^{n+1} - \rho_N^n}{\tau_n} + 2\frac{\rho_N^{n+1} - \rho_{N-1}^{n+1}}{h^2} + \theta(\rho_N^*, T_N^{n+1})\rho_N^{n+1} = \frac{2F_{N-1/2}(\rho_N^*, T_N^{n+1})}{h}, \tag{22}$$

$$\frac{\rho_0^{n+1} - \rho_0^n}{\tau_n} + 2\frac{\rho_0^{n+1} - \rho_1^{n+1}}{h^2} + \theta(\rho_0^*, T_0^{n+1})\rho_0^{n+1} = -\frac{2F_{1/2}(\rho_0^*, T_0^{n+1})}{h},$$

where

$$F_{i+1/2}(\rho_i^*, T_i^{n+1}) = \chi_{i+1/2}^+\rho_i^* - \chi_{i+1/2}^-\rho_{i+1}^*, \quad \chi_{i+1/2} = \chi\left(\frac{T_{i+1}^{n+1} + T_i^{n+1}}{2}\right)T_{x,i}^{n+1},$$

$$F_{i-1/2}(\rho_i^*, T_i^{n+1}) = \chi_{i-1/2}^+\rho_{i-1}^* - \chi_{i-1/2}^-\rho_i^*, \quad \chi_{i-1/2} = \chi\left(\frac{T_i^{n+1} + T_{i-1}^{n+1}}{2}\right)T_{\bar{x},i}^{n+1}.$$

Further, we discretize the sensitivity and adjoint problems (15), (17). For the sensitivity problem we use the following implicit-explicit finite difference scheme

$$\frac{\delta T_i^{n+1} - \delta T_i^n}{\tau} = \delta T_{\bar{x}x,i}^{n+1} + f'(T_i^{n+1})\rho_i^*\delta T_i^n + f(T_i^{n+1})\delta\rho_i^{n+1}, \quad i = 1, \ldots, N_x - 1,$$

$$\frac{\delta T_0^{n+1} - \delta T_0^n}{\tau} = \frac{2}{h}\delta T_{x,0}^{n+1} + f'(T_0^{n+1})\rho_0^*\delta T_0^n + f(T_0^{n+1})\delta\rho_0^{n+1}, \quad T_N^{n+1} = 0, \tag{23}$$

$$\delta T_i^{N_t} = 0, \quad i = 1, 2, \ldots, N_x - 1.$$

Regarding to the adjoint problem (17), since we have terminal condition, we may invert the time or advance layer by layer backward in time

$$\frac{Y_i^n - Y_i^{n+1}}{\tau} = -Y_{\bar{x}x,i}^n - f'(T_i^n)\rho_i^*Y_i^{n+1}$$
$$+ \frac{2}{\tau h}\sum_{k=1}^{K}\sum_{m=1}^{M}(T_i^n - G_{mk})\mathbf{1}_{km}(i,j), \quad i = 1, \ldots, N_x - 1,$$

$$\frac{Y_0^n - Y_0^{n+1}}{\tau} = -\frac{2}{h}Y_{x,0}^n - f'(T_0^n)\rho_0^*Y_o^{n+1} + \frac{2}{\tau h}\sum_{k=1}^{K}\sum_{m=1}^{M}(T_i^n - G_{mk})^2\mathbf{1}_{km}(i,j), \tag{24}$$

$$Y_N^n = 0, \quad Y_i^{N_t} = 0, \quad i = 1, 2, \ldots, N_x - 1,$$

where $\mathbf{1}_{km}(i,j)$ is the indicator function.

For the numerical identification algorithm we need also the approximation of the double integrals in (19) in order to compute γ_l. We apply the following well-known second order trapezoidal quadrature

$$
\int_0^{t_f}\int_0^L g(x,t)dtdx \approx \frac{h\tau}{4}\sum_{n=0}^{N_t-1}\sum_{i=0}^{N_x-1}\Big(g(x_i,t_j)+g(x_{i+1},t_j) \\
+ g(x_i,t_{j+1})+g(x_{i+1},t_{j+1})\Big).
\tag{25}
$$

The local error is $O((N_tN_x)^{-2})$, while the global error is $O((N_tN_x)^{-1})$ [5].

5   Numerical Simulations

Here we explain the implementation of the developed numerical method and discuss results from numerical experiments.

5.1   Computational Details

We consider rescaled problem (1)–(4) [2] defined in $x \in [0,1]$, $t \in [0,t_f/L^2]$ and

$$
x := x/L, \;\; t := t/L^2, \;\; \rho := L\rho, \;\; \theta_0 := L\theta_0, \;\; f := Lf, \;\; \rho_{\text{tot}} = \int_0^1 \rho(x,t)dx. \tag{26}
$$

The main reason for using this rescaling is to lessen the computational domain, since the original time and space intervals can be very large. Thus, solving the original problem numerically, taking into account that at each iteration we have to solve three parabolic problems, the method will become extremely expensive. On the other hand, dealing with the rescaled problem, we may compute the solution for very small mesh step size and obtain similar accuracy as for the original problem, but reducing the computational time significantly.

Further, we give the results and data for the original problem (1)–(4), but the computations will be performed for the rescaled model (1)–(4), (26).

We take the following set of parameters [6]

$$
T_\theta = 21, \;\; T_\chi = 25, \;\; \chi_1 = \chi_2 = 1, \;\; f_{\text{high}} = 0.6, \;\; f_{\text{low}} = 3,
$$
$$
T_f = 15, \;\; \gamma = 1, \;\; m = 10, \;\; \theta_0 = 4.10^{-3}.
$$

In order to compute the derivative $f'(T)$, we apply the smoothing

$$
f(T) = s(T)f_{\text{low}} + (1 - s(T))f_{\text{high}}, \;\; s(x) = 0.5 - 0.5\tanh\frac{T - Tf}{\epsilon}, \;\; 0 < \epsilon < 1.
$$

Algorithm 2. Two-step algorithm

First step – realization for IP

Require: ρ_0, ε

Ensure: ρ_{l+1}, $T(x, t; \rho_{l+1})$

  $l \leftarrow 0$

Calculate $J(\rho_l)$, using (12)

  while $J(\rho_l) > \varepsilon$ and $(J(\rho_{l-1}) > J(\rho_l),\, l > 0)$ **do**

    $T(x, t; \rho_l) \leftarrow$ direct problem (21)

    Calculate $J(\rho_l)$, using (12)

    $Y(x, t; \rho_l) \leftarrow$ adjoint problem (24) to $(T(x, t; \rho_l), G_{MK}, \rho_l)$

    Calculate $J'(\rho_l)$, using (16)

    Calculate α_l, using (19), (25)

    $\delta\rho^{n+1} \leftarrow \alpha_l$

    $\delta T(x, t; \rho_l) \leftarrow$ sensitivity problem (23) to $(T(x, t; \rho_l), \rho_l)$

    Calculate β_l, using (20)

    $\rho_{l+1} \leftarrow \rho_l - \alpha_l \beta_l$

    $l \leftarrow l + 1$

  end while

  $\rho^* \leftarrow \rho_{l+1}$

Second step – recompute ρ

Require: ρ^*, T

Ensure: ρ;

  $\rho^0 \leftarrow \rho_0$, $n \leftarrow 0$

  while $t < t_f$ **do**

    $\rho^{n+1} \leftarrow$ direct problem (22)

    $n \leftarrow n + 1$

  end while

  $\rho \leftarrow \rho^{n+1}$

Let $E_i^n = V(x_i, t_n) - V_i^n$, $V = \{T, \rho\}$. We will give relative errors

$$\mathcal{E} = \frac{\max\limits_{0 \le i \le N_x} \max\limits_{0 \le n \le N_t} |E_i^n|}{\max\limits_{0 \le i \le N_x} \max\limits_{0 \le n \le N_t} |V(x_i, t_n)|}, \quad \mathcal{E}_2 = \frac{\left(\sum\limits_{i=0}^{N_x} \sum\limits_{n=0}^{N_y} (E_i^n)^2 \right)^{1/2}}{\sum\limits_{i=0}^{N_x} \sum\limits_{n=0}^{N_y} (V(x_i, t_n))^2}.$$

In view of (12), the iterative procedure is stopped when $\varepsilon \ge \delta^2$, where $|(T(x_m, t_k; a) - G_{mk}| \approx \delta$.

The realization of the numerical method is described by Algorithm 2.

5.2 Numerical Results

We verify the accuracy of the proposed numerical method for the test example with the following exact solution

$$T(x, t) = e^{-t/(3t_f)}(T_a + 40(1 - (x/L)^4)), \quad \rho(x, t) = e^{-t/(t_f)}\rho^0(x),$$

where

$$\rho^0(x) = \frac{C_1\chi_1}{2\overline{f}\left(\cosh\left(0.5\chi_1\sqrt{C_1}x\right)\right)^2}, \quad C_1 = \left(\frac{2}{L\chi_1}\mathrm{acosh}\left(e^{0.5\chi_1(T_\chi - T_a)}\right)\right)^2 + 7,$$

is the steady state solution, obtained in [6], $\overline{f} = 1$. To obtain this solution, we include appropriate function in the right-hand side of the Eqs. (1) and (2).

In practice, the temperature is measured by sensors, which can send information at arbitrarily small time intervals, but the line location depends on the distance between the frames in the beehive, approximately 4–5 cm [2]. For the experiment, we use exact measurements, $M = 3$, $K = 27$, $T_a = -9\,°C$, $L = 10$ cm at $t_f = 100$ min and to generate initial density, we use perturbed data, adding a Gaussian noise to the exact solution and similarly we get the measured temperature data with perturbation.

In Table 1 we give errors of the numerical density and temperature for perturbation of the temperature (p_T) and density (p_ρ). For all runs the ratio $\tau = h^2$ is fixed and $\epsilon = 0.5$. As can be expected, the level of the perturbation significantly influence the accuracy of the solution. We observe that for small perturbation, i.e. less than 2% both for initial density and temperature measurements, the spatial order of convergence is close to two. In the case of bigger deviation for the initial density, even for exact temperature data, the convergence rate of the recovered temperature decreases, especially in maximal norm. For big perturbation. For very large perturbation (10%) the precision of the restored solutions is acceptable, but we observe a lack or very slow convergence for different norms.

Table 1. Relative errors in maximal and L_2 norms.

N_x	p_ρ	p_T	Density		Temperature	
			$\mathcal{E}$	$\mathcal{E}_2$	$\mathcal{E}$	$\mathcal{E}_2$
40	1%	0%	3.788e−2	2.753e−2	4.849e−4	2.849e−4
80	1%	0%	1.226e−2	4.139e−3	1.518e−4	7.561e−5
40	5%	0%	5.978e−2	3.029e−2	1.178e−3	4.632e−4
80	5%	0%	4.579e−2	1.435e−2	4.500e−4	1.386e−4
40	2%	2%	4.336e−2	2.826e−2	3.513e−3	1.895e−3
80	2%	2%	2.064e−2	6.499e−3	9.565e−4	5.541e−4
40	5%	5%	5.978e−2	3.066e−2	3.290e−3	1.738e−3
80	5%	5%	4.579e−2	1.456e−2	2.131e−3	1.285e−3
40	10%	3%	8.706e−2	3.823e−2	4.421e−3	2.450e−3
80	10%	3%	8.775e−2	2.829e−2	1.711e−3	8.617e−4
40	10%	10%	8.705e−2	3.858e−2	5.130e−3	2.824e−3
80	10%	10%	8.775e−2	2.814e−2	4.125e−3	2.525e−3

On Figs. 1–2 we depict exact and numerical bee density and temperature, computed by Algorithm 2, $N = 41$, $\tau = h^2$ for $p_T = p_\rho = 5\%$.

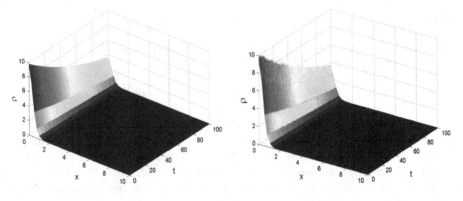

Fig. 1. Exact (*left*) and recovered (*right*) bee density, $p_T = p_\rho = 5\%$.

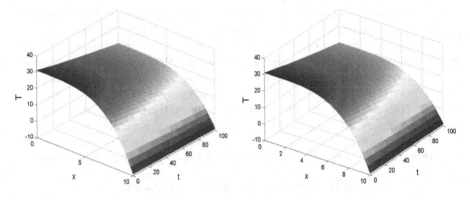

Fig. 2. Exact (*left*) and recovered (*right*) temperature, $p_T = p_\rho = 5\%$.

On Fig. 3 we plot exact and recovered solution profiles of the temperature and bee density at time $t = 31.25$ min, which belongs to a set of measured time point and $t = 75$ min, which is not point of measurements. In both cases we observe very good fitting of the numerical to the exact solution. The error of the restored density is bigger at points close to the boundary $x = 0$, while precision of the numerical temperature is fine in the whole region. This can be observed also on Figs. 1 and 2 for the whole computational domain.

6   Discussion and Future Work

In this work, we have developed a numerical optimization identification approach for describing the thermoregulation process in honey bee colonies in winter, using the Keller-Segel model [6]. We propose a two-stage numerical method based on conjugate GM: first, we recover the temperature T and the density ρ by solving a semilinear parabolic inverse problem. On the second stage we recover again the density ρ, by solving the strongly nonlinear convection-diffusion equation.

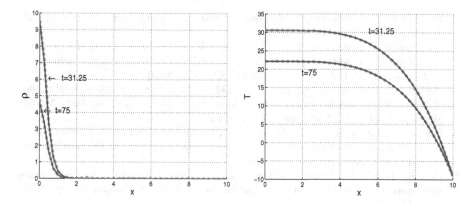

Fig. 3. Exact (*solid line*) and recovered (*dashed line*) profiles of the bee density (*left*) and temperature (*right*), $p_T = p_\rho = 5\%$.

The theoretical analysis for existence and uniqueness of quasi-solution, as well as exact (strict) solution of the IP will be subject of next study. Different versions of the conjugate GM can be found in the literature, depending on how the coefficients γ_l and β_l from (19) and (20), respectively, are computed, see e.g. [1,10,15,16,21]. The convergence analysis of the Algorithms 1,2 on this base requires additional efforts and will be done in our next work.

Additionally, our future work will be focused on development of optimization algorithm for simultaneously determination of the density and the initial temperature by temperature observations. On this base, we plan to solve an inverse problem for estimation of the parameters θ_0, m and γ that are difficult for measurements, as it is suggested in [6].

Acknowledgment. This work is supported by the Bulgarian National Science Fund under the Project KP-06-PN 46-7 "Design and research of fundamental technologies and methods for precision apiculture".

References

1. Alifanov, O.M., Artioukhine, E.A., Rumyantsev, S.V.: Extreme Methods for Solving Ill-posed problems with Applications to Inverse Heat Transfer Problems. Begell House, New York (1995)
2. Atanasov, A.Z., Koleva, M.N., Vulkov, L.G.: Numerical analysis of thermoregulation in honey bee colonies in winter based on sign-changing chemotactic coefficient model. In: Springer Proceedings in Mathematics and Statistics, submitted, July 2022
3. Atanasov, A.Z., Georgiev, S.G., Vulkov, L.G.: Reconstruction analysis of honeybee colony collapse disorder modeling. Optim. Eng. **22**(4), 2481–2503 (2021). https://doi.org/10.1007/s11081-021-09678-0
4. Bagheri, S., Mirzaie, M.: A mathematical model of honey bee colony dynamics to predict the effect of pollen on colony failure. PLoS ONE **14**(11), e0225632 (2019)

5. Bakhvalov, N.S.: Numerical Methods, p. 663. Mir Publishers, Moscow (1977). Translated from Russian to English
6. Bastaansen, R., Doelman, A., van Langevede, F., Rottschafer, V.: Modeling honey bee colonies in winter using a Keller-Segel model with a sign-changing chemotactic coefficient. SIAM J. Appl. Math. **80**(20), 839–863 (2020)
7. Bellomo, N., Bellouquid, A., Tao, Y., Winkler, M.: Toward a mathematical theory of Keller-Segel models of patern formulation in biological tissues. Math. Models Methods Appl. Sci. **25**(09), 1663–1763 (2015)
8. Calovi, M., Grozinger, C.M., Miller, D.A., Goslee, S.C.: Summer weather conditions influence winter survival of honey bees (Apis mellifera) in the northeastern United States. Sci. Rep. **11**, 1553 (2021)
9. Cao, K., Lesnic, D.: Reconstruction of the perfusion coefficient from temperature measurements using the conjugate gradient method. Int. J. Comput. Math. **95**(4), 797–814 (2018)
10. Chavent, G.: Nonlinear Least Squares for Inverse Problems: Theoretical Foundation and Step-by Guide for Applications. Springer, Cham (2009)
11. Chen, J., DeGrandi-Hoffman, G., Ratti, V., Kang, Y.: Review on mathematical modeling of honeybee population dynamics. Math. Biosci. Eng. **18**(6), 9606–9650 (2021)
12. Esch, H.: Über die körpertemperaturen und den wärmehaushalt von Apis mellifica. Z. Vergleich. Physiol. **43**, 305–335 (1960). https://doi.org/10.1007/BF00298066
13. Fakhraie, M., Shidfar, A., Garshasbi, M.: A computational procedure for estimation of an unknown coefficient in an inverse boundary value problem. Appl. Math. Comput. **187**, 1120–1125 (2007)
14. Guzman-Novoa, E., Eccles, C.Y., McGowan, J., Kelly, P.G., Correa- Bentez, A.: Varroa destructor is the main culprit for the death and reduced populations of overwintered honey bee (Apis mellifera) colonies in Ontario, Canada. Apidologie **41**, 443–450 (2010)
15. Hanke, M.: Conjugate Gradient Type Methods for Ill-Posed Problems, p. 144. Chapman and Hall/CRC, NewYork (1995)
16. Hasanov, A., Romanov, V.: Introduction to the Inverse Problems for Differential Equations. Springer, NewYork (2017). https://doi.org/10.1007/978-3-319-62797-7
17. Heinrich, B.: Energetics of honeybee swarm thermoregulation. Science **212**, 565–566 (1981)
18. Kevan, P.G., Guzman, E., Skinner A., van Engelsdorp, D.: Colony collapse disorder in Canada: do we have a problem? HiveLights, pp. 14–16, May (2007). http://hdl.handle.net/10214/2418
19. Krivorotko, O., Kabanikhin, S., Zhang, S., Kashtanova, V.: Global and local optimization in identification of parabolic systems. J. Inverse Ill-Posed Prob. **28**(6), 899–913 (2020)
20. Lemke, M., Lamprecht, A.: A model of heat production and thermoregulation in winter clusters of honey bee using differential heat conduction equations. J. Theor. Biol. **142**(2), 261–273 (1990)
21. Lesnic, D.: Inverse Problems with Applications in Science and Engineering. CRC Press, London (2020)
22. Ocko, S.A., Mahadevan, L.: Collective thermoregulation in bee clusters. J. R. Soc. Interface **11**(91), 20131033 (2014)
23. Samarskii, A.A.: The Theory of Difference Schemes. Marcel Dekker Inc, New York (2001)
24. Peters, J.M., Peleg, O., Mahadevan, L.: Collective ventilation in honeybee nest. J. R. Soc. Interface **16**(150), 20180561 (2019)

25. Ratti, V., Kevan, P.G., Eberl, H.J.: A mathematical model of forager loss in honeybee colonies infested with *varroa destructor* and the acute bee paralysis virus. Bull. Math. Biol. **79**(6), 1218–1253 (2017). https://doi.org/10.1007/s11538-017-0281-6
26. Stabentheiner, A., Kovac, H., Brodschneider, R.: Honeybee colony thermoregulation regulatory mechanisms and contribution of individuals in dependence on age, location and thermal stress. PLoS ONE **5**(1), e8967 (2010)
27. Tautz, J.: The Buzz about Bees: Biology of a Superorganism. Springer, Berlin (2008). https://doi.org/10.1007/978-3-540-78729-7
28. Wang, B., Liu, J.: Recovery of thermal conductivity in two-dimensional media with nonlinear source by optimizations. Appl. Math. Lett. **60**, 73–80 (2016)
29. Watmough, J., Camazine, S.: Self-organized thermoregulation of honeybee clusters. J. Theor. Biol. **176**(3), 391–402 (1995)

Gradient Optimization in Reconstruction of the Diffusion Coefficient in a Time Fractional Integro-Differential Equation of Pollution in Porous Media

Tihomir Gyulov[1]([✉])[ID] and Lubin Vulkov[2]

[1] Department of Mathematics, University of Ruse "Angel Kanchev",
8th Studentska Str., 7017 Ruse, Bulgaria
tgulov@uni-ruse.bg
[2] Department of Applied Mathematics and Statistics, University of Ruse
"Angel Kanchev", 8th Studentska Str., 7017 Ruse, Bulgaria
lvalkov@uni-ruse.bg

Abstract. An optimization Lagrange multiplier adjoint equation approach for an inverse problem of estimating a space-dependent diffusion coefficient in a problem of pollution in a porous media is developed. The study is reduced to the reconstruction of the highest derivative coefficient of a time fractional integro-differential equation. The well-posedness of the direct problem is discussed and energy estimates for its solution are obtained. Formula for the gradient of the least-squares cost functional is derived by Lagrange multipliers method. The conjugate gradient method for the unknown diffusion coefficient is also discussed.

Keywords: Integro-differential equation · Time-fractional derivative · Porous media · Least-squares discrepancy functional · Lagrange multiplier · Fréchet derivative · Conjugate gradient method

1 Introduction

The mathematical model studied in this paper describes dynamical processes in porous media. Such models are applied in the context of the industrial or accidental pollution of soil with liquid volatile organic pollutant, see, e.g., [4,28]. Usually, they are formed as classical integer order systems of parabolic and/or ordinary differential equations. To be more concrete, we study a simple mathematical model [16] for the rate limited mass transfer of a contaminant between the gaseous (mobile) and liquid (immobile) phase in non-equilibrium. Consider the 1D version of the system (6), (7) in [16] as follows:

$$\frac{\partial u}{\partial t} - \frac{\partial}{\partial x}\left(a(x)\frac{\partial u}{\partial x} - v(x)u \right) + r\alpha(x)(u-w) = 0, \quad \text{in } Q_\infty, \tag{1}$$

$$\frac{\partial w}{\partial t} + \alpha(x)w = \alpha(x)u, \quad \text{in } Q_\infty = \Omega \times (0, \infty), \quad \Omega = (0, l), \tag{2}$$

$$u(x, 0) = u_0(x), \quad w(x, 0) = w_0(x) \text{ on } \Omega. \tag{3}$$

Here u and v denote the concentrations of a contaminant in the corresponding phases. Generally, this system is an example of a so-called *mobile-immobile model* (MIM) [14] of solute transport in porous media. Further references can be found, for example, in [13, 24]. It is usually assumed in such models that the fluid phase is separated into a mobile or flowing part and an immobile (stagnant) one. The advection-dispersion solute transport appears only in the first one, see Eq. (1). The mass transfer between these two regions can be described by a first-order kinetics in non-equilibrium (see, e.g., the term $\alpha(x)(u - w)$). For a reference to porous media modelling in general, see, e.g., [6].

It has shown useful to characterize the anomalous diffusion observed in some particular porous media (e.g. low-permeability and heavy heterogenity [23]) via fractional derivatives due to their intrinsic memory effect and non-locality, see [1, 8, 9, 15, 35] for instance. Fractional derivatives have applications in other areas as well [25, 27]. In this paper the first-order time derivatives in the Eqs. (1) and (2) are replaced by Caputo fractional derivatives [8, 9], see Eqs. (4)–(5) below. As a result, the new system is similar to the fractal MIM system presented in [23] and [26].

Numerical investigation of the solution of the direct problem (4)–(5) is presented in [26]. A closely related system of fractional equations which models pressures in fractured porous medium is studied numerically in [31] as well. The determination of the unknown time-dependent boundary source term of a non-local boundary value problem for a time-fractional diffusion equation with Laplacian is investigated in [5]. The reconstruction of the initial data from final time measurements for a time-fractional diffusion equation with a general elliptic operator is presented in [33]. Parameter identification problems for systems of parabolic equations are studied in [22] in use of tensor train decomposition method. An inverse problem about the identification of the fractional orders of the derivatives in time of the problem (4)–(5) is investigated in [23].

Our research is focused on the determination of the unknown diffusion coefficient $a(x)$ given additional information obtained from measurements. We use the conjugate gradient method and the adjoint problem formulation for two boundary value problems in order to estimate the unknown diffusion coefficient for a time fractional integro-differential equation.

To the best of our knowledge, there are no other present studies on the estimation problem considered here in the modeling of pollution in porous media.

The rest of the paper is organized as follows. In Sect. 2, the fractional solute transport MIModel is formulated and the direct and inverse problems are presented. In Sect. 3 the well-posedness of the direct (forward) problems is discussed. In Sect. 4 the quasi-solution of the inverse problem is studied. A formula for the gradient of the cost functional is presented in Sect. 5. Finally, Sect. 6 concludes the paper by summarizing our results and proposing some further related problems.

2 Formulation of Direct and Inverse Problems for the Fractional Model

It is usually assumed in a mobile-immobile model that the solute transport through diffusion and advection is present only in the mobile zone. Then, it has been shown by many studies in last decades that it could be more suitable to use fractional derivatives in order to describe memory effect and anomalous diffusion in certain porous media, see, for example, $[1, 8, 9, 15, 23, 25, 31, 35]$. Correspondingly, the modification of the system (1)–(3) reads

$$\frac{\partial^\beta u}{\partial t^\beta} - \frac{\partial}{\partial x}\left(a(x)\frac{\partial u}{\partial x} - v(x)u\right) + r\alpha(x)(u - w) = 0, \quad \text{in } Q_T, \tag{4}$$

$$\frac{\partial^\gamma w}{\partial t^\gamma} + \alpha(x)w = \alpha(x)u, \quad \text{in } Q_T = \Omega \times (0, T), \quad \Omega = (0, l). \tag{5}$$

with initial condition (3). Here $0 < \beta, \gamma \le 1$ and

$$\frac{\partial^\theta v}{\partial t^\theta} = \frac{1}{\Gamma(1-\theta)} \int_0^t (t - s)^{-\theta} \frac{\partial v}{\partial s}(x, s)ds \tag{6}$$

is the Caputo derivative $[8, 19, 21, 29]$.

Further, we adopt classical notations of fractional calculus, see, for reference, the monographs $[21, 29]$ and the review paper $[34]$.

Denote

$$L^p(0, T) := \left\{ v : \int_0^T |v(t)|^p dt, \infty \right\}$$

with $p \ge 1$ and

$$W^{1,1}(0, T) := \left\{ v \in L^1(0, T) : \frac{dv}{dt} \in L^1(0, T) \right\},$$

where the norms are defined as usual by

$$\|v\|_{L^p(0,T)} := \left(\int_0^T |v(t)|^p \right)^{1/p}, \quad \|v\|_{W^{1,1}(0,T)} := \|v\|_{L^1(0,T)} + \left\| \frac{dv}{dt} \right\|_{L^1(0,T)}$$

The Caputo derivative (6) is well-defined for

$$v \in W^{1,1}(0, T) \quad \text{and} \quad \frac{d^\theta v}{dt^\theta} \in L^1(0, T)$$

by virtue of the Young inequality.

Recall that the left and right Riemann-Liouville fractional integrals of order θ are defined as follows

$$\left(I_{0+}^\theta u\right)(t) = \frac{1}{\Gamma(\theta)} \int_0^t \frac{u(\tau)}{(t - \tau)^{1-\theta}} d\tau, \quad \left(I_{T-}^\theta u\right)(t) = \frac{1}{\Gamma(\theta)} \int_t^T \frac{u(\tau)}{(\tau - t)^{1-\theta}} d\tau,$$

respectively. In addition, denote by

$$\left(D_{0+}^{\theta}u\right)(t) = \frac{1}{\Gamma(1-\theta)}\frac{d}{dt}\int_0^t \frac{u(\tau)}{(t-\tau)^{\theta}}d\tau = \frac{d}{dt}\left(\left(I_{0+}^{1-\theta}u\right)(t)\right),$$

$$\left(D_{T-}^{\theta}u\right)(t) = \frac{-1}{\Gamma(1-\theta)}\frac{d}{dt}\int_t^T \frac{u(\tau)}{(\tau-t)^{\theta}}d\tau = -\frac{d}{dt}\left(\left(I_{T-}^{1-\theta}u\right)(t)\right),$$

the corresponding left and right Riemann-Liouville fractional derivatives for $0 < \theta < 1$. For simplicity, we will write $I_{0+}^{\theta}u$ instead of $\left(I_{0+}^{\theta}u\right)(t)$, $D_{0+}^{\theta}u$ instead of $\left(D_{0+}^{\theta}u\right)(t)$ and etc.

We will use a formula for integration by parts, see, Proposition 2.19 from [33]. Namely, let $0 < \theta < 1$ and suppose that $w(t) \in AC[0,T]$, $w'(t) \in C_{1-\theta}[0,T]$, and $v(t) \in AC_{1-\theta}^T[0,T]$, where

$$C_{\alpha}[0,T] := \{y(t) : y(t)t^{\alpha} \in C[0,T]\}$$

$$AC_{\alpha}^T[0,T] := \{y(t) : y(t)(T-t)^{\alpha} \in AC[0,T]\}.$$

Then we have

$$\int_0^T \frac{\partial^{\theta}w}{\partial t^{\theta}}(t)v(t)dt = w(T)I_{T-}^{1-\theta}v(T) - w(0)I_{T-}^{1-\theta}v(0) + \int_0^T w(t)D_{T-}^{\theta}v(t)dt. \quad (7)$$

The following Gronwall-type inequality has been proved in [3].

Lemma 1. *Let the nonnegative absolutely continuous function $y(t)$ satisfy the inequality*

$$\frac{d^{\beta}y(t)}{\partial t^{\beta}} \leq c_1 y(t) + c_2(t), \quad 0 < \beta \leq 1 \quad (8)$$

for almost all t in $[0,T]$, where $c_1 = \text{const} > 0$ and $c_2(t)$ is an integrable non-negative function on $[0,T]$. Then

$$y(t) \leq y(0)E_{\beta}(c_1 t^{\beta}) + \Gamma(\beta)E_{\beta,\beta}(c_1 t^{\beta})I_{0+}^{\beta}c_2(t), \quad (9)$$

where

$$E_{\beta}(z) = \sum_{n=0}^{\infty} z^n/\Gamma(\beta n + 1) \quad and \quad E_{\beta,\mu}(z) = \sum_{n=0}^{\infty} z^n/\Gamma(\beta n + \mu)$$

are the Mittag-Leffler functions.

Formula (9) is based on the solution of the corresponding differential equation, namely

$$y(t) = y(0)E_{\beta}\left(c_1 t^{\beta}\right) + \int_0^t (t-\tau)^{\beta-1}E_{\beta,\beta}\left(c_1(t-\tau)^{\beta}\right)c_2(\tau)d\tau,$$

Applying this formula to Eq. (5), we find

$$w(x,t) = w_0(x)E_{\gamma}\left(-\alpha(x)t^{\gamma}\right)$$
$$+ \alpha(x)\int_0^t (t-\tau)^{\gamma-1}E_{\gamma,\gamma}\left(-\alpha(x)(t-\tau)^{\gamma}\right)u(x,\tau)d\tau, \quad (10)$$

Then, we insert (10) in (4) to obtain the integro-differential equation:

$$\frac{\partial^\beta u}{\partial t^\beta} = \frac{\partial}{\partial x}\left(a(x)\frac{\partial u}{\partial x} - v(x)u\right) - r\alpha(x)u$$

$$+ r\alpha^2(x)\int_0^t \rho(x,t,\tau)u(x,\tau)d\tau + f(x,t) \tag{11}$$

$$\equiv \mathcal{Z}u + f,$$

where

$$\rho(x,t,\tau) = (t-\tau)^{\gamma-1}E_{\gamma,\gamma}(-\alpha(x)(t-\tau)^\gamma), \tag{12}$$
$$f(x,t) = r\alpha(x)w_0(x)E_\gamma(-\alpha(x)t^\gamma).$$

The problem for the integro-differential Eq. (11) with given coefficients and right-hand side subject to the initial condition (3) and Dirichlet boundary conditions or third boundary conditions is referred as **direct** or forward problem [2, 10, 18, 20, 33].

The inverse problem studied here is about the determination of the spacewise-dependent diffusion coefficient $a(x)$ if an additional information from measurements on a finite set of space points is given, namely,

$$u(x_i, t) = g_i(t), \quad i = 1, \dots, I, \quad x_i \in (0, l). \tag{13}$$

3   Well-Posedness of the Direct Problem

In this section, we discuss the well-posedness of the direct (forward) problem and derive a priori estimates to the solutions.

3.1   The First Boundary Value Problem

Consider zero Dirichlet boundary conditions on the left and right boundary of the rectangle Q_T

$$u(0,t) = u(l,t) = 0 \tag{14}$$

and the initial conditions (3) on $\Omega = (0, l)$. We assume the smoothness conditions for the coefficients.

Theorem 1. *Suppose that*

$$0 < a_0 < a(x) \in C^1(Q_T) \tag{15}$$
$$v(x), \alpha(x) \in C(\overline{Q}_T)$$

are fulfilled and

$$u(x,t) \in C^{2,0}(Q_T) \cap C^{1,0}(\overline{Q}_T), \frac{\partial^\beta u(x,t),}{\partial t^\beta} \in C(Q_T).$$

Then the a priori estimate holds

$$\|u\|_0^2 + I_{0+}^\beta\left\|\frac{\partial u}{\partial x}\right\|_0^2 \le C\left(\|u_0\|_0^2 + I_{0+}^\beta\|f\|_0^2\right), \tag{16}$$

where C is a positive constant which does not depend on u_0 and f.

The existence and uniqueness of solution with corresponding smoothness could be studied following [21, 29, 33].

Proof. Following [3], we multiply Eq. (11) by $2u$ and integrate the result from $x = 0$ to $x = l$:

$$2\left(\frac{\partial^\beta u}{\partial t^\beta}, u\right) = -2\left(a\frac{\partial u}{\partial x}, \frac{\partial u}{\partial x}\right) + 2\left(v\frac{\partial u}{\partial x}, u\right) - 2\left(r\alpha^2 u, u\right) \tag{17}$$

$$+ 2\left(r\int_0^t \alpha^2 \rho u d\tau, u\right) + 2(f, u), \tag{18}$$

where we have denoted for arbitrary functions $\varphi, \psi \in L^2(0, l)$

$$(\varphi, \psi) = \int_0^l \varphi\psi dx \quad (\varphi, \varphi) = \|\varphi\|_0^2.$$

Using Lemma 1 in [3] we have

$$2\left(\frac{\partial^\beta u}{\partial t^\beta}, u\right) \geq \left(1, \frac{\partial^\beta}{\partial t^\beta} u^2\right) = \frac{\partial^\beta}{\partial t^\beta}\|u\|_0^2.$$

It follows from (15) the existence of positive constant $c > 0$ such that

$$\max_{\bar{Q}_T}(|v(x)|, |a(x)|, |\alpha(x), |r\alpha^2(x)\rho(x, t, \tau)/(t - \tau)^{\gamma-1}|) \leq c.$$

Then, using the ε - Cauchy inequality, we have:

$$\left(v\frac{\partial u}{\partial x}, u\right) = \int_0^l v\frac{\partial u}{\partial x} u dx \leq \varepsilon \left\|\frac{\partial u}{\partial x}\right\|_0^2 + \frac{c^2}{4\varepsilon}\|u\|_0^2$$

$$\left|\left(r\alpha^2\int_0^t \rho u d\tau, u\right)\right| = c\left|\int_0^l \left(u\int_0^t (t-\tau)^{\gamma-1} u d\tau\right) dx\right|$$

$$\leq \frac{c}{2}\left|\int_0^l \left(\int_0^t (t-\tau)^{\gamma-1}\left(u^2(x, t) + u^2(x, \tau)\right) d\tau\right) dx\right|$$

$$= \frac{c}{2}\left(\frac{t^\gamma}{\gamma}\|u\|_0^2 + \Gamma(\gamma)I_{0+}^\gamma\|u\|_0^2\right)$$

$$(f, u) = \int_0^l f u dx \leq \frac{1}{2}\|u\|_0^2 + \frac{1}{2}\|f\|_0^2.$$

Now, the above inequalities imply

$$\frac{\partial^\beta}{\partial t^\beta}\|u\|_0^2 + 2a_0\left\|\frac{\partial u}{\partial x}\right\|_0^2 \leq C_1(\varepsilon)\|u\|_0^2 + 2\varepsilon\left\|\frac{\partial u}{\partial x}\right\|_0^2 + C_2 I_{0+}^\gamma\|u\|_0^2 + \|f\|_0^2.$$

Taking $\varepsilon = \frac{a_0}{2}$, we get

$$\frac{\partial^\beta}{\partial t^\beta}\|u\|_0^2 + a_0\left\|\frac{\partial u}{\partial x}\right\|_0^2 \leq C_3\|u\|_0^2 + C_4 I_{0+}^\gamma\|u\|_0^2 + C_5\|f\|_0^2. \tag{19}$$

Applying I_{0+}^{β} to both sides of (19) we find

$$\|u\|_0^2 + I_{0+}^{\beta}\left\|\frac{\partial u}{\partial x}\right\|_0^2 \le C_7 I_{0+}^{\beta}\|u\|_0^2 + C_8 I_{0+}^{\beta+\gamma}\|u\|_0^2$$
$$+ C_9 I_{0+}^{\beta}\|f\|_0^2 + \|u_0(x)\|_0^2. \tag{20}$$

Note that

$$I_{0+}^{\beta+\gamma}\|u\|_0^2 = \frac{1}{\Gamma(\beta+\gamma)} \int_0^t (t-\tau)^{\beta+\gamma-1}\|u(\cdot,\tau)\|_0^2\, d\tau$$
$$= \frac{1}{\Gamma(\beta+\gamma)} \int_0^t (t-\tau)^{\gamma}(t-\tau)^{\beta-1}\|u(\cdot,\tau)\|_0^2\, d\tau$$
$$\le \frac{\Gamma(\beta)}{\Gamma(\beta+\gamma)} t^{\gamma} I_{0+}^{\beta}\|u\|_0^2.$$

Hence, the inequality (20) takes the form

$$\|u\|_0^2 + I_{0+}^{\beta}\left\|\frac{\partial u}{\partial x}\right\|_0^2 \le C_{10} I_{0+}^{\beta}\|u\|_0^2 + C_{11}\left(I_{0+}^{\beta}\|f\|_0^2 + \|u_0(x)\|_0^2\right) \tag{21}$$

Applying Lemma 1, we get the a priori estimate (16). □

3.2   The Third Boundary Value Problem

Now we consider the Eq. (11) with initial condition (3) and boundary conditions

$$a(0)\frac{\partial u}{\partial x}(0,t) = \beta_1(t)u(0,t) - \mu_1(t),$$
$$-a(l)\frac{\partial u}{\partial x}(l,t) = \beta_2(t)u(l,t) - \mu_2(t). \tag{22}$$

Theorem 2. *Let the conditions (15) be fulfilled, $|\beta_i(t)|$ be bounded, $i = 1, 2$, as well as*

$$u(x,t) \in C^{2,0}(Q_T) \cap C^{1,0}(\overline{Q}_T), \quad \frac{\partial^{\beta}u(x,t)}{\partial t^{\beta}} \in C(Q_T).$$

Then the a priori estimate holds

$$\|u\|_0^2 + I_{0+}^{\beta}\left\|\frac{\partial u}{\partial x}\right\|_0^2 \le C\left(\|u_0\|_0^2 + I_{0+}^{\beta}\left(\|f\|_0^2 + \mu_1^2 + \mu_2^2\right)\right), \tag{23}$$

where C is a positive constant which does not depend on u_0, μ_1, μ_2 and f.

Proof. Just as in the proof of Theorem 1, we multiply Eq. (11) by $2u$ and integrate from $x = 0$ to $x = l$ to obtain

$$\frac{\partial^{\beta}}{\partial t^{\beta}}\|u\|_0^2 + 2a_0\left\|\frac{\partial u}{\partial x}\right\|_0^2 \le 2\left. u\left(a\frac{\partial u}{\partial x} - vu\right)\right|_0^l + 2\varepsilon\left\|\frac{\partial u}{\partial x}\right\|_0^2$$
$$+ C_1(\varepsilon)\|u\|_0^2 + C_2 I_{0+}^{\gamma}\|u\|_0^2 + \|f\|_0^2. \tag{24}$$

We transform the first term of the right-hand side as follows:

$$u\left(a\frac{\partial u}{\partial x} - vu\right)\Big|_0^l = u(l,t)(\mu_2(t) - (v(l) + \beta_2(t))u(l,t))$$

$$+ u(0,t)(\mu_1(t) + (v(0) - \beta_1(t))u(0,t))$$

$$= -(v(l) + \beta_2(t))(t)u^2(l,t) + (v(0) - \beta_1(t))u^2(0,t) \quad (25)$$

$$+ \mu_2(t)u(l,t) + \mu_1(t)u(0,t)$$

$$\leq \varepsilon\left\|\frac{\partial u}{\partial x}\right\|_0^2 + C_3\|u\|_0^2 + C_4(\mu_1^2(t) + \mu_2^2(t)).$$

Now, by virtue of (25) and taking $\varepsilon = a_0/4$, the inequality (24) takes the form

$$\frac{\partial^\beta}{\partial t^\beta}\|u\|_0^2 + \left\|\frac{\partial u}{\partial x}\right\|_0^2 \leq C_5(\varepsilon)\|u\|_0^2 + C_2 I_{0+}^\gamma\|u\|_0^2 + C_0(\|f\|_0^2 + \mu_1^2(t) + \mu_2^2(t))$$

Acting on the both sides of this inequality the operator of the fractional integration I_{0+}^β and applying Lemma 1 we obtain the a priori estimate (23). □

4 Fréchet Gradient of the Discrepancy Functional

In this section we implement the Lagrange multiplier method [2,10,18] to derive the Fréchet gradient of the least-squares discrepancy functional corresponding to the quasi solution of the inverse problem (11), (13), (3) with first (14) or third boundary conditions (22). These inverse problems themselves are ill-posed [2,10,18,30] due to the fact that their solutions are unstable with respect to the data (13). In addition, the uniqueness of the solution of the inverse problem depends in general on the location of the points x_i in (13) as well as the initial and boundary conditions.

The inverse problem (11), (13), (3) and (14) or (22) can be written in operator form $\mathcal{A}(a) = g$, where $\mathcal{A} : A \to G$ is an injective operator, $a \in A = \{a_0 \leq a(x), a \in C(\overline{\Omega})$ is the admissible set, $g \in G$, G is Euclidian space of data $g = (g_1(t), \ldots, g_I(t))$. In the current paper, the inverse problems is reduced to the minimization problem

$$a^* = \operatorname*{argmin}_{a \in A} J(a), \quad J(a) = \sum_{i=1}^{I} \int_0^T (u(x_i, t; a) - g_i(t))^2 dt \quad (26)$$

subject to the condition that $u(x,t;a)$ is the solution of the problem (11), (3) and (14) or (22).

The reconstruction of the space-dependent diffusion coefficient $a(x)$ from the interior point observations (13) in the considered problem of pollution in porous media can be investigated numerically using **conjugate gradient method (CGM)**. The CGM is based on the Fréchet gradient of the least square objective functional (26). Let us recall to the reader that the functional increment $\delta J(a) = J(a + \delta a) - J(a)$ can be represented in the form

$$\delta J(a) = \langle \mathcal{U}, \delta a \rangle_H, \tag{27}$$

where $\langle \cdot, \cdot \rangle$ is the scalar product in H. Then $\nabla J(a) = \mathcal{U}$. We further consider the adjoint operator

$$\mathcal{Z}^* \Psi := \frac{\partial}{\partial x} \left(a(x) \frac{\partial \Psi}{\partial x} \right) + v(x) \frac{\partial \Psi}{\partial x} - r\alpha(x)\Psi + r\alpha^2(x) \int_t^T \rho(x, \tau, t)\Psi(x, \tau)\mathrm{d}\tau.$$

Theorem 3. *The functional $J(a)$ in (26) is Fréchet differentiable and its gradient is given by*

$$\nabla J(a) = \int_0^T \frac{\partial \Psi}{\partial x} \frac{\partial u}{\partial x} \mathrm{d}t, \tag{28}$$

where $\Psi(x, t)$ is the solution of the adjoint backward problem to the forward problem (11), (3) and (14) or (22), i.e.,

$$D_{T-}^{\beta} \Psi = \mathcal{Z}^* \Psi + \sum_{i=1}^{I} 2(u - g_i)\delta(x - x_i) \quad in \ Q_T, \tag{29}$$

$$I_{T-}^{1-\beta} \Psi(x, T) = 0, \quad x \in \Omega, \tag{30}$$

and

$$\Psi(0, t) = 0, \quad \Psi(l, t) = 0, \quad t \in (0, T), \tag{31}$$

in the case of boundary conditions (14), or

$$\begin{aligned} a(0) \frac{\partial \Psi}{\partial x}(0, t) &= (\beta_1(t) - v(0)) \, \Psi(0, t), \\ -a(l) \frac{\partial \Psi}{\partial x}(l, t) &= (\beta_2(t) + v(l)) \, \Psi(l, t). \end{aligned} \tag{32}$$

for boundary conditions (22).

Proof. We denote the dependence of the solution $u(x, t)$ from parameter $a(x)$ by $u(x, t; a) := u(x, t)$. Suppose that for $a \in A$ and $a + \delta a \in A$, $\delta u(x, t; \delta a) := u(x, t; a + \delta a) - u(x, t; a)$.

Then the deviation $\delta u := \delta u(x, t; \delta a)$ satisfies the following initial boundary value problem with an accuracy up to terms of order $o(|\delta a|)^2$, the so-called **sensitivity problem**:

$$\frac{\partial^{\beta} \delta u}{\partial t^{\beta}} = \mathcal{Z}(\delta u) + \frac{\partial}{\partial x} \left(\delta a \frac{\partial u}{\partial x} \right) \quad in \ Q_T, \tag{33}$$

$$\delta u(x, 0) = 0, \quad x \in \Omega, \tag{34}$$

$$\delta u(0, t) = 0, \quad \delta u(l, t) = 0, \quad t \in (0, T), \tag{35}$$

or

$$a(0)\frac{\partial}{\partial x}\delta u(0,t) = \beta_1(t)\delta u(0,t), \quad t \in (0,T),$$

$$-a(l)\frac{\partial}{\partial x}\delta u(l,t) = \beta_2(t)\delta u(l,t), \quad t \in (0,T),$$

(36)

for the case of boundary conditions (22).

We use the Lagrange multipliers method [2,10,18] for minimization of (26) subject to the constraints (11), (3) and boundary conditions (14) or (22). Using the Lagrange functional

$$F(u,a,\Psi) := \hat{J}(u) + \int_0^T \int_0^l \Psi \left(\frac{\partial^\beta u}{\partial t^\beta} - \mathcal{Z}u - f \right) dxdt,$$

we consider the unconstrained problem. Here $\hat{J}(u)$ is defined as the functional J in (27) considerd as a function of the state variable u, i.e., $J(a) = \hat{J}(u(x,t;a))$, $\Psi(x,t)$ is the Lagrange multiplier and the function $u(x,t)$ satisfies (3) and boundary conditions (14) or (22). Ignoring the terms of second order, we obtain the first variation of the Lagrange functional

$$\delta F(u,a,\Psi) = \sum_{i=1}^I \int_0^T 2(u(x_i,t) - g_i(t))\delta u(x_i,t)dt$$

$$+ \int_0^T \int_0^l \Psi \left(\frac{\partial^\beta \delta u}{\partial t^\beta} - \mathcal{Z}(\delta u) - \frac{\partial}{\partial x}\left(\delta a \frac{\partial u}{\partial x} \right) \right) dxdt$$

$$+ \int_0^T \int_0^l \delta\Psi \left(\frac{\partial^\beta u}{\partial t^\beta} - \mathcal{Z}u - f \right) dxdt.$$

Using integration by parts (see (7)) we obtain

$$\int_0^T \Psi \frac{\partial^\beta \delta u}{\partial t^\beta} dt = \int_0^T \delta u D_{T-}^\beta \Psi dt,$$

$$\int_0^l \Psi \frac{\partial}{\partial x}\left(a\frac{\partial \delta u}{\partial x} - v\delta u \right) dx = \int_0^l \delta u \left(\frac{\partial}{\partial x}\left(a\frac{\partial \Psi}{\partial x} \right) + v\frac{\partial \Psi}{\partial x} \right) dx,$$

$$\int_0^l \Psi \frac{\partial}{\partial x}\left(\delta a\frac{\partial u}{\partial x} \right) dx = -\int_0^l \delta a\frac{\partial u}{\partial x}\frac{\partial \Psi}{\partial x} dx,$$

$$r\int_0^T \int_0^l \Psi(x,t)\alpha^2(x)\int_0^t \rho(x,t,\tau)\delta u(x,\tau)dt$$

$$= r\int_0^l \alpha^2(x)\int_0^T \left(\int_\tau^T \Psi(x,t)\rho(x,t,\tau)dt \right) \delta u(x,\tau)d\tau dx,$$

since δu satisfies the homogeneous boundary and initial conditions. Hence, at a stationary point of F, the Lagrange multiplier Ψ satisfies the adjoint problem, u is the solution of the direct problem and

$$\int_0^T \frac{\partial \Psi}{\partial x}\frac{\partial u}{\partial x} dt = 0.$$

Further, given $a(x)$ let $\Psi(x,t)$ be the solution of the adjoint problem (29)–(32), $u(x,t)$ be the solution of the direct problem and δu is the solution of the sensitivity problem (33)–(34) and (35) or (36) corresponding to variation δa. Then, we find

$$\delta J(a) \equiv \delta F(u,a,\Psi) = \int_0^l \delta a \int_0^T \frac{\partial \Psi}{\partial x} \frac{\partial u}{\partial x} dt dx. \tag{37}$$

Comparing (27) with (37), we obtain formula (28) for the gradient of the functional $J(a)$.   □

Usually the numerical optimization with respect that the coefficient $a(x)$ is performed over a certain finite-dimensional space. Given basis functions $\varphi_i(x)$, let $a(x)$ be represented as a linear combination

$$a(x) = \sum_{j=1}^{J} a_j \varphi_j(x) \tag{38}$$

with parameters a_j, $i = 1,\ldots,J$. Then in the case of such (parametric) optimization we obtain similarly to above (see [32] as well)

$$\nabla J = \nabla J(a_1,\ldots,a_J) = \left\{ \frac{\partial J}{\partial a_j} \right\}_{j=1}^{J}, \tag{39}$$

$$\frac{\partial J}{\partial a_j} = \int_0^l \varphi_j(x) \int_0^T \frac{\partial \Psi}{\partial x} \frac{\partial u}{\partial x} dt dx. \tag{40}$$

Here $\Psi(x,t)$, $u(x,t)$ are as in (28).

5   Conjugate Gradient Method

In this section we briefly discuss how the *conjugate gradient method* (CGM) is applied in order to reconstruct the space-dependent diffusion coefficient $a(x)$ from interior concentration observations in the considered problem of pollution in porous media. The Fréchet gradient (28) is used in order to minimize the misfit between the measured and the computed concentrations defined by the least-squares objective functional (26).

In general, the following iterative procedure is performed in application of the CGM (see, e.g., [2,10,18,32])

$$a^{n+1}(x) = a^n(x) + \beta^n d^n(x), \quad n = 0,1,\ldots. \tag{41}$$

Here a^{n+1} and a^n are the successive approximations of the minimizer, n is the iteration number, β^n is the search step size and $d^n(x)$ is the search direction.

The search direction $d^n(x)$ at each step is a linear combination of the steepest descent direction at the current approximation and the search direction at the previous iterations, i.e.,

$$d^0(x) = -\nabla J[a^0(x)], \quad d^n(x) = -\nabla J[a^n(x)] + \gamma^n d^{n-1}(x), \quad i = 1,2,\ldots, \tag{42}$$

with γ^n being the *conjugation coefficient*. There are various forms of γ^n known in literature, [2, 17, 18]. For example, a common choice is the following:

$$\gamma^n = 0, \quad \gamma^n = \frac{\displaystyle\int_0^l |\nabla J[a^n(x)]|^2 \mathrm{d}x}{\displaystyle\int_0^l |\nabla J[a^{n-1}(x)]|^2 \mathrm{d}x}, \quad n = 1, 2, \ldots$$

which is suggested by Fletcher and Reeves [12]. Line search is applied in order to obtain the coefficient β^n, i.e., minimization along the specified search direction, $\beta^n = \mathrm{argmin}_\beta\, J(a^n(x) + \beta d^n(x))$.Taking into account the sensitivity problem (33), (34), (35) (or (36)), the following choice of the search step size can be derived similarly to [11] after linearization of the objective functional $J(a^n(x) + \beta d^n(x))$ with respect to β

$$\beta^n = \frac{\sum_{i=1}^{I}(u^n(x_i) - g_i)\delta u^n(x_i)}{\sum_{i=1}^{I}[\delta u^n(x_i)]^2},$$

with $u^n(x)$ and $\delta u^n(x)$ being the solutions of direct and sensitivity problems, respectively, where $a(x) = a^n(x)$ and $\delta a = d^n(x)$.

According to the discrepancy principle the iterative procedure is stopped when the difference between the measured and the estimated values is of the order of the noise level determined by the standard deviation σ of the measurements, i.e., when

$$|u(x_i, t) - g_i(t)| \approx \sigma, \quad i = 1, \ldots, I, \quad x_i \in (0, l), \quad t \in (0, T) \tag{43}$$

Substituting (43) in the definition of the functional (26) the following stopping criterion is obtained

$$J(a(x)) < \xi := \frac{1}{2}I\sigma^2. \tag{44}$$

For more details, see, e.g., [2, 10, 18].

6   Conclusions

In this paper using a quasi solution method, we studied the determination of the diffusion coefficient in a time-fractional integro-differential equation of pollution in porous media. The advection-diffusion process in the mobile region and the solute transport in the immobile region are described by a system of a time-fractional parabolic equation and a time-fractional ordinary differential equation, respectively. The study is reduced to a diffusion coefficient identification problem for a given measured data about the contaminant concentration in the mobile zone. We use the Lagrange multipliers methodology for minimization of a least squares functional whose gradient is represented in terms of the solutions of the direct and the adjoint problems.

The focus of the paper is mainly on the theoretical analysis of the problem in $\mathbb{R}^1$. We plan to extend this analysis for $\mathbb{R}^2$ and $\mathbb{R}^3$ problems. Also, we plan to apply the CGM method described in Sect. 5 in order to reconstruct numerically the unknown diffusion coefficients. As soon as the integro-differential equations can not be solved analytically in general, a finite element method (FEM) similar to the approach in [31] will be developed. The total gradient of the cost functional could then be computed by artificial neural network as in [7].

Acknowledgement. This research is supported by the Bulgarian National Science Fund partially under the Bilateral Project KP/Russia 06/12 "Numerical methods and algorithms in the theory and applications of classical hydrodynamics and multiphase fluids in porous media" from 2020, and partially by the Project FNSE-03 of the University of Ruse "Angel Kanchev".

References

1. Alaimo, G., Piccolo, V., Cutolo, A., Deseri, L., Fraldi, M., Zingales, M.: A fractional order theory of poroelasticity. Mech. Res. Commun. **100**, 103395 (2019). https://doi.org/10.1016/j.mechrescom.2019.103395
2. Alifanov, O.M., Artioukhine, E.A., Rumyantsev, S.V.: Extreme Methods for Solving Ill-Posed Problems with Applications to Inverse Heat Transfer Problems. Begell House, Danbury (1995)
3. Alikhanov, A.A.: A priori estimates for solutions of boundary value problems for fractional-order equations. Differ. Equ. **46**, 660–666 (2010). https://doi.org/10.1134/S0012266110050058
4. Armstrong, J.E., Frind, E.O., McClellan, R.D.: Nonequilibrium mass transfer between the vapor, aqueous, and solid phases in unsaturated soils during vapor extraction. Water Resour. Res. **30**(2), 355–368 (1994). https://doi.org/10.1029/93WR02481
5. Asl, N.A., Rostamy, D.: Identifying an unknown time-dependent boundary source in time-fractional diffusion equation with a non-local boundary condition. J. Comput. Appl. Math. **355**, 36–50 (2019). https://doi.org/10.1016/j.cam.2019.01.018
6. Bear, J.: Modeling Phenomena of Flow and Transport in Porous Media, 1st edn. Springer, Cham (2018). https://doi.org/10.1007/978-3-319-72826-1
7. Berg, J., Nyström, K.: Neural network augmented inverse problems for PDEs (2017). https://doi.org/10.48550/arXiv.1712.09685
8. Caputo, M.: Vibrations of an infinite viscoelastic layer with a dissipative memory. J. Acoust. Soc. Am. **56**(3), 897–904 (1974). https://doi.org/10.1121/1.1903344
9. Caputo, M., Plastino, W.: Diffusion in porous layers with memory. Geophys. J. Int. **158**(1), 385–396 (2004). https://doi.org/10.1111/j.1365-246X.2004.02290.x
10. Chavent, G.: Nonlinear Least Squares for Inverse Problems: Theoretical Foundations and Step-by-Step Guide for Applications. Scientific Computation, Springer, Dordrecht (2010). https://doi.org/10.1007/978-90-481-2785-6
11. Fakhraie, M., Shidfar, A., Garshasbi, M.: A computational procedure for estimation of an unknown coefficient in an inverse boundary value problem. Appl. Math. Comput. **187**(2), 1120–1125 (2007). https://doi.org/10.1016/j.amc.2006.09.015
12. Fletcher, R., Reeves, C.M.: Function minimization by conjugate gradients. Comput. J. **7**(2), 149–154 (1964). https://doi.org/10.1093/comjnl/7.2.149

13. Gao, G., Zhan, H., Feng, S., Fu, B., Ma, Y., Huang, G.: A new mobile-immobile model for reactive solute transport with scale-dependent dispersion. Water Resour. Res. **46**(8), 1–16 (2010). https://doi.org/10.1029/2009WR008707
14. van Genuchten, M.T., Wierenga, P.J.: Mass transfer studies in sorbing porous media I. Analytical solutions. Soil Sci. Soc. Am. J. **40**(4), 473–480 (1976). https://doi.org/10.2136/sssaj1976.03615995004000040011x
15. Goulart, A.G., Lazo, M.J., Suarez, J.M.S.: A new parameterization for the concentration flux using the fractional calculus to model the dispersion of contaminants in the planetary boundary layer. Physica A **518**, 38–49 (2019). https://doi.org/10.1016/j.physa.2018.11.064
16. Gyulov, T.B., Vulkov, L.G.: Reconstruction of the lumped water-to-air mass transfer coefficient from final time or time-averaged concentration measurement in a model porous media. Report on the 16th Annual Meeting of the Bulgarian Section of SIAM, 21–23 December 2021, Sofia, Bulgaria (2021)
17. Hanke, M.: Conjugate Gradient Type Methods for Ill-Posed Problems, 1st edn. Chapman and Hall/CRC, New York (1995). https://doi.org/10.1201/9781315140193
18. Hasanov, A.H., Romanov, V.G.: Introduction to Inverse Problems for Differential Equations. Springer, Cham (2017). https://doi.org/10.1007/978-3-319-62797-7
19. Jovanović, B.S., Vulkov, L.G., Delić, A.: Boundary value problems for fractional PDE and their numerical approximation. In: Dimov, I., Faragó, I., Vulkov, L. (eds.) NAA 2012. LNCS, vol. 8236, pp. 38–49. Springer, Heidelberg (2013). https://doi.org/10.1007/978-3-642-41515-9_4
20. Kandilarov, J., Vulkov, L.: Determination of concentration source in a fractional derivative model of atmospheric pollution. In: AIP Conference Proceedings, vol. 2333, no. 1, p. 090014 (2021). https://doi.org/10.1063/5.0042092
21. Kilbas, A.A., Srivastava, H.M., Trujillo, J.J.: Theory and Applications of Fractional Differential Equations. North-Holland Mathematics Studies, vol. 204, 1st edn. North-Holland, Amsterdam (2006)
22. Krivorotko, O., Kabanikhin, S., Zhang, S., Kashtanova, V.: Global and local optimization in identification of parabolic systems. J. Inverse ILL-Posed Probl. **28**(6), 899–913 (2020). https://doi.org/10.1515/jiip-2020-0083
23. Li, G., Jia, X., Liu, W., Li, Z.: An inverse problem of determining fractional orders in a fractal solute transport model (2021). https://doi.org/10.48550/arXiv.2111.13013
24. Li, X., Wen, Z., Zhu, Q., Jakada, H.: A mobile-immobile model for reactive solute transport in a radial two-zone confined aquifer. J. Hydrol. **580**, 124347 (2020). https://doi.org/10.1016/j.jhydrol.2019.124347
25. Liu, F.J., Li, Z.B., Zhang, S., Liu, H.Y.: He's fractional derivative for heat conduction in a fractal medium arising in silkworm cocoon hierarchy. Therm. Sci. **19**(4), 1155–1159 (2015). https://doi.org/10.2298/TSCI1504155L
26. Liu, W., Li, G., Jia, X.: Numerical simulation for a fractal MIM model for solute transport in porous media. J. Math. Res. **13**, 31–44 (2021). https://doi.org/10.5539/jmr.v13n3p31
27. Mahiuddin, M., Godhani, D., Feng, L., Liu, F., Langrish, T., Karim, M.: Application of caputo fractional rheological model to determine the viscoelastic and mechanical properties of fruit and vegetables. Postharvest Biol. Technol. **163**, 111147 (2020). https://doi.org/10.1016/j.postharvbio.2020.111147
28. Mendoza, C.A., Frind, E.O.: Advective-dispersive transport of dense organic vapors in the unsaturated zone: 1. Model development. Water Resour. Res. **26**(3), 379–387 (1990). https://doi.org/10.1029/WR026i003p00379

29. Podlubny, I.: Fractional Differential Equations. Mathematics in Science and Engineering, vol. 198. Elsevier, Amsterdam (1999)

30. Tikhonov, A., Arsenin, V.: Solutions of ILL-Posed Problems. Winston, Washington (1977)

31. Tyrylgin, A., Vasilyeva, M., Alikhanov, A., Sheen, D.: A computational macroscale model for the time fractional poroelasticity problem in fractured and heterogeneous media (2022). https://doi.org/10.48550/arXiv.2201.07638

32. Vabishchevich, P.N., Denisenko, A.Y.: Numerical methods for solving the coefficient inverse problem. Comput. Math. Model. **3**(3), 261–267 (1992). https://doi.org/10.1007/BF01133895

33. Wei, T., Xian, J.: Variational method for a backward problem for a time-fractional diffusion equation. ESAIM: M2AN **53**(4), 1223–1244 (2019). https://doi.org/10.1051/m2an/2019019

34. Yamamoto, M.: Fractional calculus and time-fractional differential equations: revisit and construction of a theory. Mathematics **10**(5), 698 (2022). https://doi.org/10.3390/math10050698

35. Zhou, H., Yang, S., Zhang, S.: Modeling non-darcian flow and solute transport in porous media with the Caputo-Fabrizio derivative. Appl. Math. Model. **68**, 603–615 (2019). https://doi.org/10.1016/j.apm.2018.09.042

Flash Flood Simulation Between Slănic and Vărbilău Rivers in Vărbilău Village, Prahova County, Romania, Using Hydraulic Modeling and GIS Techniques

Cristian Popescu[1]([✉]) and Alina Bărbulescu[2]

[1] Technical University of Civil Engineering Bucharest, 124 Lacul Tei Boulevard, Bucharest, Romania
cristiannicolae.popescu@gmail.com

[2] Transilvania University of Brașov, 5 Turnului Street, Brașov, Romania

Abstract. Flash flooding occurs in different hilly regions with circular shaped catchments across Europe, including Romania, when heavy rains or rapid melting of the snow are encountered. Vărbilău is a village in the Subcarpathian hills, susceptible to flash floods. In some situations, roads and buildings are affected. This study aims to analyze the flash flood from the 22nd of July, 2018, compare it with the same phenomena formed at higher and lower flow rates, and determine the impact on settlements and roads. The research is done based on field observations, hydrological data from the Romanian National Institute of Hydrology, Geographic Information System (GIS) database, as well as by processing information in GIS environment (ArcMap 10.2.2) and hydraulic modeling in HEC-RAS 6.2. Data is imported in GIS and prepared for the flooding simulation in HEC-RAS. After the model is obtained based on flow values, the affected buildings and roads are quantified in GIS. The area is prone to flooding. Upstream from the confluence the Slănic valley is much more affected than the valley of Vărbilău because the buildings are in the proximity of the river and the valley is narrow, while downstream from the confluence only two bridges and one county road are affected. The utility of protective levees is presented in the paper, while road accessibility for the emergency intervention authorities registers higher values in the northern part.

Keywords: Flood · Hydraulic modeling · Settlements · Confluence · Rivers

1 Introduction

Hydrological phenomena are part of the hazards that are quite unpredictable and can affect large areas, including inhabited ones. The challenge comes when hazards occur,

GIS database from European Environment Agency (eea.europa.eu).
Microsoft Worldwide building footprints derived from satellite imagery.
Hydrological flow data from the Romanian National Institute of Hydrology.

© The Author(s), under exclusive license to Springer Nature Switzerland AG 2023
D. Simian and L. F. Stoica (Eds.): MDIS 2022, CCIS 1761, pp. 309–327, 2023.
https://doi.org/10.1007/978-3-031-27034-5_21

when the authorities must intervene first of all to evacuate the population, save all lives and belongings as much as possible. The subsequent challenge is the temporary relocation of the victims in the situation where they can no longer live in their own households as a result of their destruction.

Floods, flash floods, torrents, hurricanes, tsunami waves, landslides are produced by extreme hydrological phenomena all over the Globe [1]. In the area of the Ganges-Brahmaputra delta in Bangladesh and India, catastrophic floods occur as a result of the presence of cyclones. Cyclones are formed as a result of the differential heating between the Indian subcontinent and the homonymous ocean. Also, the orographic barrier in the north leads to the fall of extremely high precipitation in the south. Hurricanes are found in the Caribbean, South America, East Asia etc. These lead to catastrophic floods that cause significant casualties and material losses [2]. Tsunami waves occur as a result of underwater earthquakes. They can be tens of meters high and travel at speeds between 300 and 700 km/h. When they reach the shore area, they have an impressive power of destruction. As a result of impressive rainfall, flash floods occur in areas with steep slopes. Water accumulation may occur downstream. If the rains take place over a long period in flat areas with elongated valleys, floods may occur. Torrents are a temporary watercourse with a huge flow, which can also transport materials, resulting from heavy rains, melting snow or in some situations, as a result of the breaking of anthropogenic water accumulations (artificial dam lakes). They create relief microforms such as ravines, and by depositing materials at the base of the torrent, they can also block communication routes or destroy houses [2]. Precipitation can cause landslides, especially in deforested areas, in areas with undermined slopes. Landslides can cover communication routes; they can destroy houses if they are located on them.

In Europe, the rich precipitation is recorded in the western area as a result of the oceanic influence, decreasing gradually towards the eastern side [3]. Also, they increase in altitude. Floods have occurred in recent years in countries such as Germany, Belgium, Switzerland and the Netherlands. Moreover, the latter has areas located below sea level. Thus, if the level of the planetary ocean were to increase significantly, these areas could be strongly affected [3]. The important supply of water from countries such as Germany, Slovakia, Austria, Hungary can lead to an increase in the level of the Danube, which thus floods areas downstream in Romania. Floods also occurred in Central Europe in countries such as Poland and the Czech Republic. In the hilly and mountainous areas of Europe (Alps, Carpathians, Italian Piedmont, Romanian Subcarpathians) where the slopes are higher, flash floods can occur [3]. Factors such as deforestation, soils with low permeability, showers, the circular shape of hydrographic basins favor flash floods. They flow towards the downstream areas where, if the bed of the riverbed is flat, the valleys are elongated and bordered by slopes, thus making it impossible to drain in other directions, accumulations of water can occur (floods). At the opposite pole we have prolonged droughts as a result of the absence of precipitation [4].

In Romania, precipitation decreases from west to east and from north to south. The specific climate is temperate-continental, with oceanic influences in the west and center, Mediterranean in the southwest, continental in the east and south [5], Pontic in the coastal area and Baltic Scandinavian in the area of Suceava. Precipitation may fall in significant amounts, sometimes in torrential form, in the months of May, June and the beginning of

July. In these months, there is a risk of flooding and flash floods. Also, as a result of the rapid melting of the snow, floods can occur in the spring. Floods (water accumulation) occur on large rivers, such as Siret, Mureş, Olt, Danube and their tributaries in the plain or large meadow areas. In the areas with steep slopes, especially in the Subcarpathians and Carpathians, flash floods occur.

Drought takes place especially in the summer, starting in June [5].

Floods have a negative impact if they occur in inhabited areas. In Romania there are many settlements located near rivers. Even if the houses are not destroyed, the effects are unpleasant as a result of the deposition of alluvium in the yards or over the roads. The groundwater can be affected and it cannot be used for a while. When houses are destroyed, there is a need to relocate and find new shelters for people, sometimes permanently [6].

Mătreaţă and Mătreaţă (2010) provide a useful perspective on flooding in small basins. They are characterized by their unpredictable manifestation [7].

Sanyal and Lu (2005) show a correlation between water depth in affected rural areas and higher elevations that provide temporary shelter. This ingenious solution can be extrapolated and improved at the level of study basins as a temporary measure for affected locals: identifying non-flooding proximity areas where camping tents can be set up for shelter [8].

Skentos (2018) identifies geological importance in modeling terrain configuration through GIS techniques and the topographic positioning index. Thus, the author was able to spatialize areas with a steep or smooth slope according to this index, practically determining the importance of the slope factor in rapid runoff. He also presented a simple geological map where rock types influencing erosion or runoff could be seen [9].

Costache, Popescu and Bărbulescu (2022) talk about the extent of flooding in an area of Buzău County, Romania. HEC-RAS software for hydraulic modeling for different overflow probabilities is used, while GIS techniques are applied to quantify the number of buildings and roads affected [10].

Iosub, Minea and Romanescu (2015) talk about the use of HEC-RAS modelling in flood risk analysis in the middle and lower sector of a river. They manage to quantify the damage for a 1% occurrence flood based on the simulation. This kind of analysis can be applied for a 0.1%, 2%, 5% or 10% flood occurrence or any other important river flows [11].

It is very important to identify the areas with flood risk. Thus, the authorities can take measures or at least reduce the impact of the phenomena when they take place.

2 Study Area

The village of Vărbilau is chosen because floods occurred here in the past, including in 2018 based on the data from the Romanian National Institute of Hydrology, but also because here is the confluence of the Slănic with Vărbilău. The village of Vărbilău is located between two rivers. This could be an additional risk in the occurrence of flooding and flash-flooding.

This article will analyze the event that took place on July 22nd 2018, when an impressive amount of precipitation fell in a very short time. Thus, runoffs occurred on the slopes, resulting in flash floods on the Slănic and Vărbilău rivers. This indication is given, for example, by the fact that the flow increased from 0.49 m³/s, at 6:00 a.m., to 98.6 m³/s, at 3:30–4:00 p.m. At 5:00 p.m. the flow suddenly dropped to 19.2 m³/s, so that the following day at 6:00 a.m. it would be 1.02 m³/s. This indicates that there was a flash flood in Vărbilău. If accumulation had occurred, then the flow rate would have decreased more slowly. Moreover, this area is more susceptible to flash floods as a result of the steeper slopes and the narrower valleys. The accumulation part (flooding) can occur downstream, after the confluence, where the valleys are wider and the slope is smoother. On Vărbilau, values of almost 200 m³/s were reached.

Also, a comparison will be presented between the flow recorded in 2018 and two other hypothetical flow values, one 50% lower, the other 50% higher. It should be mentioned that much higher values can be recorded in these areas, an indication being another river also called Slănic from Buzău county which registered much higher flow values [10]. It is located near the area analyzed in this paper and has a similar extent to that of the Vărbilău river. There will also be flow simulations that will include two protection levees and an accessibility map for the emergency intervention teams.

Another reason why this village is chosen is because the location of human settlements in the vicinity of rivers (some houses are less than 30 m away from the rivers), but also because the difference in altitude between the thalweg and some settlements is small, under 5 m.

The flooding phenomena can be repeated. Hance, hydraulic simulation will be presented in order to show the impact.

Vărbilău is located in Romania in the south-central part (Fig. 1) of the country.

It can be noticed that it is very close to the parallel of 45° northern latitude.

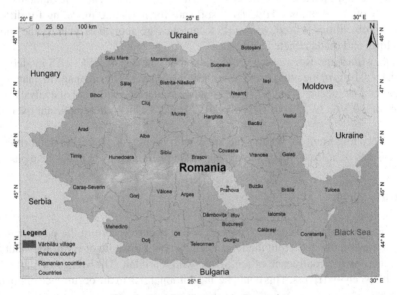

Fig. 1. Varbilău location in Romania.

Vărbilău is a commune in Prahova county (Fig. 2), consisting of the villages of Coțofenești, Livadea, Podu Ursului, Poiana Vărbilău and Vărbilău (residence). The analysis will concentrate on the residence [12].

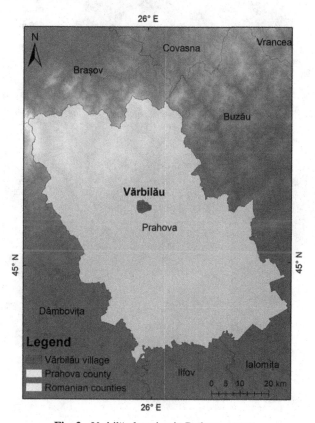

Fig. 2. Varbilău location in Prahova county.

Vărbilău is part of the the Subcarpathian Curved hills. The alternation of hills with depressions, valley lanes, witnesses of erosion can be observed. From an altimetric point of view, the analyzed area (Fig. 3) is located between 308 m in the southern part (Vărbilău riverbed) and 626 m in the northeastern part (the hilly area). The studied area is located on the sinking ends of the Paleogene flysch deposit, which constitutes the two spurs, the Văleni spur and the Homorâciu-Prăjani spur [5].

Vărbilău commune is located in the hilly transition zone between the high mountainous area (cooler) and the plain area (warmer), so the climate will be of high hills (over 500 m) [3]. This type of climate corresponds to an average annual temperature of 9 to 10 °C (the annual average in Câmpina is 9.6°), with average annual thermal amplitudes of 22 to 24 °C (in January the average temperature is -2 to -3 °C, and in July of 20 to 21 °C). The average annual amount of precipitation is 600–700 mm/m^2/year. The dominant direction of the winds is SE and NW, average speed is 2–3 m/s [6].

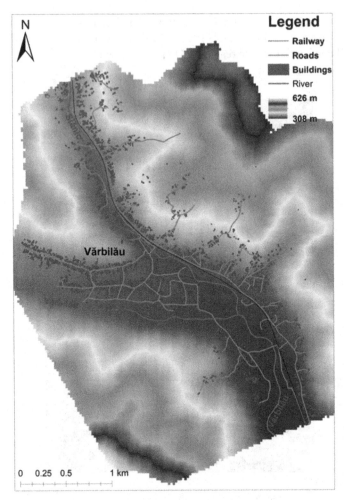

Fig. 3. Varbilău village.

Most of the commune is covered with forest vegetation and secondary with meadows installed on the site of former forests (24%). Following the distribution of the vegetation, the influence of the morphometric elements can be distinguished, which imprint the vertical zonation, but also the influence of the morphological elements (pools, micro depressions), which produce deviations from the general rule [1].

The soils are argiluvisols, cambisols class, hydromorphic soils class, undeveloped soils [1].

Regarding the communication routes, Vărbilau village is crossed by the Ploiesti-Buda-Slănic (Prahova) railway, while by road it is crossed by the county roads DJ102 (Poiana Vărbilau-Slănic) and DJ 101T (Ostrovu-Vărbilău). The population of the whole municipality is approximately 6000 inhabitants. Thus, the density of 136 inhabitants/km^2 is above the national average, of 84 inhabitans/km^2 [12].

3 Methodology

ArcMap 10.2.2 is a GIS tool from ESRI used primarily to prepare data for flood simulation in HEC-RAS 6.2 hydraulic modeling software. The flood simulation is based on the flow data recorded on July 22nd, 2018. Also, the quantification of the flood impact was also determined with the help of GIS (Geographic Information System) [13].

The following GIS databases are added to ArcMap in order to start the processing: Digital Terrain Model at 25 m (DTM) eea.europa.eu, Romanian road network, Romanian building network from Microsoft worldwide building footprints, railway network, river network, county and administration units. The coordinate system used is Stereo 70 (31700) [14].

The first step is the creation of the location maps of the village of Vărbilău. Thus, the map of Romania is constructed by adding a Digital Terrain Model (DTM), a raster (a raster image file is a rectangular array of regularly sampled values known as pixels. In this case, the pixel cell size is of 25 m), a polygon database of the counties and countries which Romania borders. The Black Sea is also been added. The latter are shapefiles, that is a simple, nontopological format for storing geometric location and attribute information of geographic features. Later, through "raster clip" (raster processing) from the Data Management Tools tool, the relief of Romania is cut. The counties are added over the relief. The country's neighbors and the Black Sea are added outside of it. Finally, the location of Vărbilău commune is drawn, and the legend, latitude and longitude are added. The map is exported from the GIS application. The map of Prahova county is built in the same way, only that here the only shapefiles needed are the counties and the Vărbilău village polygon.

The next part is the preparation of the location for the simulation. Thus, the buildings, roads, rivers and railways are added for the analyzed area, namely the village of Vărbilău, with the help of the "clip" from the Geoprocessing tab. Also, the DTM is reduced to the locality level with the help of the same "raster clip". Afterwards, the HEC-GeoRAS extension must be activated or, as the case may be, installed. Thus, accessing the extension, the thalweg (centerline) of the Slănic and Vărbilău rivers, banks (left and right), flow paths (left, right and channel) and transverse profiles (cross section lines over the valley) are drawn. These are important for the simulation in HEC-RAS. After the data are validated, they are exported as RAS data in HEC-RAS where metric system is used.

Before the simulations are done, some settings must be made in HEC-RAS. It is important that when processing, the confluence is first represented as an intersection in HEC-GeoRAS. The settings will be made for the upper course of the Vărbilău (before the confluence), for the Slănic course and for the lower course of the Vărbilău (after the confluence). Thus, in "Steady flow data" under the Edit tab, the flow is added for each of the courses specified above in cubic meters, while in Boundary Conditions "critical depth" is set for the lower course of the Vărbilău river.

Another important step is to add in the "Manning's n or k" the "n" roughness coefficient. Thus, it is very important to set the correct values. A high value of the "n" value will lead to a decrease in the velocity of water flowing on a surface [15]. High values represent favorability in the accumulation of water (production of floods). In Table 1, the "n" value can be observed for each cross Sects. (28 cross lines) as per Fig. 4. N #1 represents the coefficient for the left bank, n #2 for the main channel and n #3 for the right bank. The values for the main channel are lower because it is finer (earth channel - gravelly, the "n" value is 0.025), while the banks are rougher (weedy, the "n" value is 0.03). The numbers in Table 1 are based on a list of roughness coefficient values for different surfaces [15].

Table 1. "n" value for Manning's coefficient.

River station (equivalent to cross section)	n #1	n #2	n #3
1–28	0.03	0.025	0.03

Manning's equation, used in HEC-RAS for steady flows, is introduced by the engineer Robert Manning in 1889 as an empirical equation that applies to uniform flow in open channels. It represents a function of the channel velocity, flow area and channel slope [16].

$$Q = vA = \left(\frac{1.00}{n}\right)AR^{\frac{2}{3}}\sqrt{S}$$

where:

Q = Flow rate (m^3/s);
v = Velocity (m/s);
A = Flow area (m^3);
n = Manning's Roughness Coefficient – setup by the HEC-RAS user (Table 1);
R = Hydraulic Radius, (m);
S = Channel slope (m/m).

The simulation can be run from Run tab, "steady flow analysis". HEC-RAS 6.2 uses and equation for each cross section (1D steady flow water surface profiles) in order to obtain the 2D hydraulic model. There are 28 cross sections in total [17].

Based on the documentation from HEC-RAS 6.2 the following equation is used for the profile calculation:

$$Z_2 + Y_2 + \frac{a_2 V_2^2}{2g} = Z_1 + Y_1 + \frac{a_1 V_1^2}{2g} + h_e$$

where:

Z_1, Z_2 = elevation of the main channel inverts;
Y_1, Y_2 = depth of water at cross sections;
V_1, V_2 = average velocities (total discharge/total flow area);
a_1, a_2 = velocity weighting coefficients;
g = gravitational acceleration
h_e = energy head loss

Water surface profiles are computed from one cross section to the following (Fig. 4).

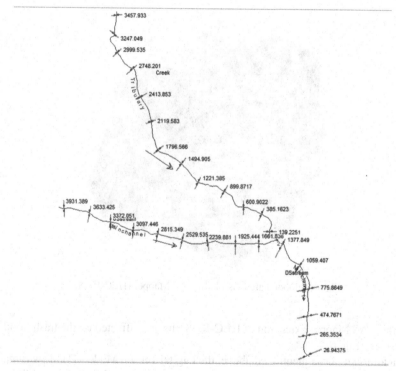

Fig. 4. All cross sections of the two rivers (caption from HEC-RAS).

After computing, the hydraulic model is obtained taking into account the DTM, surface roughness, flow values and cross sections.

In Fig. 5, we can observe the water level in a cross section.

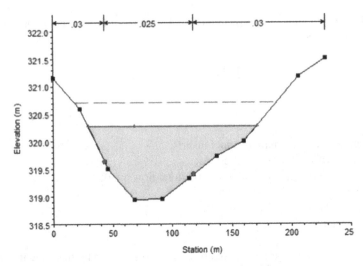

Fig. 5. Cross section from HEC-RAS.

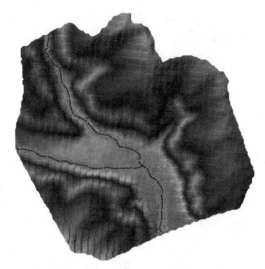

Fig. 6. Flooded area visible in RAS Mapper (HEC-RAS).

In the RAS Mapper extension of HEC-RAS, the area affected by the flash flood can be observed (Fig. 6).

In addition to those mentioned above, the Digital Terrain Model (DTM) was manipulated to add two 6 m high levees intended to protect a number of houses. The new model containing the levees is used in HEC-RAS to see the inundation band and further to compare the two situations: floods without and with levees (Fig. 7).

After verifying the correctness of the data, the data raster obtained in HEC-RAS is exported to be added back to ArcMap to quantify the impact of the flood.

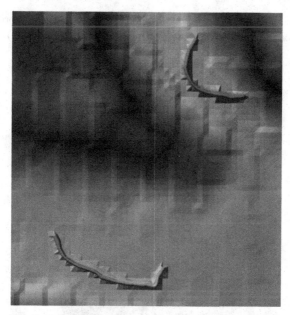

Fig. 7. Levees added in the Digital Terrain Model (DTM).

In GIS, the newly added raster is reclassified using "reclass" in 3D Analyst to obtain a uniform raster. Later, it is transformed into a polygon shapefile using "raster to polygon" from Conversion Tools [18].

The next step is to transform the buildings into points with the help of "feature to point" (features) from Data Management Tools. The obtained points are counted with the help of "spatial join" (overlay) from Analysis Tools. Thus, we determine the total number of buildings in the analyzed area based on the polygon of the administrative unit of Vărbilău village, as well as the number of flooded buildings based on the inundation band obtained in HEC-RAS [19].

The flooded roads are obtained with the help of "clip' from Geoprocessing tab. In the attribute table of the roads a new column called length is added. With the help of "calculate geometry" [19], the length of each road section is determined. Moving forward, the total length of the roads and flooded roads is determined using "summary statistics" (statistics) from Analysis Tools. By quantifying the flooded roads and buildings, analyzes can be made. Because the railway is not affected, no action is needed in GIS [20].

The total surface of the flooded area is calculated with the help of "calculate geometry" in the attribute table [19].

In addition, an accessibility map with time intervals for the emergency intervention teams (Fig. 8) is created in GIS based on the road network, taking into account the travel speed (from 5 to 50 km/h) on different road segments. The better the quality of the road, the higher the speed. The segments are located between two intersections. The starting point for the accessibility analysis is located in the north of Vărbilău village because the intervention teams enter through that area. The nearest center for emergency situations is located north of Vărbilău, in the town of Slănic, 7 km away. The authority responsible for this action is called the Inspectorate for Emergency Situations (ISU).

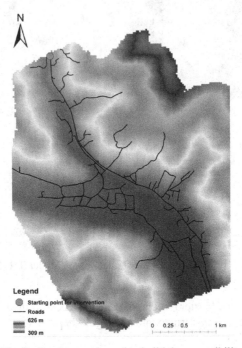

Fig. 8. Road network used to build the accessibility.

At the end, the maps are exported into JPG format. The obtained data will be presented and commented in the following chapter of the study.

4 Results and Discussions

According to the flow data presented previously (98.6 m^3/s on Slănic, 199 m^3/s on Vărbilău before the confluence and 297.6 m^3/s after the confluence), the period of the flash flood is not long. However, as a result of the heavy rainfall, roads and houses are flooded. The phenomenon manifested itself in the form of a flash flood as a result of the narrower valleys upstream of the confluence between Vărbilău and Slănic, the circular shape of the hydrographic basins, as well as the larger slopes. Thus, the speed of water movement is high.

According to Fig. 9, it can be seen that the most affected houses are in Slănic valley because they are located quite close to the river, but also because the difference in level between the houses and the river is quite small. Very few houses, 9 in number, are affected by the Vărbilău river upstream of the confluence. The valley in this area is wider, and the distance from the houses to the river is greater. Downstream of the confluence, no building was flooded. Regarding the communication routes, the most affected are also in the Slănic valley, while in the Vărbilău valley two bridges and a communal road are affected. The railway is not flooded at all.

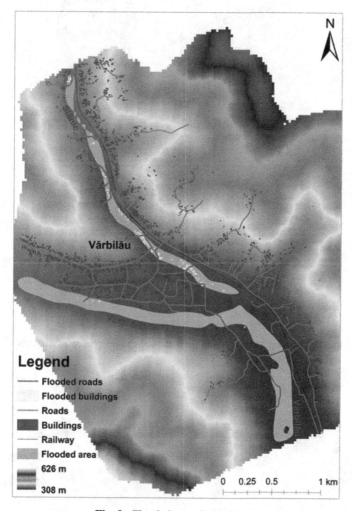

Fig. 9. Flooded areas in Vărbilău.

In Fig. 10, the affected buildings and roads in the Slănic valley can be seen. The most affected are on the right bank in the direction of the river flow.

By comparison, for flows approximately 50% lower (50 m³/s for Slănic and 100 m³/s for Vărbilău before the confluence) than those shown on both rivers above (98.6 m³/s on Slănic, 199 m³/s on Vărbilău before the confluence), it can be observed in Fig. 11, that the surface decreased, but not significantly. This is specific to flash floods. For flows 50% higher (150 m³/s for Slănic and 300 m³/s for Vărbilău), the flooded surface expanded very little as shown in Fig. 12.

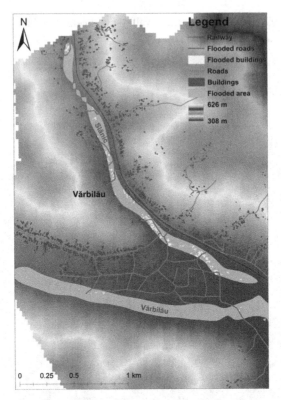

Fig. 10. Flooded areas in Vărbilău in the center and northwestern parts.

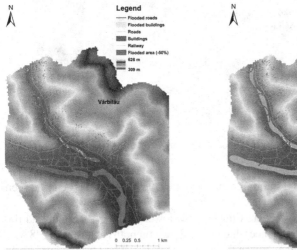

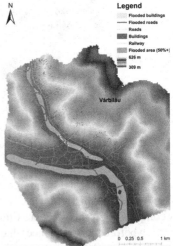

Fig. 11. Flooded areas (flow −50%).   **Fig. 12.** Flooded areas (flow +50%).

According to Table 2, for a registered flow of 98.6 m^3/s for Slănic and 199 m^3/s for Vărbilău, out of a total of 28398 m of roads, 2029 m represent the affected ones, that is 7.14%. Regarding the affected buildings, 139 buildings were flooded out of a total of 2963, that is 4.69%. The total flooded area was 0.75 km^2. The flood zone is wider after the confluence area.

For a flow of 50 m^3/s for Slănic and 100 m^3/s for Vărbilău, 5.61% out of the total roads are flooded, while 3.75% represent the affected buildings. The area covered with water is 0.62 km^2.

For a value of 150 m^3/s for Slănic and 300 m^3/s for Vărbilău, 2274 m of roads are affected, while 151 buildings are flooded. The surface covered with water is of 0.81 km^2. It can be seen that there are no big differences regarding the surfaces covered with water.

Table 2. Values of affected roads and buildings.

Flow (m^3/s)	Quantified values		Flooded surface (km^2)
	Length of flooded roads (meters)	Number of flooded buildings	
98.6 m^3/s for Slănic & 199 m^3/s for Vărbilău	2029 (7.14%)	139 (4.69%)	0.75
50 m^3/s for Slănic & 100 m^3/s for Vărbilău	1595 (5.61%)	111 (3.75%)	0.62
150 m^3/s for Slănic & 300 m^3/s for Vărbilău	2274 (8%)	151 (5.09%)	0.81
Total length of roads *meters		Total number of buildings	
28398		2963	

One of the solutions that could protect households from floods would be the construction of levees. The analysis includes two protection levees in areas where houses are affected. Flow simulation was done with values of 98.6 m^3/s for Slănic and 199 m^3/s for Vărbilău rivers for two situations, one including and the other excluding the levees, in order to see how they protect the settlements. In Fig. 13, we can see the two proposed levees, 1 and 2. The flood area has a different shape with the application of the levees.

In Fig. 14, the flood zone with and without the dike is represented (Levee 1). Without it, the water would flood at least 11 settlements. In the Fig. 15, the same can be observed (Levee 2). Without the dike, no less than 6 settlements would be affected. Thus, the two dikes protect the settlements in those areas, and the water bypasses the houses. The added levees can protect the houses at flows up to 300 m^3/s for Slănic and 500 m^3/s for Vărbilău.

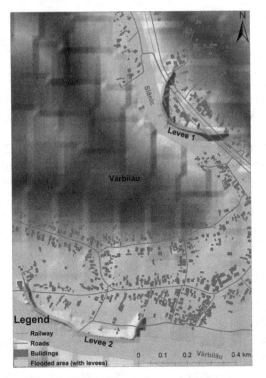

Fig. 13. Flooded areas after applying levees.

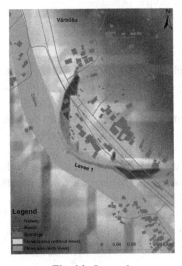

Fig. 14. Levee 1.

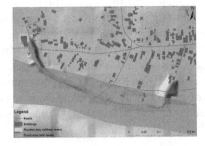

Fig. 15. Levee 2.

Any hydrological hazard requires the intervention of the authorities to save the population and properties. In Fig. 16, the accessibility in the village of Vărbilău can be observed, depending on the road network. Thus, accessibility is greatest in the northern area (less than 2 min travel time from the entry point in Vărbilău) because the teams for emergency situations come from that direction. We can observe that in the central area the movement of the emergency teams takes between 2–4 min from the entrance to the village, the density and quality of the roads being higher. In the south, accessibility is average (4–6 min), the reason being the greater distance from the entry point to the village. In the outskirts of Vărbilău, accessibility is low and very low (it takes at least 6 min to reach the houses), the reason being the poor quality and density of the roads, the long distance from the entrance and the houses located in remote areas.

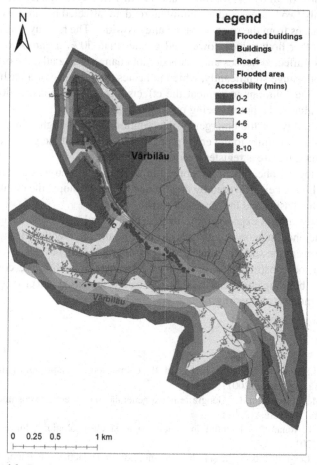

Fig. 16. Road accessibility in Vărbilău village for emergency units (ISU).

Flooding can occur frequently in this area. The situation presented above is comparable to that of a probability of overtaking of 10%. Thus, higher flows can be recorded that can affect the area more.

5 Conclusion

The area presented in the article is exposed to flash floods. Upstream of the confluence, the buildings and roads located in the Slănic valley are more strongly affected compared to those in the Vărbilău valley because the buildings are located at a smaller distance from the river, the difference in level between the buildings and the river is small and valley is narrower. Moreover, only 9 buildings were flooded in the Vărbilău valley, the rest being in Slănic at a flow of 98.6 m^3/s for Slănic and 199 m^3/s for Vărbilău. Downstream of the confluence, only two bridges and a communal road are affected because the houses are at greater distances from the river, and the valley is wider. The railway was not flooded. Compared to other flow values, lower and greater than 50%, it can be observed that the numbers for affected houses and roads do not change drastically. Also, the flooded surfaces do not change significantly, which indicates the presence of a flash flood.

Regarding the solutions to combat the effects hazards, the construction of levees proves its usefulness, the houses being protected by flash floods.

The accessibility for the emergency intervention teams is higher in the north and center due to the proximity to the entry point to the village and the good quality of the roads, while marginal areas register a reduced accessibility.

The hydrological phenomenon recorded in 2018 in Vărbilău was a flash flood that lasted several hours, being favored by the drainage from the slope, the circular shape of the hydrographic basins and the abundant rains.

The analyzed area is exposed more to flash floods and less to the accumulation of water, thus, the phenomenon occurs quickly, making the intervention of the authorities difficult. To prevent the sudden intake of water from upstream, besides levees, artificial dam lakes can be constructed in order to control the flow, while the course can be regularized. By building channels, the flow rate can be increased in a cross section. This would reduce the risk of flooding.

References

1. Ielenicz, M.: Romania. Geografie fizică, vol. II – Climă, ape, vegetaţie, soluri, mediu, Editura Universitară, Bucharest (2007)
2. Ielenicz, M., Comanescu, L.: Geografie fizică generală cu elemente de cosmologie, Editura Universitară, Bucharest (2009)
3. Ciulache, S., Ionac, N.: Esenţial în meteorologie şi climatologie: Editura Universitară, Bucharest (2011)
4. Drobot, R.: Lecţii de hidrologie şi hidrogeologie, Editura Didactica si Pedagogică, Bucharest (2020)
5. Roşu, A.: The physical geography of Romania, Editura Didactică şi Pedagogică, Bucharest (1980)
6. Velcea, V., Savu, A.: Geografia Carpaţilor şi a Subcarpaţilor Românești, Editura Didactică şi Pedagogică, Bucharest (1982)

7. Mătreață, M., Mătreață, S.: Metodologie de estimare a potențialului de producere de viituri rapide în bazine hidrografice mici. Comunicări de Geografie, XIV, Editura Universității din București. Bucharest (2010)

8. Sanyal, J., Lu, X.X.: Remote sensing and GIS-based flood vulnerability assessment of human settlements: a case study of Gangetic West Bengal India. Hydrol. Process. **19**(18), 3699–3716 (2005)

9. Skentos, A.: Topographic position index based landform analysis of Messaria (Ikaria Island, Greece). Acta Geobalcanica **4**, 7–15 (2018)

10. Costache, R., Popescu, C., Barbulescu, A.: Assessing the vulnerability of buildings to floods in the lower sector of Slănic River. Case Study of Cernătești village, Buzău County, Romania. In: IOP Conference Series Materials Science and Engineering, vol. 1242, no. 1, p. 012011 (2022)

11. Iosub, M., et al.: The use of HEC-RAS modelling in flood risk analysis. In: Air Water Conference Journal, Environment Components, Cluj-Napoca, pp. 315–322 (2015)

12. Ghinea, D.: Enciclopedia Geografia a Romaniei. Enciclopedica Publishing, Bucharest (2000)

13. Costache, R., Prăvălie, R., Mitof, I., Popescu, C.: Flood vulnerability assessment in the low sector of Sărățel catchment Case study: Joseni village. Carpathian J. Earth Environ. Sci. **10**(1), 161–169 (2015)

14. Costache, R.: Flash-flood potential index mapping using weights of evidence, decision trees models and their novel hybrid integration. Stochast. Environ. Res. Risk Assess. **33**, 1375–1402 (2019)

15. Zhe, L., Juntao, Z.: Calculation of field Manning's roughness coefficient. Agric. Water Manag. **49**(2), 153–216 (2001)

16. Al-Husseini, T.R.: Estimation of Manning's roughness coefficient for Al-Diwaniya river. J. Eng. Dev. **19**, 4 (2015)

17. US Army Corps of Engineers: Hydraulic Reference Manual v. 5.0, HEC-RAS River Analysis System, Hydrologic Engineering Center (2016)

18. Gatto, et al.: On the use of MATLAB to import and manipulate geographic data: a tool for landslide susceptibility assessment. Geographies **2**(2), 341–353 (2022)

19. Guiwen, L., et al.: Optimizing multi-way spatial joins of web feature services. SPRS Int. J. GeoInf. **6**(4), 123 (2017)

20. Bulai, M., Ursu, A.: Creating, testing and applying a GIS road travel cost model for Romania. Geographia Technica **15**(1), 8–18 (2012)

Author Index

A

Alexiev, Kiril 89
Antonijevic, Milos 104, 188
Arya, Monika 3
Atanasov, Atanas Z. 279

B

Bacanin, Nebojsa 104, 188
Bǎrbulescu, Alina 309
Bhowmik, Kowshik 122
Borissova, Daniela 42
Bukmira, Milos 188

C

Carstea, Claudia-Georgeta 73
Cocha Toabanda, Edwin 18
Cutulab, Ana-Casandra 73

D

Dezert, Jean 221
Dimitrov, Vasil 42
Dimitrova, Zornitsa 42

E

Enache-David, Nicoleta 73
Erazo, María Cristina 18

G

Georgiev, Slavi 247, 264
Gyulov, Tihomir 294

H

Hanumat Sastry, G. 3
Husac, Felix 173

J

Jothi, J. Angel Arul 136
Jovanovic, Dijana 188

K

Kanishkha, J. 136
Kapp, David 234
Khan, Mohammed Abdul Hafeez 136
Kirchhoff, Michael 54
Koleva, Miglena N. 279

L

Lytvyn, Vasyl 161

M

Marcu, Stefan-Bogdan 148
Mediakov, Oleksandr 161
Messay-Kebede, Temesguen 234
Mi, Yanlin 148
Mikhov, Rossen 42
Musgrave, John 234

N

Nau, Johannes 54

P

Peleshchak, Ivan 161
Peleshchak, Roman 161
Petrovic, Aleksandar 104
Popescu, Cristian 309
Prasad, A. M. Deva 136

R

Ralescu, Anca 122, 234
Rezaei, Ahmad 54
Richter, Johannes 54

S

Sabnis, Hrishikesh 136
Sangeorzan, Livia 73
Sarac, Marko 104
Simian, Dana 173
Slavcheva, Nevena 89
Smarandache, Florentin 221

D. Simian and L. F. Stoica (Eds.): MDIS 2022, CCIS 1761, pp. 329–330, 2023.
https://doi.org/10.1007/978-3-031-27034-5

Stankovic, Marko   188
Streitferdt, Detlef   54
Strumberger, Ivana   104, 188
Surdu, Sabina   204

T
Tabirca, Sabin   148
Tangney, Mark   148
Tchamova, Albena   221
Todorov, Venelin   247, 264
Tuba, Eva   104

V
Vulkov, Lubin   279, 294

Y
Yallapragada, Venkata V. B.   148
Yoo, Sang Guun   18

Z
Zivkovic, Miodrag   104, 188

Printed in the United States
by Baker & Taylor Publisher Services